ISNM 108:
International Series of Numerical Mathematics
Internationale Schriftenreihe zur Numerischen Mathematik
Série Internationale d'Analyse Numérique
Vol. 108

Edited by
K.-H. Hoffmann, München; H. D. Mittelmann, Tempe;
J. Todd, Pasadena

Springer Basel AG

H. Antes   P. D. Panagiotopoulos

The Boundary Integral Approach to Static and Dynamic Contact Problems

Equality and Inequality Methods

Springer Basel AG

Authors

Prof. Dr. Heinz Antes
Institut für Angewandte
Mechanik
T.U. Braunschweig
D–W–3300 Braunschweig
Germany

Prof. Dr. Panagiotis D. Panagiotopoulos
Dept. of Civil Engineering
Aristotle University
GR–54006 Thessaloniki
Greece

and

Fakultät für Mathematik und Physik
RWTH Aachen
D–W–5100 Aachen
Germany

A CIP catalogue record for this book is available from the Library of Congress,
Washington D.C., USA

Deutsche Bibliothek Cataloging-in-Publication Data

Antes, Heinz:
The boundary integral approach to static and dynamic contact
problems: equality and inequality methods / H. Antes; P. D.
Panagiotopoulos. – Basel ; Boston ; Berlin : Birkhäuser, 1992
  (International series of numerical mathematics ; Vol. 108)
  ISBN 978-3-0348-9716-7   ISBN 978-3-0348-8650-5 (eBook)
  DOI 10.1007/978-3-0348-8650-5
NE: Panagiotopoulos, Panagiotis D.:; GT

© 1992 Springer Basel AG
Originally published by Birkhäuser Verlag Basel in 1992
Softcover reprint of the hardcover 1st edition 1992

Printed from the authors' camera-ready manuscript on acid-free paper
ISBN 978-3-0348-9716-7

Dedicated to the Memory of Professor Carl Heinz
a Teacher, Friend and Colleague
greatly missed

Contents

Chapter 10 Boundary Integral Formulations for the Monotone Multivalued Boundary Conditions

Chapter 11 Elastodynamic Unilateral Problems. A B.I.E. Approach

Chapter 12 Nonconvex Unilateral Contact Problems

Chapter 13 Miscellanea

Preface

The fields of boundary integral equations and of inequality problems, or more generally, of nonsmooth mechanics, have seen, in a remarkably short time, a considerable development in mathematics and in theoretical and applied mechanics. The engineering sciences have also benefited from these developments in that open problems have been attacked succesfully and entirely new methodologies have been developed. The contact problems of elasticity is a class of problems which has offered many open questions to deal with, both to the research workers working on the theory of boundary integral equations and to those working on the theory of inequality problems. Indeed, the area of static and dynamic contact problems could be considered as the testing workbench of the new developments in both the inequality problems and in the boundary integral equations. This book is a first attempt to formulate and study the boundary integral equations arising in inequality contact problems.

The present book is a result of more than two decades of research and teaching activity of the first author on boundary integral equations and, of the second author, on inequality problems, as well as the outgrowth of seminars and courses for a variety of audiences in the Technical University of Aachen, the Aristotle University of Thessaloniki, the Universities of Bochum, of Hamburg and Braunschweig, the Pontificia Univ. Catolica in Rio de Janeiro etc.

The book is intended for a wide spectrum of readers, mathematicians and engineers alike. Chapters 2 to 6 concern the equality contact problems and have been written by the first author, whereas Chapters 7 to 13 concern the inequality contact problems and have been written by the second author, who also has the responsibility for sections $1.2 \div 1.5$ of Chapter 1.

We wish to acknowledge the great assistance we received from Dr. Zervas who prepared in a very diligent way the final text with the LaTeX program, from Dr. E. Koltsakis, Dr. E. Mistakidis, Dr. O. Panagouli for the programming of the numerical applications of Chapters 7, 9, 12 and 13 as well as for the preparation of the final figures, and from Prof. Z. Naniewicz for critically proofreading some parts of the book. We would also like to thank the editors of the ISNM series for including this volume in their series, our editors in Birkhäuser Verlag for their cooperation and all those who contributed to the emergence of the present volume.

Also, we must apologise to those whose work has been inadvertently neglected in compiling the references of this book. Moreover, we shall welcome all comments and corrections from readers.

H. Antes, P.D. Panagiotopoulos
Thessaloniki, June 1992

Introduction

The aim of the present book is the formulation and study of boundary integral equation methods for static and dynamic contact problems. Both the cases of equality and inequality contact constraints are examined. Those of the first kind lead to classical bilateral problems expressed as variational equalities, whereas the others lead to unilateral problems expressed as variational inequalities in the case of monotone constraints, or as hemivariational inequalities in the case of nonmonotone constraints. The variational equalities give rise to classical boundary integral equations, whereas the variational inequalities and the hemivariational inequalities lead to multivalued boundary integral equations. In parallel to the formulation and the mathematical study, numerical examples help illustrate the presented theories.

In the first Chapter, after some historical information concerning the evolution of the boundary integral equation methods, we give some elements of nonsmooth convex and nonconvex analysis and certain propositions concerning minimization problems, variational inequalities and hemivariational inequalities. Moreover, the equality and inequality contact conditions are defined with respect to both the convex and the nonconvex constraint superpotentials.

In the second Chapter, direct and indirect boundary integral equation methods are presented for the case of equality constraints. Attention is paid to the method of weighted residuals, the use of reciprocal theorems and the singularity method.

In the third Chapter the above methods are illustrated with respect to certain problems from the theory of bars and of beams, and from the theory of Kirchhoff and Reissner plates.

The fourth Chapter deals with the numerical implementation of the boundary integral equations for equality problems. We pay attention to the point collocation method and to the Galerkin boundary element method.

Chapter five extends the results of the previous Chapters to dynamic equality problems. Both steady state and harmonic problems are treated and several numerical applications concerning transient problems and wave propagation problems are given.

Chapter six is the last Chapter dealing with equality problems. Here, certain dynamic interaction problems as the fluid-structure interaction problem and the unilateral contact problem, treated with trial and error methods, are presented.

The aim of Chapter seven is to smoothly introduce the reader into the area of inequality problems. In this Chapter we deal with the Signorini-Fichera B.V.P. We derive two multivalued integral equations holding on the boundary of the body. A numerical application illustrates the theory.

In the next Chapter the multivalued boundary integral equations derived in the previous Chapter are studied concerning the existence and uniqueness of their solutions.

Chapter nine deals with the frictional unilateral contact B.V.P. First the problem of given friction is studied, the corresponding boundary integral expressions are derived and are subsequently studied. Then Coulomb's friction problem is formulated. Numerical examples illustrate the theory.

In Chapter ten certain general contact boundary conditions of monotone, possibly multivalued type, are considered. For them, two types of multivalued boundary integral equations are derived and studied. Here we deal both with the coercive and the semicoercive problem. For the latter, certain new necessary conditions and certain new sufficient conditions are obtained.

Chapter eleven deals with dynamic inequality problems. Using time discretization, a time difference boundary integral equation is obtained. Numerical applications concerning dynamic inequality problems illustrate the theory.

Chapter twelve deals with certain general contact boundary conditions of nonmonotone, possibly multivalued type. For these conditions, which are derived from nonconvex superpotentials, two multivalued boundary integral equations are formulated and studied. Existence results are obtained for both the coercive and the semicoercive case.

Chapter thirteen is the last Chapter of the book. Here we have included certain highly innovative topics. In the first section we deal with cracks having unilateral contact and friction interface conditions, in the second section with adhesively bonded cracks, in the third section we consider fractal interfaces subjected to unilateral contact and friction interface conditions and in the last two sections we deal with the treatment of the multivalued boundary integral equations, obtained in the previous Chapters for several inequality contact problems, in a neurocomputing environment.

Guidelines for the Reader. Abbreviations

The choice of the material of Chapter 1 is governed by the requirements of the subsequent Chapters. We expect the reader to have some knowledge on basic functional analysis especially concerning the norms and certain elementary properties of Sobolev spaces, the Lax-Milgram theorem and the trace theorem. However this functional analysis is needed only for the Chapters 8, 9 and for some parts of Chapters 10 and 12. All propositions of nonsmooth analysis (convex and nonconvex) needed in this book are given in Ch.1. We intentionally include the numerical applications within each chapter in order to permit the reader to read seperately the chapters. The book is equally well accessible to a reader unfamiliar with functional analysis who is interested only in mechanics and in applications. In this case, proofs should be disregarded and the reader should understand the variational expressions in the "usual engineering sense" assuming that spaces $[H^1(\Omega)]^3, [H^{1/2}(\Gamma)]^3$ etc. have simple three-dimensional spaces and the duality pairings $\langle \cdot, \cdot \rangle$ denote inner products.

Certain notations and abbreviations used throughout the text are listed here. All notations defined in the text are not given here. Throughout the book the summation convention with respect to a repeated index is employed. unless otherwise stated. Bold face letters denote vectors and matrices of discretized problems.

$\hat{=}$	Definition
DOFs	Degrees of Freedom
B.V.P.	Boundary Value Problem
F.E.M.	Finite Element Method
B.E.M.	Boundary Element Method
B.I.E.	Boundary Integral Equation
V.I.	Variational Inequality
H.V.I.	Hemivariational Inequality
L.C.P.	Linear Complementarity Problem
Q.P.P.	Quadratic Programming Problem
C.P.P.	Convex Programming Problem
$\fint$	Cauchy principal value
$\bar{A}$	Closure of a set A
δ_{ij}	$\delta_{ij} = \{0 \text{ if } i \neq j\}$ Cronecker's delta

Due to the diversity of the results presented in this book, i.e. of theories of purely mechanical or mathematical origin, it was in some cases impossible to achieve a uniformity of notations. Therefore, we have tried to be as meticulus as we could in the definitions in order to avoid ambiguities.

Chapter 1
Introductory Material

1.1 On the Evolution of the B.I.E.M./B.E.M.

The Integral Equation Method has a long history that goes back to 1903 when Fredholm [Fre03] published his rigorous work on integral equations encoutered in potential theory. The classical works of Tricomi [Tri28] in 1928 on integral equations, of Kellog [Kel] in 1929 on potential theory and especially that of Muskhelishvili [Mus53a,b] in 1953 and Kupradze [Kup63;68] in 1965 on elastostatics, present important applications of the integral equation techniques and show for the first time their possibilities. The notable works of the authors Tricomi [Tri57] in 1957, Mikhlin [Mik57;65a,b], Kalandiya [Kal], Smirnov [Smi64], Gakhov [Gak66], Pogorzelski [Pog66] and Ivanov [Iva76] provided a wealth of information on integral equations, their properties and their approximate solutions. However, these methods were not very popular to engineers, and, moreover, only few attempts for their numerical solution are encountered during this period before the appearance or even before the extended use of digital computers.

The methods for solving integral equations in potential fluid flow theory developed by Trefftz [Tre17] and Prager [Pra28] in 1917 and 1928, respectively, might be considered as the precursors of modern boundary integral techniques, even though such methods are really impractical without the use of a computer. The integral equations methods as practical, efficient and general computational tools, started to emerge during the 60's, a period characterized by the commencement of a wider use of computers, showed a considerable expansion and development during the 70's, in parallel to the rapid development of the computational facilities, and attained a level of maturity during the first half of the 80's.

The term "Boundary Element Method" (B.E.M.), which first appears in the literature in 1977 in the works of Banerjee and Butterfield [Ban81] and Brebbia and Dominguez [Bre77b] and indicates the surface discretization character of the method, finally prevailed over the term "Boundary Integral Equation Method" (B.I.E.M.) first introduced by Cruse [Cru73] in 1973 in conjunction with the direct version of the method. However the term B.I.E.M. is still being used by several authors. It was shown in 1977-78 by Brebbia and Dominguez [Bre77a,b;78b] that the B.E.M., like the Finite Element Method (F.E.M.) and many other numerical methods, can be considered as special cases of the general weighted residuals "principle". This established connection between the various existing numerical techniques thus providing an additional justification for prefering the term B.E.M. instead of B.I.E.M. During the same period, Zienkiewicz et al [Zie77a,b] and Atluri and Grannell [Atl78] obtained the same result on the basis of variational methods and described ways of coupling together the F.E.M. and the B.E.M. for the efficient solution of various complex problems. Note,

also, at this point that the aforementioned weighted residual "principle" is a special case of the principle of virtual work as the later is presented in the famous but difficult book on Theoretical Mechanics by Hamel [Ha].
There are actually two boundary integral equation formulations and, although they can be related to each other, as it has been shown by Brebbia and Butterfield [Bre78a] in 1978, they have really distinct roots. The form currently known as the "indirect" method is in many ways the more direct of the two approaches, since it is based on the physical evidence involved in replacing a physical boundary by a surface of "singularities", e.g. in elasticity, of forces or dislocations distributions whose intensities are adjusted to yield the same results as the original problem. Note moreover, that the singularities must not be placed on the original boundary, but can be located elsewhere in the field, as is the case for what is now called the "embedding integral equation method".

The second form, now referred to as the "direct" method, is an extension of a standard mathematical approach to the solution of partial differential equations, the Green's function approach. If the Green's function is known for a given equation, domain and type of boundary conditions, then the solution of such a problem with arbitrary boundary values for the prescribed type of boundary conditions, is reduced to a standard quadrature of known functions. Much of the mathematical physics of the last century and of the early years of the current was devoted to the quest for such Green's functions. The breakthrough came when it was recognized that the use of a Green's function which satisfied only the physics of the problem (or equivalently the differential equations of the problem), without necessarily satisfying any boundary conditions on any particular boundary (for instance a point source solution), could reduce the dimensionality of the problem (for the homogeneous case), even though this procedure gave rise to an integral rather than to a differential equation. In this sense the "direct" method can be characterized as the result of merging a physical with a mathematical approach.

The indirect method is much older than the direct one and possesses a considerable number of variants. In 1906 Fredholm [Fre06] and Lauricella [Lau07], already, used this "singularity method" to obtain a formulation of the elasticity boundary value problem (B.V.P.) as an integral equation which results if the displacements are prescribed on the boundary. Later Kupradze [Kup65] solved in the same way the "stress" problem and the "mixed" problem, for homogeneous and for non-homogeneous bodies. Lauricella found the fundamental solution for the half plane but gave no numerical example. In 1926 Miche [Mic26] obtained the same solution independently but he only roughly described the method with respect to a particular case. Massonnet [Mas49] was the first to solve the plane stress problem numerically, using a vectorial integrator of his invention (1949) and later an electronic computer. All these researchers used the fundamental solution of the half plane. From 1962 and on, G. Rieder [Rie62;68] and later his coworkers, e.g. U. Heise, T. Kermanidis, H. Grüters, H. Antes , U. Pahnke, U. Zastrow [Hei69,75;78a,b, Ker70;73;75;76, Grü, An72;73, Pah, Zas82;85], solved the plane elasticity problem using the fundamental solution for the full plane. In this context we mention also Wendland [Wen65a,b;68]. A fundamental result was obtained in 1968 by R. Arantes e Oliveira [Oli68], who distributed the singularities on a contour external to the contour of the body under consideration. Hartmann [Har81] gives the most general presentation of the Indirect Method, based on the vector potentials of the first and second kind. In contrast to several statements found in the literature, there is no lack of physical meaning in the indirect B.E.M. A very enlightening case is Massonnet's derivation who solved the plane problem of elasticity using as fundamental solution the "Flamant distribution", i.e. the radial distribution of stresses due to a concentrated force applied to a half-plane. The physical meaning of the indirect B.E.M., as first presented by Rieder [Rie68], is also described by Jaswon [Jas84].

Among those who first developed indirect boundary element techniques one can also mention here Smith and Pierce [Smi58], Hess [Hes62], Hess and Smith [Hes64], Jaswon [Jas63a], Symm [Sym63], Jaswon and Ponter [Jas63b], Harrington et al. [Has69], Mc-Donald and Wexler [McD72] and Silvester and Hsieh [Sil71] for their work on potential theory and Massonnet [Mas65], Rim and Henry [Rim67], Jaswon et al. [Jas67], Oliveira [Oli68], Butterfield and Banerjee [But70;71], Benjumea and Sikarskie [Ben72], Banerjee [Ban76], Banerjee and Butterfield [Ban77], Crouch [Cro73;76] and Glahn [Gla79] for their work in elastostatics.

The direct B.E.M. was first introduced in an explicit and general form by Jaswon [Jas63a] in 1963 in connection with the potential theory and by Rizzo [Riz67] in 1967 in connection with elastostatics, even though one can observe here that the works of Friedman and Shaw [Fri62, Sha62a] in acoustics, Banaugh and Goldsmith [Bah63a,b] in acoustics and steady elastic waves and Cruse [Cru69] on three-dimensional elastostatics are essentially based on the same idea as the direct B.E.M. In this context we also mention Christiasnsen [Chr75a,b].

Cruse and Rizzo [Cru68a], and Cruse [Cru68b] were the first to achieve numerical solutions of general transient problems in elastodynamics via the B.I.E.M. More precisely, those researchers, following the work of Rizzo [Riz67] in elastostatics, employed a direct approach with displacements and tractions appearing as the explicit unknowns and in conjunction with the Laplace transformation were able to solve a transient wave propagation problem in the half-plane. Manolis and Beskos [Man80;81;83a] improved this methodology even further and solved transient wave scattering problems. Niwa et al. [Niw85;86] and Kobayashi and Nishimura [Kob82a;83b] obtained steady-state, two-dimensional wave scattering type solutions which they synthesized with the aid of the Fourier transform to obtain the transient response. The work of Dominguez [Dom78a,b] is a first attempt to study the dynamic response of two- and three-dimensional, rigid surface and embedded foundations by the B.E.M. in the frequency domain.

Among the first papers on free vibration analysis are those of Tai and Shaw [Tai74], De May [Dem76], and Hutchinson [Hut78] in connection with the scalar wave equation governing both acoustics and anti-plane strain elastodynamics. Integral equation approaches were also presented for the plate vibration problems by Vivoli [Viv72], Vivoli and Filippi [Viv74], Niwa et al. [Niw81a] and for free plane elastodynamic problems by Niwa et al. [Niw81a,b;82a,b,c].

Time-domain based boundary integral equation formulations were first introduced by Friedman and Shaw [Fri62] in acoustics. The work of Cole et al. [Col78] in elastodynamics, in spite of its generality, is restricted from an application point of view to anti-plane strain cases. Among the first to develop general boundary element methodologies in the time domain for two and three dimensions were Niwa et al. [Niw80], Manolis [Man83b], Mansur [Ma83a], Antes [Ant85], and Karabalis and Beskos [Kar84;86], respectively. A wealth of information on the B.E.M. for problems in elastodynamics can be found in review articles such as those of Dominguez and Alarcon [Dom81], Geers [Gee83], Dominguez [Dom85], Kobayashi [Kob85;87], Karabalis and Beskos [Kar87a,b], Dominguez and Abascal [Dom87], Banerjee et al. [Ban87], Dravinski [Dra88], Tassoulas [Tas88], and especially Beskos [Bes87b;92]. Some material can also be found in the textbooks of Banerjee and Butterfield [Ban81], Brebbia et al. [Bre84a] and Beskos [Bes87a;88].

Applications of the B.E.M. abound in many areas of engineering mechanics such as in general potential theory [Jas63a, Sym63], potential fluid flow [Smi58, Hes62;64], acoustics [Fri62, Sha62a,b, Ban63a], torsion of shafts [Jas63b, Mas65, Men73, Ker76], electric and magnetic field theory [Har69, McD72, Sil71], elastostatics [Mas65, Jas67, Rim67, Riz67, Oli68, Cru69, But70;71, Ben72, Cro73, Cru73, Ban76, Cro76], elastodynamics [Ban63b, Cru68a, Mai91b], plates and shells [Jas68, New68, For69], transient heat conduction [Riz70], viscoelasticity [Riz71], plasticity [Swe71, Ric73, Men73, Kum77, Ban78, Muk78, Mai87, Pol88], water waves [Jas63a, Sha70, Lee71], viscous fluid flow [Wu73;74], ground water flow [New76, Lig77] and thermoelasticity [Riz77]. This list of references is by no means complete containing only titles of some of the early papers on each area.

Of special interest is the evolution of the integral equation approach to fracture problems due to the singularities caused by the cracks. The mathematical and numerical methods developed in this area of mechanics contributed a great deal to the further development of the B.I.E.M. Here we give certain references on the numerical methods only and we mention the works of Erdogan et al. [Erd72;73], Theocaris [The76;77a,b,c; 78a,b;79a,b,c,d], Cruse et al. [Cru71, Sny75], Krenk [Kre75;81] etc. The most systematic research on this area is due to Theocaris [The80a,b;81a,b,c,d;82a,b,c,d,e; 83a,b,c,d; 84a,b] and his coworkers Ioakimidis [Ioa76;77a,b; 79a,b,c;80a,b,c;82], Tsamasphyros [Tsa77;79; 80;81a,b; 82; 83a,b], Bartzokas [Bar89a,b], Kazantzakis [Kaz] et al. who have studied in detail both for the mathematical and the numerical point of view many types of integral equations correcting several misconceptions of the earlier researchers thus offering a reliable mathematical base to those that are to follow future research in this area. Concerning the applications to crack problems we refer to the review article by Theocaris [The81c,d].

As the first books concerning the mechanical aspects of the B.I.E.M. can be considered the monograph by Rieder [Rie62] and the monograph by Kupradze et al. [Kup65]. The first applied book on the B.E.M. and its various numerical applications appeared in 1975 as a collection of review-type articles edited by Cruse and Rizzo [Cru75]. Jaswon and Symm [Jas77] in 1977 published the first and Brebbia [Bre78b] in 1978 the second book on the method as it is applied to potential theory and elastostatics, while Brebbia and Walker [Bre80a] in 1980 revised and expanded it by adding a chapter on time-dependent and nonlinear problems. The book of Banerjee and Butterfield [Ban81], published in 1981, represents the first comprehensive work on the B.E.M. and its applications in various fields of engineering sciences. The more recent (1984) book of Brebbia, Telles and Wrobel [Bre84a] is another general work on the same subject. The boundary elements literature is recently growing up and is enriched with books devoted to specialized subjects, such as those of Mukherjee [Muk82] on creep and fracture, Parton and Perlin [Par82] and Crouch and Starfield [Cro83] on solid mechanics, Liggett and Liu [Lig83] on porous media flow, Telles [Tel83] on inelastic problems, Venturini [Ven83] on geomechanics, Hromadka [Hro84], and Ingham and Kelmanson [Ing84] on potential theory, Balas and J.&V. Sladek [Bal89] on stress analysis, Ciskowski and Brebbia [Cis91] on acoustics, Manolis and Beskos [Man88] on elastodynamics, Beskos [Bes91] on plates and shells, Kitahara[Kit] on eigenvalue problems in elastodynamics and thin plates, and Takahashi [Tak] on classical contact problems. In an effort to provide information to scientists and engineers on the most recent developments in the field of boundary elements in the form of review-type articles, publication of two series of volumes has started and is continuing under the editorship of Banerjee et al. [Ban78;82;84;86] and Brebbia [Bre81b;83a;84c;85]. Other sources of information on the B.E.M. and its various applications, besides the numerous scientific and technical journal articles, are the proceedings of a number of international conferences that have been edited by Brebbia et al. [Bre78b;80b;81a;82;83b;84b;87;88;90;91a,b].

The above list of publications is by no means exhaustive and concerns mainly the evolution of the B.E.M. from the standpoint of mechanics and engineering. Concerning the evolution of the related mathematical questions we refer to [Kr] [Prö] [Del74;85] [Ruo88;89] and to the references given there, and concerning the relation to the theory of wavelets to [Dav] and to the references given there.

Until very recently the B.I.E.M / B.E.M was applied only to equality or bilateral problems, i.e. to problems which do not have inequality constraints or any multivaluedness (i.e. complete vertical branches) in the stress-strain and/or reaction- displacement laws. We refer to the next Section and to [Pan85, Mbi88a,b] for a more precise definition of the term bilateral problems and the opposite term to it of "unilateral problems". All the aforementioned references on the B.E.M. concern only the treatment of bilateral or equality problems through the B.E.M., which leads to integral equation formulations on the boundary of the body. Exception is only the linear complementarity (L.C.P.) approach to plasticity [Com] which has certain affinity with the problems treated here. The B.I.E.M/B.E.M. when extended to unilateral or inequality problems leads to multivalued boundary integral equations or equivalently to variational or hemivariational inequalities on the boundary. We refer in this context to [Pan83b;85;87;91b, Pp87;89]. The present work is a first attempt to present how the B.I.E.M. is extended to treat inequality or unilateral problems. We focus our efforts to a special but very important class of inequality problems, the contact problems of deformable bodies.

1.2 Elements of Nonsmooth Analysis

1.2.1 Elements of Nonsmooth-Convex Analysis

We deal here with functionals taking values in the extended real line $\bar{\mathbb{R}} = \mathbb{R} \cup \{\pm\infty\}$ $= [-\infty, +\infty]$. Let us consider a convex set K, subset of a Hilbert space X, and a convex real valued functional $f: K \to \mathbb{R}$. We assume that the definition of a convex set and of a convex or strictly convex real valued functional are known. Then we can extend f to all of X by setting $\bar{f}(x) = \{f(x) \text{ for } x \in K, \infty \text{ for } x \notin K\}$ and thus we can limit our attention to functionals defined on all of X. To every convex set K we can associate a functional $I_K: X \to \bar{\mathbb{R}}$, called the indicator of K, which is defined by

$$I_K(x) = \begin{cases} 0 & \text{for } x \in K \\ \infty & \text{for } x \notin K. \end{cases} \tag{1.1}$$

With respect to a functional $f: X \to \bar{\mathbb{R}}$ we may define the epigraph set epi f by the expression

$$\text{epi } f = \{(x, \lambda) | f(x) \leq \lambda, \ \lambda \in \mathbb{R}, \ x \in X\} \tag{1.2}$$

By definition a functional f taking values on the extended real line is convex, if and only if the epi f is convex on $X \times \mathbb{R}$. Obviously the indicator I_K is a convex functional. The effective domain of the convex functional f on X is defined by

$$D(f) = \{x | x \in X, \ f(x) < \infty\} \tag{1.3}$$

A functional f is called proper if $f: X \to (-\infty, +\infty]$ and $f \not\equiv \infty$. Note that for f_1 and f_2 convex, $f_1 + f_2$ is convex as well. (Here we define that $(f_1 + f_2)(x) = \infty$ for $f_1(x) = -f_2(x) = \pm\infty$ cf. [Eke] p.67). A functional $f: X \to \bar{\mathbb{R}}$ is called lower semicontinuous (l.s.c) on X if and only if epi f is a closed subset of $X \times \mathbb{R}$. For f l.s.c, $-f$ is upper semicontinuous and conversely.

Suppose now that X is generally a locally convex Hausdorff topological vector space (L.C.H.T.V.S) and T a topology on X compatible with the duality between X and its dual space X'. Then a convex subset of (resp. a convex functional on) X which is closed (resp. l.s.c) with respect to the topology T is closed (resp. l.s.c) with respect to every other topology compatible with the duality.

In a Hilbert space (H-space) X a convex, l.s.c functional $f: X \to \bar{\mathbb{R}}$ is continuous in the $\text{int} D(f)$ (interior of $D(f)$). For the proof of these results and of the forthcoming results we refer to the classical treatises on Convex Analysis [Rock, Eke, Mor67]. Let us give now some other useful definitions. If A is a subset of X then the set of all linear combinations $\sum_i \lambda_i x_i$, $x_i \in A$ with $\sum \lambda_i = 1$, $i = 1, \ldots, n$, is called the "affine hull" of A. If additionally $\lambda_i \geq 0$ the set is called the convex hull of A and is denoted by $\text{co} A$. A point $x \in A$ is said to be a relative interior point of A if it is an interior point of A when A is regarded topologically as a subset of its affine hull. The set of all relative interior points of A is denoted as $\text{relint} A$.

Convexity is of importance in the study of optimization problems, as will become clear with the following results.

We consider the minimization problem of a convex functional on a convex set $K \subset X$, where X is a H-space. We seek a point x_0 which is a solution to the problem

$$f(x_0) = \inf\{f(x) | \ x \in K\} \text{ or } f(x_0) = \inf_K f(x). \tag{1.4}$$

If f achieves its infinum on K for $x = x_0 \in K$, we can write

$$f(x_0) = \min\{f(x) | \ x \in K\} \text{ or } f(x_0) = \min_K f(x). \tag{1.5}$$

Every x_0 satisfying (1.5) is a solution of the optimization problem. Let K be a nonempty convex, closed subset of X and f a convex, l.s.c proper functional $f\colon K \to \bar{\mathbb{R}}$. It is obvious that f can be extended to all of X, and hence the solution of (1.5) is sought in X. The following proposition concerns the existence of a minimum over K.

Proposition 1.1 Let X be a H-space with the norm $\|\cdot\|$ and

$$\lim f(x) = \infty \text{ when } \|x\| \to \infty, \quad x \in K \subset X, \tag{1.6}$$

or let

$$K \text{ be bounded.} \tag{1.7}$$

Then problem (1.5) admits at least one solution. If f is strictly convex, the solution is unique.

The solutions of problem (1.5) constitute a convex closed subset of X, as can be easily proved by considering the convexity and the lower semicontinuity of f. This subset is nonempty if Prop. 1.1 is valid. Some variational inequalities equivalent to problem (1.5) will now be obtained. X is a H-space and $\langle \cdot, \cdot \rangle$ denotes the duality pairing with the dual space X'.

Proposition 1.2 Let $f = f_1 + f_2$ be a proper functional, where f_1 and f_2 are convex, l.s.c. functionals on K and suppose, that grad f_1 exists on X. For $x_0 \in K$, the following conditions are equivalent to each other:

$$f(x_0) = \inf_K f(x); \tag{1.8}$$

$$\langle \operatorname{grad} f_1(x_0), x - x_0 \rangle + f_2(x) - f_2(x_0) \geq 0, \quad \forall x \in K; \tag{1.9}$$

and

$$\langle \operatorname{grad} f_1(x), x - x_0 \rangle + f_2(x) - f_2(x_0) \geq 0, \quad \forall x \in K. \tag{1.10}$$

Convex functionals $f\colon X \to \bar{\mathbb{R}}$ are not necessarily differentiable. Then the supporting hyperplanes to the epi f describe the differential properties of f. This leads to the notion of subdifferential. The vector $x' \in X'$, for which

$$f(x_1) - f(x) \geq \langle x', x_1 - x \rangle, \quad \forall x_1 \in X \tag{1.11}$$

holds, where $f(x)$ is finite at $x \in X$, is called the subgradient of f at x. The set of all $x' \in X'$ satisfying (1.11) is called the subdifferential of f at x and is denoted by $\partial f(x)$. We then write

$$x' \in \partial f(x). \tag{1.12}$$

The set $\{x | \partial f(x) = \emptyset\}$ is denoted by $D(\partial f)$, and is called the domain of ∂f.

The mapping $\partial f\colon X \to X'$ is multivalued and is called the subdifferential of f. If $\partial f(x) \neq \emptyset$, f is said to be subdifferentiable at x, $\partial f(x) = \emptyset$ for $x \notin D(f)$ and $f \not\equiv \infty$. From (1.11), it follows readily that a necessary and sufficient condition in order that x_0 minimize f on X is that

$$0 \in \partial f(x_0). \tag{1.13}$$

This condition indicates the role of subdifferentials in optimization theory. The existence of subdifferentials is ensured in the case of convex functionals by means of the following result.

Proposition 1.3 Let $f\colon X \to \bar{\mathbb{R}}$ be convex, and suppose that f is finite and continuous at $x_0 \in X$. Then $\partial f(x_0) \neq \emptyset$. Moreover $\partial f(x)$ is nonempty for every $x \in \mathrm{int}\, D(f)$.

The special case $f = I_K$, where K is a nonempty convex subset of X, is very important. Then

$$\partial I_K(x) = \{x' | I_K(x_1) - I_K(x) \geq \langle x', x_1 - x \rangle, \ \forall x_1 \in X\} \tag{1.14}$$

or, equivalently,

$$\partial I_K(x) = \{x' | \langle x', x_1 - x \rangle \leq 0, \ \forall x_1 \in K, \ \text{for } x \in K\}. \tag{1.15}$$

The geometrical meaning of the variational inequality

$$\langle x', x_1 - x \rangle \leq 0, \quad \forall x_1 \in K, \quad x \in K \tag{1.16}$$

is that x' is an outward normal vector to K at x. In general, the set of all vectors x' satisfying (1.16) forms an outward normal cone to K at x. This cone (a) is empty provided $x \notin K$ (b) has at least the zero element if $x \in K$, and (c) has only the zero element if $x \in \mathrm{relint}\, K$.

Subdifferentiability is closely related to the notion of "one-sided Gâteaux- differentiability". This provides a method for the "construction" of the subdifferential for a given functional.

A functional $f\colon X \to \bar{\mathbb{R}}$, where X is a H-space is said to be one-sided directional Gâteaux-differentiable at x_0 if there exists $\tilde{f}'(x_0, h)$ such that

$$\lim_{\mu \to 0_+} \frac{f(x_0 + \mu h) - f(x_0)}{\mu} = \tilde{f}'(x_0, h), \quad \forall h \in X. \tag{1.17}$$

It should be noted that $+\infty$ and $-\infty$ are allowed as limits in (1.17). Functional $h \to \tilde{f}'(x_0, h)$ is the one-sided directional Gâteaux-differential of f at x_0 with respect to the direction h. If $\tilde{f}'(x_0, h) = -\tilde{f}'(x_0, -h)\, \forall h \in X$, then f is Gâteaux-differentiable at x_0. It can be readily shown that $\tilde{f}'(x_0, \cdot)$ is a convex, positively homogeneous function of h. One important property of convex functionals is their one-sided directional Gâteaux-differentiability.

Proposition 1.4 Assume that $f\colon X \to \bar{\mathbb{R}}$ is convex. Then f is one-sided directional Gâteaux-differentiable at every $x \in X$ with $f(x) \neq \pm\infty$. Moreover

$$f(x_1) - f(x) \geq \tilde{f}'(x, x_1 - x), \quad \forall x_1 \in X \tag{1.18}$$

and

$$\tilde{f}'(x, x_1 - x) \geq -\tilde{f}'\big(x, -(x_1 - x)\big), \quad \forall x_1 \in X. \tag{1.19}$$

Suppose further that f is bounded on a neighborhood of $x_0 \in X$. Then

$$\tilde{f}'(x_0, h) = \max\{\langle x', h \rangle | x' \in \partial f(x_0)\}, \quad \forall h \in X. \tag{1.20}$$

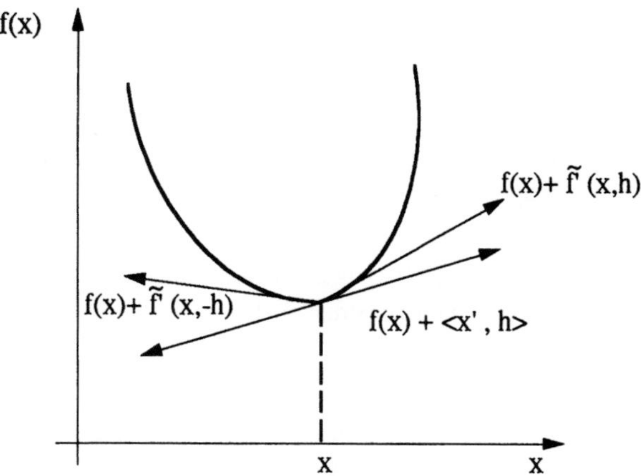

Fig. 1.1: The geometrical meaning of (1.18) and (1.19)

This last proposition permits a simple construction (fig. 1.1) of the set $\partial f(x)$. As we can see, if f maps $\mathbb{R}$ into $\bar{\mathbb{R}}$ then the subgradients x' are the slopes of the nonvertical lines through $(x, f(x))$, which have no point in common with intepi f. From $\tilde{f}'(x, 1) = f'_+(x)$ and $\tilde{f}'(x, -1) = -f'_-(x)$ (right and left derivatives), it follows that $f'_-(x) \leq x' \leq f'_+(x)$. Assume that f is a convex, l.s.c., proper functional on $\mathbb{R}$. In this case the right and left derivatives f'_+ and f'_- can be extended, when $x \notin D(f)$, by setting $f'_+ = f'_- = \infty$ (resp. $f'_+ = f'_- = -\infty$) for points lying to the right (resp. to the left) of $D(f)$. We may then write for $f: \mathbb{R} \to \bar{\mathbb{R}}$

$$\partial f(x) = \{x' \in \mathbb{R} \,|\, f'_-(x) \leq x' \leq f'_+(x)\}. \tag{1.21}$$

Proposition 1.5 Let $f: X \to \bar{\mathbb{R}}$ be convex and suppose that grad $f(x)$ exists at x. Then $\partial f(x) = \{\text{grad } f(x)\}$. Conversely, if f is finite and continuous at x and if $\partial f(x)$ has only one element, then grad f exists at x and $\partial f(x) = \{\text{grad } f(x)\}$.

Concerning simple examples illustrating the notion of the subdifferential we refer to [Rock, Pan85]. Now the addition rule for subdifferentials is given.

Proposition 1.6 Let $f_1: X \to \bar{\mathbb{R}}$ and $f_2: X \to \bar{\mathbb{R}}$. Then for every $x \in D(\partial f_1) \cap D(\partial f_2)$

$$\partial f_1(x) + \partial f_2(x) \subset \partial(f_1 + f_2)(x). \tag{1.22}$$

Assume further that grad f_2 exists at the point x. Then if $x \in D(\partial f_1)$, (1.22) holds as a set equality.

The following example illustrates the addition rule. Let f be a convex functional on X such that grad f exists everywhere, and let $K \neq \emptyset$ be a convex subset of X. The minimization problem of f over K is equivalent to the minimization of $f + I_K$ over all of X. For x_0 to be a solution of this problem, it is necessary and sufficient (cf. (1.13)) that

$$0 \in \partial(f(x_0) + I_K(x_0)).$$

which is equivalent (Prop. 1.6) to the relation

$$-\operatorname{grad} f(x_0) \in \partial I_K(x_0). \tag{1.23}$$

For further results of the subdifferential calculus we refer to [Rock, Eke].

We denote by $R(\partial f)$ the range of ∂f, i.e., the set $\{x'|x' \in \partial f(x), x' \in X', x \in X\}$. Further we shall give a proposition on the relative interior of $R(\partial f)$, due to Schatzman [Scha]

Proposition 1.7 (a) Suppose that $\mathcal{G}$ is a finite dimensional space and f is convex, l.s.c., proper functional on $\mathcal{G}$. Then in order that

$$0 \in \operatorname{relint} R(\partial f), \tag{1.24}$$

the following conditions are necessary and sufficient: that $\mathcal{G}$ be the direct sum of the vector spaces $\mathcal{G}_1$ and $\mathcal{G}_2$, that f be the invariant on $\mathcal{G}_1$ and that there exist constants $c_1 > 0$ and c_2 such that

$$f(q) \geq c_1|q|_{\mathcal{G}} + c_2, \quad \forall q \in \mathcal{G}_2. \tag{1.25}$$

(b) Let X be a H-space , $\mathcal{G}$ a finite dimensional subspace of X, $\mathcal{G}'$ its dual $x' \in X'$, and $x'|_{\mathcal{G}} \in \mathcal{G}'$. Further, let Φ be a convex, l.s.c. and proper functional on X. Then in order that

$$x'|_{\mathcal{G}} \in \operatorname{relint} R(\partial \Phi_{\bar{x}}), \tag{1.26}$$

where $\Phi_{\bar{x}}$ is a convex, l.s.c. and proper functional on $\mathcal{G}$ defined by

$$\Phi_{\bar{x}}(q) = \Phi(\bar{x} + q) \quad \text{for } \bar{x} \in D(\Phi), \tag{1.27}$$

it is necessary and sufficient that $\mathcal{G}$ be the direct sum of the vector spaces $\mathcal{G}_1$ and $\mathcal{G}_2$ as in a), such that the function $F: x \rightarrow \Phi(x) - \langle x', x \rangle$ is invariant on $\mathcal{G}_1$, and that there exist constants $c_3 > 0$ and c_1, c_2 such that for any $\tilde{x} \in \tilde{\mathcal{G}}$ where $\tilde{\mathcal{G}}$ is closed and $X = \mathcal{G} \oplus \tilde{\mathcal{G}}$, the inequality

$$\Phi(\tilde{x} + q) - \langle x', q \rangle_X \geq c_1 - c_2\|\tilde{x}\|_X + c_3\|q\|_X, \quad \forall q \in \mathcal{G}_2 \tag{1.28}$$

holds.

Note that a consequence of this proposition is that the sets $\operatorname{relint} R(\partial \Phi_x)$ and $\overline{R(\partial \Phi_{\bar{x}})}$ are independent of $\bar{x}$, if $\bar{x} \in D(\Phi)$. Moreover, by taking into account that the closure of $R(\partial \Phi)$ is identical to the closure of $\operatorname{relint} R(\partial \Phi)$ [Rock], we find that the same is true for $\overline{R(\partial \Phi_{\bar{x}})}$.

Let $\{X, X'\}$ be a dual pair of H-spaces and $\langle x', x \rangle$ the duality pairing. For a functional $f: X \rightarrow \bar{\mathbb{R}}$ there arises the question as to whether an affine continuous function $\langle x', \cdot \rangle - \mu, \mu \in \mathbb{R}$, exists which is a minorant of f, i.e.

$$f(x) \geq \langle x', x \rangle - \mu, \quad \forall x \in X.$$

A necessary and sufficient condition is that

$$\mu \geq \sup_{x \in X} \left(\langle x', x \rangle - f(x) \right).$$

This relation introduces the conjugate functional f^c, which is defined on X' by the relation

$$f^c(x') = \sup_{x \in X} \left(\langle x', x \rangle - f(x) \right). \tag{1.29}$$

f^c can be regarded as the pointwise supremum of the family of affine continuous functionals $g(\cdot) = \langle \cdot, x \rangle - \mu$ with $(x, \mu) \in \text{epi} f$. Let us denote by $\Gamma(X)$ the set of functions $f: X \to \bar{\mathbb{R}}$, which are the pointwise suprema of a family of affine continuous functionals $\langle x', \cdot \rangle + \alpha$, $\alpha \in \mathbb{R}$ on X. The following propositions hold.

Proposition 1.8 The class $\Gamma(X)$ consists exactly of the convex, l.s.c., proper functionals $f: X \to \bar{\mathbb{R}}$ and of the constants $+\infty$ and $-\infty$.

Further we denote by $\Gamma_0(X)$ the set of functionals $f \in \Gamma(X)$ such that $f \not\equiv \pm\infty$. It is obvious that $\Gamma_0(X)$ consists precisely of the convex, l.s.c. and proper functionals on X.

Proposition 1.9 Let f be a convex functional on X. Then f^c is a convex, l.s.c. functional on X'. If, moreover, f is proper, then f^c is proper as well, and conversely. Moreover $f^{cc} = f$.

The conjugacy operation $f \to f^c$ can be considered as a one-to-one correspondence between $\Gamma_0(X)$ and $\Gamma_0(X')$ and is called Fenchel transformation (also Fenchel-Young or Legendre-Fenchel or polarity transformation).

The following proposition exhibits the relation between ∂f and ∂f^c.

Proposition 1.10 Assume that f is a convex, proper functional on X. The following conditions are equivalent to one another.

(i) $x' \in \partial f(x);$ (1.30)

(ii) $\sup_{z \in X} \left(\langle x', z \rangle - f(z) \right)$ is achieved at $z = x;$ (1.31)

(iii) $f(x) + f^c(x') \leq \langle x', x \rangle;$ and   (1.32)

(iv) $f(x) + f^c(x') = \langle x', x \rangle.$ (1.33)

     If, additionally, f is l.s.c., these conditions are equivalent to:

(v) $x \in \partial f^c(x');$ and   (1.34)

(vi) $\sup_{z' \in X'} \left(\langle z', x \rangle - f^c(z') \right)$ is achieved at $z' = x'$.   (1.35)

Suppose that $f = I_K$, where K is a nonempty, convex subset of X. Then I_K^c is given on X' by

$$I_K^c(x') = \sup_{x \in K} \langle x', x \rangle. \tag{1.35a}$$

I_K^c is called the support function of K. Now let K be a linear subspace M of $\mathbb{R}^n$. Then the supremum is ∞, unless $\langle x', x \rangle = 0$, $\forall x \in M$. Accordingly, $I_K^c = I_{M^\perp}$, where $M^\perp$ denotes the orthogonal complement of M.

Let X and X' be two vector spaces and A a mapping from X into the power set $\mathcal{P}(X')$ of X'. The mapping A is called a multivalued operator or multivalued mapping or multifunction. In this case, from $A(x_1) = x_1'$ and $A(x_1) = y_1'$ it does not follow that $x_1' = y_1'$, as happens with single-valued functions. Considering A as a subset of $X \times X'$, we can write $A(x) = \{y \in X' | (x, y) \in A\}$. The set $D(A) = \{x | x \in X, A(x) \neq \emptyset\}$ is called the domain of A and the set $R(A) = \bigcup_x A(x), x \in X$, the range of A. Because

A is multivalued, we will write $y \in A(x)$, where $x \in D(A)$ and $y \in X'$. If A and B are two multivalued operators on X, then $\lambda A + \mu B$, $\lambda, \mu \in \mathbb{R}$, is a multivalued operator mapping x into $\lambda A(x) + \mu B(x) = \{\lambda y + \mu z | y \in A(x), z \in B(x)\}$. Moreover, $D(\lambda A + \mu B) = D(A) \cap D(B)$. Suppose further that X and X' are dual H-spaces with duality pairing $\langle x', x \rangle$ for $x \in X, x' \in X'$. The multivalued mapping $A \colon X \to \mathcal{P}(X')$ is said to be monotone if

$$\langle y_1 - y_2, x_1 - x_2 \rangle \geq 0, \tag{1.36}$$

$$\forall x_1, x_2 \in D(A), \quad \forall y_1 \in A(x_1), \quad \forall y_2 \in A(x_2).$$

If $\geq$ is replaced by $>$, then A is said to be strictly monotone.

Let f be a convex proper functional on X. Then it can be shown that ∂f is a monotone multivalued function from X into $\mathcal{P}(X')$.

The graph of the multivalued operator $A \colon X \to \mathcal{P}(X')$ is a set $\mathcal{Q}(A) = \{(x,y)|(x,y) \in D(A) \times X', y \in A(x)\}$. Then $\mathcal{Q}(A_1) \subset \mathcal{Q}(A_2)$, if and only if $A_1(x) \subset A_2(x) \forall x \in X$. The set $\mathcal{A}$ of the monotone operators from X into $\mathcal{P}(X')$ can be partially ordered by graph inclusion. It can be shown, furthermore, that every totally ordered subset of $\mathcal{A}$ has an upper bound. Then by means of the Zorn Lemma $\mathcal{A}$ contains at least one maximal element, which is called a maximal monotone operator. Accordingly, a monotone operator $A \colon X \to \mathcal{P}(X')$ is called maximal monotone if and only if $\mathcal{Q}(A) \subset \mathcal{Q}(B)$ implies that $A = B$, where $B \colon X \to \mathcal{P}(X')$ is an arbitrary monotone operator, i.e., if and only if $\mathcal{Q}(A)$ is not properly contained in any other monotone subset of $X \times X'$. From the above we obtain equivalently that an operator $A \colon X \to \mathcal{P}(X')$ is called maximal monotone if and only if i) A is monotone and ii) for every $x \in X$ and $y \in X'$ such that

$$\langle y - y_1, x - x_1 \rangle \geq 0, \quad \forall x_1 \in D(A), \quad \forall y_1 \in A(x_1) \tag{1.37}$$

the relation

$$y \in A(x) \tag{1.38}$$

holds.

The following proposition relates the theory of maximal monotone operators to subdifferentiation.

Proposition 1.11 The subdifferential ∂f of a convex, l.s.c., proper functional f on X, where X is a H-space, is a maximal monotone operator.

The class of the monotone operators $\beta \colon \mathbb{R} \to \mathcal{P}(\mathbb{R})$ is subsequently considered. A complete nondecreasing curve in $\mathbb{R}^2$ is the graph $\mathcal{Q}(\beta)$ of a maximal monotone mapping $\beta \colon \mathbb{R} \to \mathcal{P}(\mathbb{R})$. In a Cartesian coordinate system such a graph is similar to the graph of a continuous nondecreasing function, with the difference that it may contain vertical segments as well. The maximal monotone graphs in $\mathbb{R}^2$ are used for the formulation of unilateral boundary conditions. A proposition now follows relating the complete nondecreasing curves in $\mathbb{R}^2$ and the subdifferentials ∂f of convex, l.s.c and proper functionals on $\mathbb{R}$.

Proposition 1.12 Let $\beta \colon \mathbb{R} \to \mathcal{P}(\mathbb{R})$ be a maximal monotone mapping. A convex, l.s.c., proper functional $f \colon \mathbb{R} \to \bar{\mathbb{R}}$ can be determined up to an additive constant such that

$$\beta = \partial f. \tag{1.39}$$

Accordingly the graphs of the subdifferentials ∂f, where $f \in \Gamma_0(\mathbb{R})$, are precisely
the complete nondecreasing curves of $\mathbb{R}^2$. Note that Prop. 1.12 gives a method for
the determination of f, when β is given. There exists also another simple method
which is called regularization method: Let β be a maximal monotone operator. For
the determination of f, we define a continuous and single-valued function β_ε (called a
regularized operator), which depends on $\varepsilon > 0$. For $\varepsilon \to 0_+$ the graph $\mathcal{Q}(\beta_\varepsilon)$ "coincides"
with the graph $\mathcal{Q}(\beta)$. f_ε results from β_ε through integration (fig. 1.2). From f_ε we
obtain f as $e \to 0_+$.

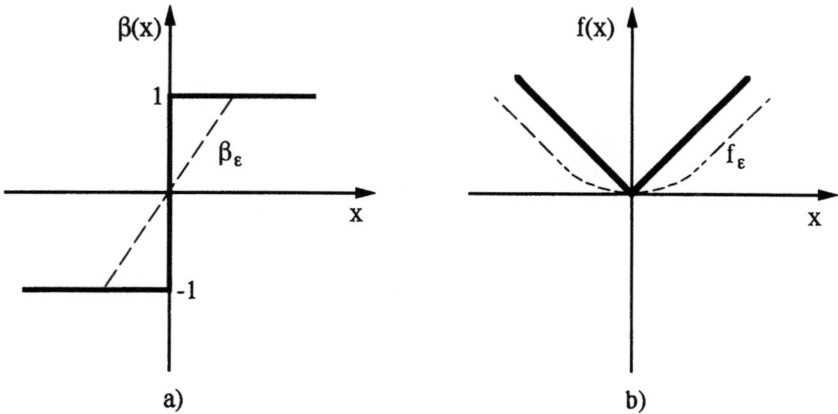

Fig. 1.2: Regularization of $f(x) = |x|, x \in \mathbb{R}$. The functions β and β_ε

1.2.2 Elements of Nonsmooth-Nonconvex Analysis

The notion of subdifferential ∂f has been extended by F.H. Clarke [Clar73;75,81;83]
and by R.T. Rockafellar [Rock79;80] to nonconvex functionals. This new concept is
called "generalized gradient", is denoted here by $\bar{\partial} f$ and has been applied by the second
author to define in Mechanics a new class of variational expressions, the hemivariational
inequalities [Pan81;82;83a,85,91a]. In this context cf. also [Au84;90;91][Mor88a,b].

Let X be a H-space, X' its dual, $\langle \cdot, \cdot \rangle$ the duality pairing between X and X' and
S a subset of a H-space. Suppose generally that $A: S \to X$ is a multivalued operator.
Then as $\tilde{s} \to s$, the set $\liminf A(\tilde{s})$ consists of all $x \in X$ such that for every $Y \in F_x$
there exists $U \in F_s$, and for every $\tilde{s} \in U, Y \cap A(\tilde{s}) \neq \emptyset$. We denote here by F_x (resp. F_s)
a filter of neighborhoods of x (resp. of s). If $\liminf A(\tilde{s}) = A(s)$ for all $s \in S$, then the
multivalued mapping A is called l.s.c. The definition of $\limsup A(\tilde{s})$ is obtained in the
same way. For a function $g: S \times X \to [-\infty, +\infty]$, we define the expression "limsupinf"
as

$$h(s, x) = \limsup_{\tilde{s} \to s} \inf_{\tilde{x} \to x} g(\tilde{s}, \tilde{x}) = \sup_{Y \in F_x} \inf_{U \in F_s} \sup_{\tilde{s} \in S} \inf_{\tilde{x} \in Y} g(\tilde{s}, \tilde{x}). \tag{1.40}$$

For a set $C \in X$ and for $x \in X$, the tangent cone $T_C(x)$ to C at x is by definition

$$T_C(x) = \liminf_{\substack{\tilde{x} \to_C x \\ \mu \to 0_+}} \frac{1}{\mu}(C - \{\tilde{x}\}); \tag{1.41}$$

here $\tilde{x} \to_C x$ means that $\tilde{x} \to x$ with $\tilde{x} \in C$. Equivalently,

$$\begin{aligned}
T_C(x) \;=\; &\{y | y \in X, \text{ for } \mu_n \to 0_+, \text{ and } x_n \to_C x, \\
&\text{there exists } y_n \to y \text{ with } y_n + \mu_n x_n \in C\}.
\end{aligned}$$

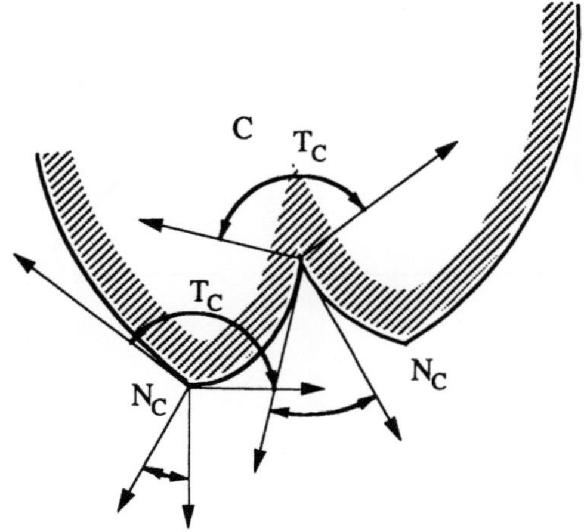

Fig. 1.3: Tangent and normal cones

It may be proved that $T_C(x)$ is a convex closed cone which contains 0. The normal cone $N_C(x)$ to C at x is defined as

$$N_C(x) = \{x' | x' \in X', \langle z, x' \rangle \le 0, \quad \forall z \in T_C(x)\}. \tag{1.42}$$

Obviously,

$$T_C(x) = \{y | y \in X, \langle y, z' \rangle \le 0, \quad \forall z' \in N_C(x)\}. \tag{1.43}$$

If C is convex, the definition (1.42) coincides with the definition (1.15) of the normal cone to a convex set. Let f be a function on X with values in $[-\infty, +\infty]$. The upper subdifferential $f^{\uparrow}(x, y)$ of f at $x (f(x)$ finite) with respect to y is defined by

$$f^{\uparrow}(x, y) = \text{limsupinf} \frac{f(\tilde{x} + \mu \tilde{y}) - \tilde{\alpha}}{\mu} \tag{1.44}$$

as in (1.40). Here S is epi $f \times [0, \infty)$ and

$$g(\tilde{x}, \tilde{\alpha}, \mu, \tilde{y}) = \begin{cases} (f(\tilde{x} + \mu \tilde{y}) - \tilde{\alpha})/\mu & \text{if } \mu > 0 \\ -\infty & \text{if } \mu = 0 \end{cases} \cdot$$

Moreover the "limsupinf" is formed as $(\tilde{x}, \tilde{\alpha}) \to (x, f(x))$ with $f(\tilde{x}) \leq \tilde{\alpha}, \mu \to 0_+$ and $\tilde{y} \to y$. If f is l.s.c. at x, then (1.44) is simplified and the limsupinf is formed as $\tilde{x} \to x, f(\tilde{x}) \to f(x), \tilde{y} \to y$ and $\mu \to 0_+$. Function $g: y \to f^{\uparrow}(x, y)$ is convex, l.s.c. and positively homogeneous. Moreover

$$\text{epi } g = T_{\text{epi } f}(x, f(x)). \tag{1.45}$$

If f is convex, then

$$f^{\uparrow}(x, y) = \liminf_{\tilde{y}' \to y} \tilde{f}'(x, \tilde{y}), \quad \forall y \in X, \tag{1.46}$$

where $\tilde{f}'(x, y)$ is the one-sided directional Gâteaux-differential.

Analogously to $f^{\uparrow}(x, y)$, the lower subdifferential $y \to f^{\downarrow}(x, y)$ is again defined by (1.44) but now with the "limsupinf" replaced by "liminfsup" which is formed as $(\tilde{x}, \tilde{a}) \to (x, f(x))$ with $\tilde{a} \leq f(\tilde{x}), \mu \to 0_+$ and $\tilde{y} \to y$. If f is Lipschitzian on a neighborhood of x, then

$$f^{\uparrow}(x, y) = -f^{\downarrow}(x, -y) = f^0(x, y), \quad \forall y \in X \tag{1.47}$$

where $y \to f^0(x, y)$ is the directional differential of Clarke at x in the direction y and is

$$f^0(x, y) = \limsup \frac{f(\tilde{x} + \mu y) - \tilde{a}}{\mu}. \tag{1.48}$$

Here the "limsup" is formed as $(\tilde{x}, \tilde{a}) \to (x, f(x))$ with $\tilde{a} \geq f(\tilde{x})$ and $\mu \to 0_+$. We recall that f is said to be Lipschitzian at x if a neighborhood U of x exists such that f is finite on U, and for some continuous seminorm p of X

$$|f(x_1) - f(x_2)| \leq c\|x_1 - x_2\|, \quad c = c(U) \text{ const. } > 0, \quad \forall x_1, x_2 \in U. \tag{1.49}$$

f is Lipschitzian at x if it is continuously differentiable at x, or if it is convex (or concave) and finite at x, or if f is the linear combination of Lipschitzian functions at x.

We shall give two equivalent definitions of the generalized gradient $\bar{\partial} f(x)$ of $f: X \to [-\infty, +\infty]$ at $x \in X$ for $f(x)$ finite:

(i) $\bar{\partial} f(x) = \{x' | x' \in X', f^{\uparrow}(x, x_1 - x) \geq \langle x', x_1 - x \rangle, \; \forall x_1 \in X\} \tag{1.50}$

and

(ii) $\bar{\partial} f(x) = \{x' | x' \in X', (x', -1) \in N_{\text{epi} f}(x, f(x))\}. \tag{1.51}$

Note that $\bar{\partial} f(x) = \emptyset$ if $f^{\uparrow}(x, 0) = -\infty$, otherwise $\bar{\partial} f(x) \neq \emptyset$ and for every $y \in X$

$$f^{\uparrow}(x, y) = \sup\{\langle y, x' \rangle | x' \in \bar{\partial} f(x)\}. \tag{1.52}$$

Generally, $\bar{\partial} f(x)$ is a convex closed subset of X' for the $\Sigma(X', X)$-topology. If, moreover, $f^{\uparrow}(x, y)$ is finite for every y, then $\bar{\partial} f(x)$ is a nonempty $\Sigma(X', X)$-compact subset of X', and conversely. Then (1.52) is valid with "sup" replaced by "max". If I_C denotes the indicator of the set C, then it may easily be verified that

$$\bar{\partial} I_C(x) = N_C(x) \tag{1.53}$$

and

$$I_C^{\uparrow}(x,y) = I_{T_C(x)}(y). \tag{1.54}$$

If f is convex (resp. concave and bounded below on a neighborhood of x), then

$$\bar{\partial} f(x) = \partial f(x) \tag{1.55}$$

resp.

$$\bar{\partial} f(x) = -\partial(-f)(x) \tag{1.56}$$

at every x where f is finite. For grad $f(\cdot)$ continuous at x,

$$\bar{\partial} f(x) = \{\text{grad } f(x)\}. \tag{1.57}$$

It is easily verified through (1.44) and (1.50) that if f has at x a local minimum, then

$$0 \in \bar{\partial} f(x). \tag{1.58}$$

f is called substationary at x if (1.58) holds [Rock79]. Local minima and a large class of local maxima are substationarity points, but the converse is not always true. We say that x is a substationarity point of f with respect to a closed set K if $f + I_K$ is substationary at x.

Suppose now that f is a maximum-type function, i.e., $f = \max\{\varphi_1, \ldots, \varphi_m\}$ where $\varphi_i = \varphi_i(x)$ $i = 1, \ldots, m$, $x \in \mathbb{R}^n$ are smooth functions. We denote the sets $\{x|\varphi_i = f\}$ by A_i. It is easy to verify that

$$\bar{\partial} f(x) = \{\text{grad } \varphi_i(x)\} \quad \text{if } x \in A_i,$$

$$\bar{\partial} f(x) = \text{co}\{\text{grad } \varphi_i(x), \text{grad } \varphi_j(x)\} \quad \text{if } x \in A_i \cap A_j$$

and

$$\bar{\partial} f(x) = \text{co}\{\text{grad } \varphi_i(x), \text{grad } \varphi_j(x), \text{grad } \varphi_k(x)\} \quad \text{if } x \in (A_i \cap A_j) \cap A_k, \quad \text{etc.}$$

If

$$f^{\uparrow}(x,y) = \tilde{f}'(x,y) \tag{1.59}$$

for every $y \in X$, f is called $\bar{\partial}$-regular. This is the case if f is convex or a maximum-type function. We have for f, g Lipschitz functions on X

$$\bar{\partial}(f + g)(x) \subset \bar{\partial} f(x) + \bar{\partial} g(x) \tag{1.60}$$

Relation (1.60) holds as a set equality if f and g are $\bar{\partial}$-regular.

Suppose finally that $C = \{x \in \mathbb{R}^n \,|\, f(x) \leq 0\}$. Then at a point x_0 with $f(x_0) = 0$

$$N_C(x_0) \subset \{\lambda x'|x' \in \mathbb{R}^n, \lambda \geq 0, x' \in \bar{\partial} f(x_0)\}, \tag{1.61}$$

whenever f is Lipschitzian on a neighborhood of x_0 and $0 \notin \bar{\partial} f(x_0)$. If, moreover, f is $\bar{\partial}$-regular at x_0, then (1.61) holds as an equality.

Suppose that $\beta : \mathbb{R} \to \mathbb{R}$ is a function such that $\beta \in L^{\infty}_{\text{loc}}(\mathbb{R})$. For any $\rho > 0$ and $\xi \in \mathbb{R}$ let us define

$$\bar{\beta}_{\rho}(\xi) = \operatorname*{ess\,inf}_{|\xi_1 - \xi| \le \rho} \beta(\xi_1) \quad \text{and} \quad \bar{\bar{\beta}}_{\rho}(\xi) = \operatorname*{ess\,sup}_{|\xi_1 - \xi| \le \rho} \beta(\xi_1) \tag{1.62}$$

Obviously the monotonicity properties of $\rho \to \bar{\beta}_{\rho}(\xi)$ and $\rho \to \bar{\bar{\beta}}_{\rho}(\xi)$ imply that

$$\bar{\beta}(\xi) = \lim_{\rho \to 0_+} \bar{\beta}_{\rho}(\xi) \quad \text{and} \quad \bar{\bar{\beta}}(\xi) = \lim_{\rho \to 0_+} \bar{\bar{\beta}}_{\rho}(\xi) \tag{1.63}$$

exist. Let us define the multivalued function

$$\tilde{\beta}(\xi) = [\bar{\beta}(\xi), \bar{\bar{\beta}}(\xi)] \tag{1.64}$$

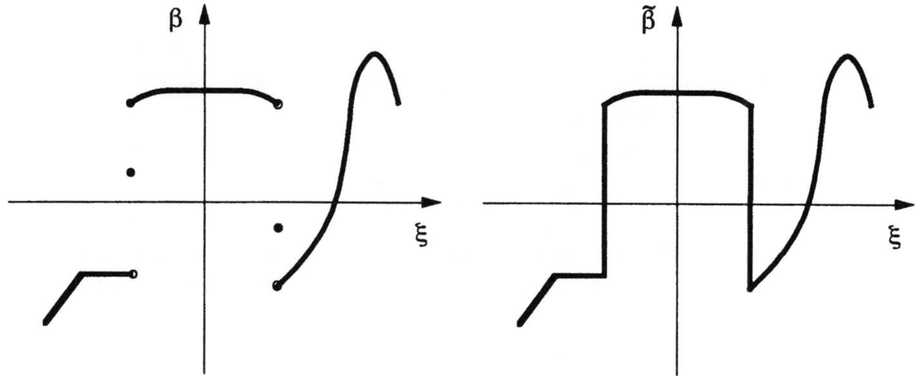

Fig. 1.4: On the definition of β and $\tilde{\beta}$

where $[\cdot, \cdot]$ denotes simply the interval. Roughly speaking (fig. 1.4) $\tilde{\beta}$ results from the generally discontinuous function β by "filling in the gaps". For instance if at ξ $\beta(\xi_+) > \beta(\xi_-)$ (resp. $\beta(\xi_+) < \beta(\xi_-)$) then $\tilde{\beta}(\xi) = [\beta(\xi_-), \beta(\xi_+)]$ (resp. $\tilde{\beta}(\xi) = [\beta(\xi_+), \beta(\xi_-)]$). It was proved by Chang [Cha] that a locally Lipschitz function $\tilde{j}$ can be determined up to an additive constant by the relation

$$\tilde{j}(\xi) = \int_0^{\xi} \beta(\xi_1) d\xi_1 \tag{1.65}$$

such that $\bar{\partial}\tilde{j}(\xi) \subset \tilde{\beta}(\xi)$. If moreover $\beta(\xi_{\pm})$ exist for each $\xi \in \mathbb{R}$ then

$$\bar{\partial}\tilde{j}(\xi) = \tilde{\beta}(\xi). \tag{1.66}$$

1.3 Contact Problems

1.3.1 Monotone Multivalued Boundary and Interface Conditions. Pointwise Formulations

Using the definitions given in Sect. 1.2 we may define subdifferential boundary conditions in a deformable body. These boundary conditions include as special cases the classical boundary conditions of mechanics. We denote by Ω an open bounded subset of $\mathbb{R}^3$ which is occupied by a deformable body. The boundary of Ω is denoted by Γ. The points $x \in \Omega, x = \{x_i\}, i = 1,2,3$, are referred to a Cartesian coordinate system. We denote by $S = \{S_i\}$ the stress vector on Γ. $S_i = \sigma_{ij}n_j$, where $\sigma = \{\sigma_{ij}\}$ is an appropriately defined stress tensor and $n = \{n_i\}$ is the outward unit normal vector on Γ. The vector S may be decomposed into a normal component S_N and a tangential component S_T with respect to Γ

$$S_N = \sigma_{ij}n_jn_i \quad \text{and} \quad S_{T_i} = \sigma_{ij}n_j - (\sigma_{ij}n_in_j)n_i. \tag{1.67}$$

Analogously[1] to S_N and S_T, u_N and u_T denote the normal and the tangential components of the displacement vector u with respect to Γ. S_N and u_N are considered as positive if they are parallel to n.

A maximal monotone operator $\beta_i \colon \mathbb{R} \to \mathcal{P}(\mathbb{R})$ is introduced and a boundary condition of the form

$$-S_i \in \beta_i(u_i) \tag{1.68}$$

is considered in the i-th direction. Then (Prop. 1.12) a convex, l.s.c and proper functional j_i on $\mathbb{R}$ may be determined up to an additive constant such that

$$\beta_i = \partial j_i. \tag{1.69}$$

Then (1.68) is written as

$$-S_i \in \partial j_i(u_i). \tag{1.70}$$

This relation is a subdifferential boundary condition and is understood pointwise, i.e., as a relation between $-S_i(x) \in \mathbb{R}$ and $u_i(x) \in \mathbb{R}$ at every point $x \in \Gamma$. Obviously, (1.70) may also be written in the inverse form

$$u_i \in \partial j_i^c(-S_i) \tag{1.71}$$

and

$$u_i \in \beta_i^c(-S_i), \tag{1.72}$$

where $\beta_i^c = \partial j_i^c$ is again a maximal monotone operator on $\mathbb{R}$ and is the inverse operator of β_i. The graph of β_i, referred to a Cartesian system Oxy, is a complete nondecreasing curve in $\mathbb{R}^2$ which is generally multivalued; thus the graph may include segments parallel to both coordinate axes. "Superpotential" j_i (resp. j_i^c) is a local superpotential (resp. conjugate superpotential) and expresses the potential (resp. complementary

---

[1]In the Chapters 2÷6, which deal with the equality contact problems, we use for technical reasons (many indices in the same letter) the indices n and t instead of N and T.

energy) of the contact constraint [Mor68]. The constraint (1.68) may be considered as the material law of a fictive spring of zero length at x in the ith-direction.

Analogously to (1.68), a boundary condition of the form

$$-S_N \in \beta_N(u_N) = \partial j_N(u_N) \tag{1.73}$$

may be defined. Assume that j is a convex, l.s.c., proper functional on $\mathbb{R}^3$. Then a contact relation of the form

$$-S \in \partial j(u) \tag{1.74a}$$

is defined pointwise on Γ, i.e., as a monotone relation between $S(x)$ and $u(x)$. Equivalently to (1.74), we may write

$$u \in \partial j^c(-S) \tag{1.74b}$$

and

$$j(u) + j^c(-S) = -u_i S_i. \tag{1.74c}$$

Similarly to (1.74), a subdifferential law

$$-S_T \in \partial j_T(u_T) \tag{1.75}$$

may be considered.

In some classes of mechanical problems, similar boundary conditions may be defined between S and the partial time derivative of the displacement $\partial u/\partial t$, or the velocity v. We give some examples to illustrate these boundary conditions.

(i) The classical boundary conditions $u_i = 0$ can be put in the form (1.68) through the operator

$$\beta_i(u_i) = \begin{cases} \mathbb{R} & \text{if } u_i = 0 \\ \emptyset & \text{otherwise} \end{cases},$$

or through the functional $j_i(u_i) = \{0 \text{ if } u_i = 0 \text{ and } \infty \text{ otherwise}\}$. The boundary conditions $S_i = C_i$ is written in the form (1.68) or (1.70) with $\beta_i(u_i) = -C_i$ (C_i given) or $j_i(u_i) = -C_i u_i$ (no summation) for every $u_i \in \mathbb{R}$.

(ii) The boundary condition

$$-S_N = k u_N, \ k \text{ const} > 0 \tag{1.76}$$

may be written in the form (1.73) by setting

$$\beta_N(u_N) = k u_N, \quad j_N(u_N) = \frac{1}{2} k u_N^2.$$

This is the Winkler law, which describes in a simplified manner the interaction between a deformable body and the soil. This law is used in practical civil engineering.

(iii) The foregoing boundary condition does not describe the case in which the body loses contact with the support [Pan75]. To do so we should consider the following law:

$$\text{if } \ u_N < 0, \quad \text{then} \quad S_N = 0; \tag{1.77a}$$

$$\text{if } \ u_N \geq 0, \quad \text{then} \quad S_N + k u_N = 0; \quad k \text{ const} > 0. \tag{1.77b}$$

Relation (1.77a) corresponds to the case of noncontact and (1.77b) to the case of contact. The regions of contact and noncontact are not known a priori; thus we have a free B.V.P. The respective operator β_N (resp. j_N) is given by

$$\beta_N(u_N) = \begin{cases} ku_N & \text{if } u_N \geq 0 \\ 0 & \text{if } u_N < 0 \end{cases}$$

and

$$j_N(u_N) = \begin{cases} \frac{1}{2}ku_N^2 & \text{if } u_N \geq 0 \\ 0 & \text{if } u_N < 0 \end{cases}.$$

$j_N(u_N)$ can be written compactly as $\frac{1}{2}ku_{N+}^2$, where u_{N+} denotes the positive part of u_N, i.e., $u_{N+} = \sup\{0, u_N\}$. Relations (1.77) are called conditions of unilateral contact for a linear Winkler law, whereas (1.76) is the condition of bilateral contact.

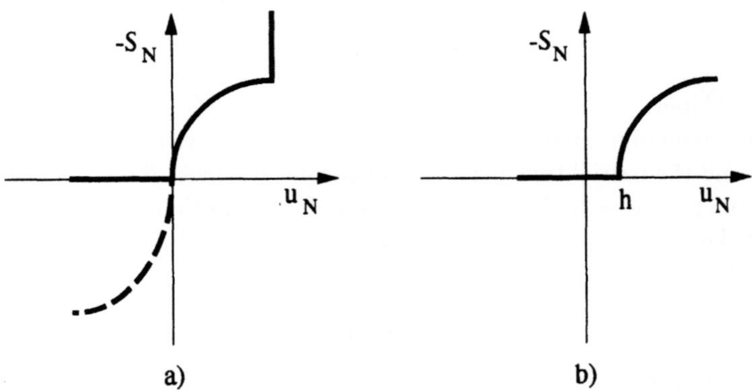

Fig. 1.5: Unilateral contact boundary conditions

We can consider generally the operators

$$\beta_N(u_N) = \begin{cases} \beta_1(u_N - h) & \text{if } u_N \geq h \\ 0 & \text{if } u_N < h \end{cases} \tag{1.78}$$

Here β_1 is assumed to be a maximal monotone operator on $\mathbb{R}$ such that $0 \in \beta_1(0)$. Eq. (1.78) leads to unilateral contact boundary conditions, but with a nonlinear Winkler law and a support at a given distance $h = h(x)$ from the body under consideration (fig. 1.5). The relations are not sufficient to formulate a B.V.P., but they must be combined with a boundary condition concerning S_T or u_T or both, e.g. $S_T = C_T$, where $C_T = C_T(x)$ is given, or $u_T = 0$, or, more generally, (1.75). It is also possible for β_N to change from point to point, in which case $\beta_N = \beta_N(u_N(x), x)$. Note that the uncoupling of the contact conditions in the tangential and in the normal directions is a considerable simplification of the mechanical problem and can be avoided.

(iv) If only the body is deformable while the support is not, then the boundary conditions of Signorini hold [Fich63;64;72, Duv72]. They read (fig. 1.6)

$$\text{if } u_N < 0, \quad \text{then } S_N = 0;$$
$$\text{if } u_N = 0, \quad \text{then } S_N \leq 0, \tag{1.79}$$

or equivalently

$$S_N \leq 0, \quad u_N \leq 0, \quad \text{and} \quad S_N u_N = 0. \tag{1.80}$$

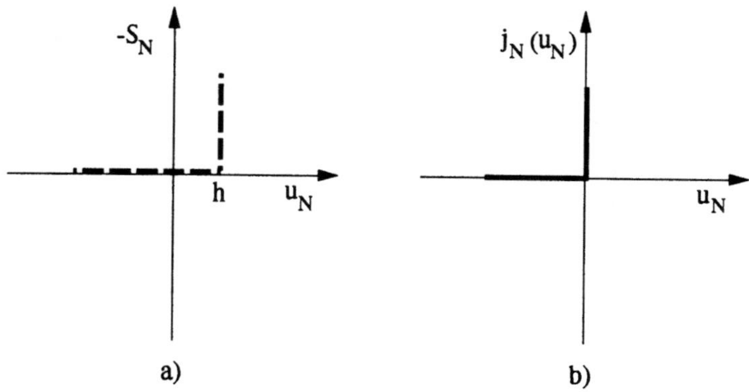

Fig. 1.6: The Signorini contact boundary condition

The respective operator β_N is

$$\beta_N(u_N) = \begin{cases} 0 & \text{if } u_N < 0 \\ [0, +\infty) & \text{if } u_N = 0 \\ \emptyset & \text{if } u_N > 0 \end{cases},$$

and the corresponding superpotential

$$j_N(u_N) = \begin{cases} 0 & \text{if } u_N \leq 0 \\ \infty & \text{if } u_N > 0 \end{cases}$$

To describe the contact with the possibility of debonding (or detachment) between two deformable bodies we consider a boundary condition analogous to (1.79) on the simplifying assumption that the boundary displacements are sufficiently small. As the two bodies cannot penetrate one another, we assume that the sum of the displacements $u_N^{(1)}$ and $u_N^{(2)}$ of the two bodies and of the existing normal distance between them $h = h(x)$ must be greater than, or equal to the approach u^0 of the two bodies in the normal direction due to a rigid body displacement. We denote by $\bar{u}_N$ the quantity

$$u_N^{(1)} + u_N^{(2)} + h - u^0,$$

and let R_N be the respective contact force. The contact conditions read:

$$\text{if } \bar{u}_N > 0, \quad \text{then } R_N = 0;$$
$$\text{if } \bar{u}_N = 0, \quad \text{then } R_N \geq 0. \tag{1.81}$$

(v) The next example concerns the friction boundary conditions [Duv71, Mor86]. We consider the following boundary conditions:

$$\text{if } |S_T| < \mu|S_N|, \quad \text{then} \quad u_{T_i} = 0, \quad i = 1,2,3 \tag{1.82a}$$

$$\text{if } |S_T| = \mu|S_N|, \text{ then there exists } \lambda \geq 0 \text{ such that } u_{T_i} = -\lambda S_{T_i}, \; i = 1,2,3. \tag{1.82b}$$

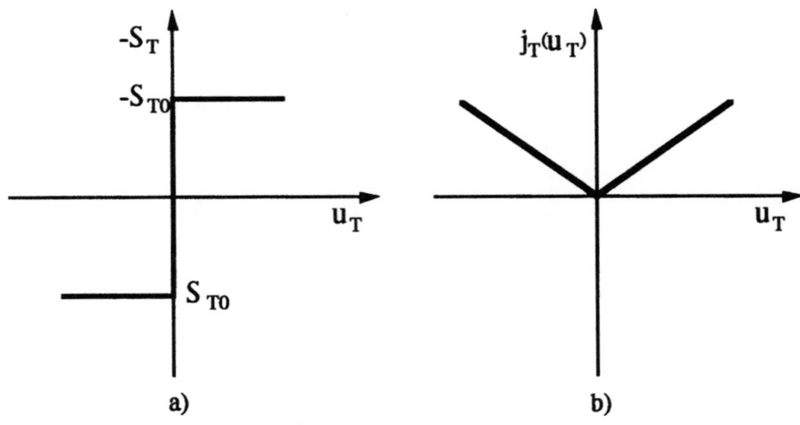

Fig. 1.7: The friction boundary condition

Here $\mu = \mu(x) > 0$ denotes the coefficient of friction and $|\cdot|$ the usual $\mathbf{R}^3$-norm. If Ω is a two-dimensional body, Γ is a curve, and thus S_T, u_T may be referred to a local right-handed coordinate system (n, τ) on Γ where τ denotes the unit vector tangential to Γ. Then (1.82a,b) can be put in the form

$$-S_T \in \beta_T(u_T), \tag{1.83}$$

where (fig. 1.7)

$$\beta_T(u_T) = \begin{cases} [-\mu|S_N|, +\mu|S_N|] & \text{if } u_T = 0 \\ \mu|S_N| & \text{if } u_T > 0 \\ -\mu|S_N| & \text{if } u_T < 0 \end{cases} \tag{1.84}$$

Assume further that $S_N = C_N$, where C_N is given, and denote $\mu|C_N|$ by S_{T_0}. Then

$$\beta_T(u_T) = \partial(S_{T_0}|u_T|). \tag{1.85}$$

If Ω is a three-dimensional body, then (1.82a,b) can be put only in the form (1.75) with

$$j_T(u_T) = S_{T_0}|u_T|. \tag{1.86}$$

We can verify that

$$j_T^c(-S_T) = \begin{cases} 0 & \text{if } |S_T| \leq S_{T_0} \\ \infty & \text{otherwise} \end{cases}, \tag{1.87}$$

and thus (1.82a, b) are equivalently written in the form

$$u_T \in \partial j_T^c(-S_T). \tag{1.88}$$

Note that the two dual subdifferential formulations give rise, for $u_T, S_T \in \mathbb{R}^3$ to the variational inequality

$$j_T(u_T^\star) - j_T(u_T) \geq -S_{T_i}(u_{T_i}^\star - u_{T_i}), \quad \forall u_T^\star \in \mathbb{R}^3 \tag{1.89}$$

and to

$$-u_{T_i}(S_{T_i}^0 - S_{T_i}) \leq 0, \quad \forall S_T^0 \in \mathbb{R}^3 \text{ such that } |S_T^0| \leq S_{T_0} \tag{1.90}$$

for $|S_T| \leq S_{T_0}$.

In dynamic problems, a friction law of the form

$$-S_T \in \partial j_T(v_T) = \partial(S_{T_0}|v_T|) \tag{1.91}$$

can be considered (Coulomb's law of friction). Here v_T denotes the tangential velocity which is equal to $\partial u_T/\partial t$ if the displacements are sufficiently small.

It is possible to combine the friction boundary condition with the unilateral contact boundary condition. We then obtain the following relations:

$$\begin{aligned} \text{if } u_N < 0, \quad &\text{then } S_N = 0, \quad S_{T_i} = 0, \quad i = 1, 2, 3 \\ \text{if } u_N \geq 0, \quad &\text{then } S_N + k u_N = 0, \end{aligned} \tag{1.92}$$

where k is a constant > 0 and $(1.82a, b)$ hold. In this case, however, it is not possible to write the boundary conditions in the subdifferential form (1.74). A generalization of (1.91) is obtained if the superpotential j_T is

$$j_T(u_T) = |C_N|(\mu_a^2 u_{Ta}^2 + \mu_b^2 u_{Tb}^2)^{1/2}. \tag{1.93}$$

Here a and b are two orthogonal directions termed orthotropy directions, which are defined at every point on the surface of the body, u_{Ta} and u_{Tb} are the components of the displacement u_T with respect to a local coordinate system (a, b), and μ_a and μ_b are the two corresponding friction coefficients. The resulting friction law is called orthotropic friction law (cf. in this context also [Pan85]). Note that if we have two deformable bodies in contact, the interface friction condition can be described by the same laws given above with the only difference that the tangential displacement u_T must be replaced by the relative tangential displacement $[u_T]$.

Further, we give some subdifferential boundary conditions arising in the theory of plates. Ω is here an open bounded subset $\mathbb{R}^2$ defined by the middle surface of the plate. Γ denotes the boundary of Ω. The points of Ω are referred to a fixed Cartesian coordinate system $Ox_1 x_2 x_3$. The x_1- and x_2-axes coincide with the middle surface of the plate, and the x_3-axis with the direction of the normal to the middle surface. The positive direction of the x_3-axis is upwards. The displacements of the plate in its plane are denoted by u_1, u_2 and vertical to its plane by w. By M_n and K_n we denote respectively the bending moment and the total or Kirchhoff shearing force [Gir] on the boundary of the plate, and we introduce boundary conditions of the form

$$M_n \in \beta_1\left(\frac{\partial w}{\partial n}\right) = \partial j_1\left(\frac{\partial w}{\partial n}\right) \tag{1.94}$$

$$-K_n \in \beta_2(w) = \partial j_2(w) \tag{1.95}$$

Concerning (1.95) the same contact problems are included in this formalism as for (1.73) and (1.75). With respect to (1.94) we can mention here the rotational friction boundary condition which is identical to the plastic hinge boundary conditions. It reads

$$\text{if } |M_n| < M_0 \quad \text{then } \frac{\partial w}{\partial n} = 0$$
$$\text{if } |M_n| = M_0 \quad \text{then there exists } \lambda \geq 0 \text{ such that } M_n = \lambda\frac{\partial w}{\partial n} \tag{1.96}$$

and can be put in the form (1.94), where β_1 has the form

$$\beta_1\left(\frac{\partial w}{\partial n}\right) = \begin{cases} [-M_0, M_0] & \text{if } \frac{\partial w}{\partial n} = 0 \\ M_0 & \text{if } \frac{\partial w}{\partial n} > 0 \\ -M_0 & \text{if } \frac{\partial w}{\partial n} < 0 \end{cases} \tag{1.97}$$

and M_0 is a prescribed bending moment.

In the theory of plates, another type of subdifferential relation can be formulated. Assume that the load vector f at every point $x \in \Omega_0 \subset \Omega$ consists of a part $\bar{f}$ which is given and of another part $\bar{\bar{f}}$ related to the displacement of that point by a relation of the form

$$- \bar{\bar{f}} \in \beta_3(w) = \partial j_3(w). \tag{1.98}$$

Here β_3 and j_3 have the same properties as β_i and j_i in (1.94)(1.95). As an application let us consider a plate which at points $x_0 \in \Omega_0 \subset \Omega$ $\bar{\Omega}_0 \cap \Gamma = \emptyset$, is at a distance $h = h(x)$ from a deformable support. It is assumed that the support causes a reaction force which is proportional to its deformation (Winkler support). We may then write the relation

$$- \bar{\bar{f}} \in \beta(w) \quad \text{in} \quad \Omega_0 \subset \Omega,$$

and

$$\bar{\bar{f}} = 0 \quad \text{in} \quad \Omega - \Omega_0.$$

β is a maximal monotone operator defined by

$$\beta(w) = \begin{cases} k(w - h)\, k \text{ const} > 0 & \text{if } w \geq h \\ 0 & \text{if } w < h. \end{cases} \tag{1.99}$$

For other types of subdifferential laws for deformable bodies we refer to [Pan85]. Note here that besides the boundary conditions (1.73) and (1.75), where the actions normally and tangentially to the boundary are considered separately, laws of the form

$$-S_N \in \partial j_N(u_N; S_T) \tag{1.100}$$

$$-S_T \in \partial j_T(u_T; S_N) \tag{1.101}$$

can be also considered. Of such type is e.g. the unilateral contact with friction boundary condition. The numerical treatment of such boundary conditions is made possible by means of a multistep decomposition technique introduced in [Pan75] for the unilateral contact problem with friction. We assume in the first step that $S_T = S_T^{(1)}$, where $S_T^{(1)}$ is given and we solve the problem with the boundary conditions $-S_N \in \partial j_N(u_N; S_T^{(1)})$, $S_T = S_T^{(1)}$. The solution of this problem yields the values for S_N, say $S_N^{(2)}$. Then the second step problem is considered with the boundary conditions $-S_T \in \partial j_T(u_T; S_N^{(2)})$, $S_N = S_N^{(2)}$. The numerical solution of this problem offers the value of S_T, say $S_T^{(3)}$. This procedure continues until the differences $|S_N^{(i+1)} - S_N^{(i)}|$, $|S_T^{(i+1)} - S_T^{(i)}|$, are made sufficiently small.

Finally let us recall that general contact laws of the form (1.74a) or equivalently (1.74b) can be considered. In this context we refer the reader to the general subdifferential material law $\sigma \in \partial w(\varepsilon)$, where $\sigma = \{\sigma_{ij}\}$, (resp. $\varepsilon = \{\varepsilon_{ij}\}$) the stress (resp. strain) tensor, which is in detail explained in [Pan85].

1.3.2 Extension of the Monotone Multivalued Boundary Conditions to Function Spaces

Until now we have considered subdifferential boundary conditions of a pointwise nature. The superpotentials define a relation, at every point of the boundary Γ, between the value of S_N and u_N etc. It is however necessary for the formulation and the study of B.V.Ps connected with subdifferential boundary conditions to consider extensions of these boundary conditions to function spaces. Roughly speaking, we are concerned with the following important question: if $j(u(x)) \in \Gamma_0(\mathbb{R}^3)$, what will the properties of $\int_\Gamma j(u(x))d\Gamma$ be?

Assume generally that $(\Gamma, \mathcal{B}, \mu)$ is a positive measure space with $\mu(\Gamma) < \infty$. If $A: H \to \mathcal{P}(H)$ is a multivalued operator and H is a Hilbert space identified with its dual, then we can define the operator $\bar{A}$ on $L^2(\Gamma, H)$ (extension of A to $L^2(\Gamma, H)$) by setting

$$f \in \bar{A}(u) \Leftrightarrow f(x) \in A(u(x)) \quad \mu - \text{a.e. on } \Gamma \tag{1.102}$$

It is shown in [Brez73] that if A is maximal monotone, $\bar{A}$ is too. The following proposition holds

Proposition 1.13 Let $A = \partial\varphi$, where φ is a convex, l.s.c., proper functional on H. For $u \in L^2(\Gamma, H)$, we define the functional

$$\Phi(u) = \begin{cases} \int_\Gamma \varphi(u(x))d\mu & \text{if } \varphi(u) \in L^1(\Gamma) \\ \infty & \text{otherwise} \end{cases}, \tag{1.103}$$

Then Φ is convex, l.s.c., and proper on $L^2(\Gamma, H)$, and $\bar{A} = \partial\Phi$, i.e., $\partial\Phi$ is the extension of $\partial\varphi$ to $L^2(\Gamma, H)$.

For the proof of this proposition we refer to [Brez72, Pan85]. At this point we mention only that in order to prove that $\bar{A} = \partial\Phi$, we show the equivalence of the following two conditions: for $u, f \in L^2(\Gamma, H)$,

(i) $\quad \Phi(v) - \Phi(u) \geq \displaystyle\int_\Gamma (f, v - u)_H d\mu \quad \forall v \in L^2(\Gamma, H);$ $\tag{1.104}$

(ii) $\quad \varphi(v(x)) - \varphi(u(x)) \geq \big(f(x), v(x) - u(x)\big)_H \, \mu-\text{a.e. on } \Gamma, \, \forall v \in L^2(\Gamma, H).$ (1.105)

Moreover we can show that each of the variational inequalities (1.104), (1.105) is equivalent to the following condition:

(iii) there exists $A \subset \Gamma$ with $\mu(A) = 0$ such that

$$\varphi(v) - \varphi\big(u(x)\big) \geq \big(f(x), v - u(x)\big)_H, \quad \forall x \in \Gamma - A, \ \forall v \in H. \tag{1.106}$$

In order to extend the subdifferential contact boundary conditions we apply Prop. 1.13 for the usual Lebesgue measure and $H = \mathbb{R}^3$, if for instance (1.74a) is considered, or $H = \mathbb{R}$, in the case of (1.73) etc. Here we shall construct the extension of $-S(x) \in \partial j(u(x))$ to $L^2(\Gamma)$. A method similar to what follows could be used for the other types of subdifferential boundary conditions. The functional

$$\Phi(u) = \begin{cases} \int_\Gamma j\big(u(x)\big)d\Gamma & \text{if } j(u) \in L^1(\Gamma) \\ \infty & \text{otherwise} \end{cases}, \tag{1.107}$$

is convex, l.s.c. and proper on $[L^2(\Gamma)]^3$. For $u, S \in [L^2(\Gamma)]^3$ the relation

$$-S \in \partial\Phi(u) \tag{1.108}$$

holds if and only if

$$-S(x) \in \partial j\big(u(x)\big) \quad \text{a.e. on } \Gamma. \tag{1.109}$$

These two relations are obviously equivalent to the inverse relations

$$u \in \partial\Phi^c(-S) \tag{1.110}$$

and

$$u(x) \in \partial j^c\big(-S(x)\big) \quad \text{a.e. on } \Gamma, \tag{1.111}$$

as they are to the relations

$$\Phi(u) + \Phi^c(-S) = -\int_\Gamma u_i(x)S_i(x)d\Gamma \tag{1.112}$$

and

$$j\big(u(x)\big) + j^c\big(-S(x)\big) = -u_i(x)S_i(x) \quad \text{a.e. on } \Gamma, \tag{1.113}$$

where

$$\Phi^c(-S) = \begin{cases} \int_\Gamma j^c\big(-S(x)\big)d\Gamma & \text{if } j^c(-S) \in L^1(\Gamma) \\ \infty & \text{otherwise} \end{cases}. \tag{1.114}$$

We shall often have in deformable bodies that the displacement field $u \in [H^1(\Omega)]^3$. Then $u \to \Phi(u|_\Gamma)$ is a convex, l.s.c. and proper functional on $[H^1(\Omega)]^3$ due to the continuity of the trace application. Let us denote by $\bar\Phi$ the restriction $\Phi|_{[H^{1/2}(\Gamma)]^3}$ of Φ to $[H^{1/2}(\Gamma)]^3$, assuming that $\bar\Phi$ on $H^{1/2}(\Gamma)$ is not identically equal to $+\infty$. Then $\partial\Phi$ and $\partial\bar\Phi$ define maximal monotone graphs on

$$[L^2(\Gamma)]^3 \times [L^2(\Gamma)]^3 \quad \text{and} \quad [H^{1/2}(\Gamma)]^3 \times [H^{-1/2}(\Gamma)]^3$$

respectively. In many B.V.Ps we shall encounter the boundary condition written for $u \in [H^1(\Omega)]^3$ and $S \in [H^{-1/2}(\Gamma)]^3$ in the form

$$\Phi(\gamma v) - \Phi(\gamma u) \geq \langle -S, \gamma v - \gamma u \rangle, \quad \forall v \in [H^1(\Omega)]^3. \tag{1.115a}$$

Since the trace application $v \to \gamma v$ is surjective from $[H^1(\Omega)]^3$ onto $[H^{1/2}(\Gamma)]^3$, (1,115a) yields that

$$\bar{\Phi}(v) - \bar{\Phi}(\gamma u) \geq \langle -S, v - \gamma u \rangle, \quad \forall v \in [H^{1/2}(\Gamma)]^3. \tag{1.115b}$$

which is equivalent to

$$-S \in \partial\bar{\Phi}(\gamma u) \quad \text{or} \quad \gamma u \in \partial\bar{\Phi}^c(-S). \tag{1.115c}$$

Recall that $\langle \cdot, \cdot \rangle$ denotes the duality pairing between $[H^{1/2}(\Gamma)]^3$ and $[H^{-1/2}(\Gamma)]^3$ which coincides with the integral $\int_\Gamma S_i(v_i - u_i)d\Gamma$ if $S \in [L^2(\Gamma)]^3$. Henceforth, the bar on Φ and the γ will be omitted if no ambiguity occurs. The expression (1.115c) is a weak formulation of the boundary condition (1.74a) since only if $S \in [L^2(\Gamma)]^3$ the complete equivalence holds.

Further the weak formulations of some contact boundary conditions will be studied. We denote by H_T the space

$$H_T = \{v | v \in [H^{1/2}(\Gamma)]^3, \ v_i n_i = 0 \text{ a.e. on } \Gamma\} \tag{1.116}$$

and we recall ([Pan85] p.32) that if $a = \{a_i\} \in [H^{1/2}(\Gamma)]^3$, and $a_N = a_i n_i$, $a_T = \{a_{T_i}\}$ where $a_{T_i} = a_i - a_N n_i$, then the mapping $a \to \{a_N, a_T\}$ is an isomorphism from $[H^{1/2}(\Gamma)]^3$ onto $H^{1/2}(\Gamma) \times H_T$. In the dual spaces a'_N and a'_T are uniquely determined by the relation

$$\langle a', a \rangle = \langle a'_N, a_N \rangle_{1/2} + \langle a'_T, a_T \rangle_{H_T} \quad \forall a \in [H^{1/2}(\Gamma)]^3, \tag{1.117}$$

where $\langle \cdot, \cdot \rangle_{1/2}$ and $\langle \cdot, \cdot \rangle_{H_T}$ denote the duality pairings on $H^{1/2}(\Gamma) \times H^{-1/2}(\Gamma)$ and $H'_T \times H_T$. Obviously $a' \to \{a'_N, a'_T\}$ is again an isomorphism from $([H^{1/2}(\Gamma)]^3)'$ onto $H^{-1/2}(\Gamma) \times H'_T$. For all the above it is sufficient that Γ be $C^{1,1}$-regular ($C^{0,1}$-regularity, i.e a Lipschitz boundary is also possible [Has82] with minor modifications).

Suppose now that Φ is a proper functional on $[H^{1/2}(\Gamma)]^3$. Φ is decomposable if two proper functionals Φ_N on $H^{1/2}(\Gamma)$ and Φ_T on $H_T(\Gamma)$ can be determined such that

$$\Phi(v) = \Phi_N(v_N) + \Phi_T(v_T), \quad \forall v \in [H^{1/2}(\Gamma)]^3, \tag{1.118}$$

where

$$v = \{v_i\}, \quad v_i = v_N n_i + v_{T_i}.$$

Proposition 1.14 Suppose that Γ is $C^{1,1}$-regular and that the functional Φ is proper and decomposable on $[H^{1/2}(\Gamma)]^3$. If Φ is convex (resp. l.s.c.), then Φ_N and Φ_T are convex (resp. l.s.c.) as well, and conversely. Moreover,

$$\Phi^c(T) = \Phi_N^c(T_N) + \Phi_T^c(T_T), \quad \forall T \in [H^{-1/2}(\Gamma)]^3, \tag{1.119}$$

where

$$\langle T, v \rangle = \langle T_N, v_N \rangle_{1/2} + \langle T_T, v_T \rangle_{H_T}, \quad \forall v \in [H^{1/2}(\Gamma)]^3. \tag{1.120}$$

For the proof see [Pan85] p.109 and [Hün].

(i) Suppose now that $C_T \in H'_T$ and $c_N \in H^{1/2}(\Gamma)$ are given and let us consider the boundary conditions

$$u_N = c_N \quad \text{and} \quad S_T = C_T \tag{1.121}$$

We define a functional Φ on $[H^{1/2}(\Gamma)]^3$ by (cf. 1.118)

$$\Phi_T(v_T) = -\langle C_T, v_T \rangle_{H_T} \quad \text{for } v_T \in H_T \tag{1.122a}$$

and

$$\Phi_N(v_N) = \begin{cases} 0 & \text{if } v_N = c_N \\ \infty & \text{if } v_N \neq c_N \end{cases}, \tag{1.122b}$$

for $v_N \in H^{1/2}(\Gamma)$. Then Φ belongs to the class $\Gamma_0([H^{1/2}(\Gamma)]^3)$ and is decomposable. By Prop. 1.14, (1.119) holds, and since

$$\Phi_T(T_T) = \begin{cases} 0 & \text{if } T_T = -C_T \\ \infty & \text{otherwise} \end{cases}, \tag{1.122c}$$

for $T_T \in H'_T$ and

$$\Phi_N^c(T_N) = \langle T_N, c_N \rangle_{1/2} \quad \text{for} \quad T_N \in H^{-1/2}(\Gamma) \tag{1.122d}$$

it results that

$$\Phi^c(T) = \begin{cases} \langle T_N, c_N \rangle_{1/2} & \text{if } T_T = -C_T \\ \infty & \text{otherwise} \end{cases}, \tag{1.122e}$$

If $C_{T_i} \in L^2(\Gamma)$, then the conditions (1.121) are equivalent to the pointwise boundary conditions

$$u_N(x) = c_N(x) \quad \text{and} \quad S_{T_i}(x) = C_{T_i}(x) \quad \text{a.e. on } \Gamma. \tag{1.123}$$

(ii) Let $C_T \in H'_T$ be given. We study the Signorini boundary condition

$$u_N \leq 0 \quad \text{a.e. on} \quad \Gamma, \quad \langle S_N, u_N \rangle_{1/2} = 0 \quad S_N \leq 0 \quad \text{in} \quad H^{-1/2}(\Gamma) \tag{1.124}$$

and

$$S_T = C_T. \tag{1.125}$$

It should be noted that the last of (1.124) is by the definition of a non-negative distribution (which is a measure) equivalent to

$$\langle S_N, v_N \rangle_{1/2} \geq 0 \quad \forall v_N \in H^{1/2}(\Gamma) \quad v_N \leq 0 \quad \text{a.e. on } \Gamma. \tag{1.126}$$

Further, the functional Φ is defined on $[H^{1/2}(\Gamma)]^3$ by (1.118), where for $v_N \in H^{1/2}(\Gamma)$

$$\Phi_N(v_N) = \begin{cases} 0 & \text{if } v_N \leq 0 \quad \text{a.e. on } \Gamma \\ \infty & \text{if } v_N > 0 \quad \text{on } \Gamma_1 \subset \Gamma \text{ with mes } \Gamma_1 > 0 \end{cases} \tag{1.127}$$

and for $v_T \in H_T$

$$\Phi_T(v_T) = -\langle C_T, v_T \rangle_{H_T}. \tag{1.128}$$

Then we have

$$\Phi_N^c(-T_N) = \begin{cases} 0 & \text{if } T_N \leq 0 \\ \infty & \text{otherwise} \end{cases} \tag{1.129}$$

for $T_N \in H^{-1/2}(\Gamma)$ and

$$\Phi_T^c(T_T) = \begin{cases} 0 & \text{if } T_T = -C_T \\ \infty & \text{otherwise} \end{cases} \tag{1.130}$$

for T_T in H_T'. Then (1.119) yields $\Phi^c(T)$. If $S_N \in [L^2(\Gamma)]^3$, (1.124) is equivalent to the pointwise condition (1.79) or (1.80).

(iii) Let $S_{T_0} \in L^\infty(\Gamma)$ with essinf $S_{T_0} > 0$. We denote by A, A', A_T and A_T' the injections

$$A\colon H^{1/2}(\Gamma) \to L^2(\Gamma), \quad A'\colon L^2(\Gamma) \to H^{-1/2}(\Gamma),$$

$$A_T\colon H_T \to L_T = \{v \in [L^2(\Gamma)]^3, \quad v_i n_i = 0 \quad \text{a.e. on } \Gamma\}$$

$$\text{and} \quad A_T'\colon L_T \to H_T'.$$

Obviously, $v \to \{v_N, v_T\}$ is an isometry from $[L^2(\Gamma)]^3$ onto $L^2(\Gamma) \times L_T$. Let us now consider the following boundary conditions:

$$\text{if} \quad S_T \in H_T', \quad S_T = A_T' \tilde{S}_T, \quad \tilde{S}_T \in L_T \text{ and } |\tilde{S}_T| < S_{T_0} \tag{1.131}$$

a.e. on Γ, then $u_T = 0$, with $u_T \in H_T$;

$$\text{if} \quad S_T \in H_T', \quad S_T = A_T' \tilde{S}_T, \quad \tilde{S}_T \in L_T \text{ and } |\tilde{S}_T| = S_{T_0}$$

a.e. on Γ, then there exists $\lambda \geq 0$ such that

$$u_T = -\lambda \tilde{S}_T, \quad u_T \in H_T; \tag{1.132}$$

$$S_N = C_N \text{ in } H^{-1/2}(\Gamma). \tag{1.133}$$

We can easily verify that (1.131)(1.132) are equivalent to

$$-\tilde{S}_T \in \partial F_T(A_T u_T), \quad S_T = A_T' \tilde{S}_T, \tag{1.134}$$

where

$$F_T(v) = \int_\Gamma S_{T_0} |v| d\Gamma \quad \text{for} \quad v \in L_T. \tag{1.135}$$

An equivalent formulation to (1.134) reads:

$$\begin{aligned} S_T &= A_T' \tilde{S}_T, \ |\tilde{S}_T| \leq S_{T_0} \quad \text{a.e. on } \Gamma, \\ S_{T_0} |u_T| + \tilde{S}_{T_i} u_{T_i} &= 0 \qquad \text{a.e. on } \Gamma. \end{aligned} \tag{1.136}$$

The functional F_T^c is defined on L_T by

$$F_T^c(T) = \begin{cases} 0 & \text{if } \tilde{T} \in L_T, \quad |\tilde{T}| \leq S_{T_0} \text{ a.e. on } \Gamma \\ \infty & \text{otherwise} \end{cases}. \tag{1.137}$$

In order to write (1.131)(1.132) in the form (1.115c), we set $\Phi_T = F_T \circ A_T$; then

$$\Phi_T^c(T) = \begin{cases} 0 & \text{if } T \in H_T', \text{ with } \cdot T = A_T'\tilde{T}, \ \tilde{T} \in L_T \\ \text{and} & |\tilde{T}_T| \le S_{T_0} \quad \text{a.e on } \Gamma, \\ \infty & \text{otherwise} \end{cases} \tag{1.138}$$

and thus (1.131)(1.132) takes the form

$$-S_T \in \partial\Phi_T(u_T) \quad \text{or} \quad u_T \in \partial\Phi_T^c(-S_T) \tag{1.139}$$

on $H_T \times H_T'$ for $S_T = A_T'\tilde{S}_T$.

In order to show (1.139) we use of formula for $\partial(f \circ l)(x)$ where f is a convex and l.s.c. functional and l is a linear functional (see [Eke] p.27]. Here using the special structure of the superpotential we can derive certain regularity results. Thus we can obtain from (1.139) that

$$\frac{S_T}{S_{T_0}} \in [L^\infty(\Gamma)]^3 \quad \text{and} \quad \left|\frac{S_T}{S_{T_0}}\right|_{L^\infty} \le 1. \tag{1.140}$$

even if $S_T \notin$ range of A_T'. For the proof see [Duv72, Pan85].

1.3.3 Nonmonotone Multivalued Boundary Conditions

There exist several contact problems which must be expressed through nonmonotone contact boundary conditions between $-S_N$ and u_N, between $-S_T$ and u_T, or generally between S and u. Let us give certain applications first

i) In fig. 1.8a a normal contact law between a deformable body and a support constructed by a granular material or concrete is depicted. If debonding takes place the stresses are zero.

ii) In fig. 1.8b the same contact law but without debonding is presented. Here the support is assumed to be a reinforced concrete support which in tension obeys Scanlon's zig-zag law [Flo].

iii) The diagram of fig. 1.8c concerns the adhesive contact problem. The adhesive material between body and support may sustain some small tensile force. Then a debonding takes place which may obey the brittle type diagram BOACD or the semibrittle diagram BOAED. Note that the vertical branches (i.e. the multivaluedness) in both the laws of fig. 1.9b,c are complete, i.e. for an appropriate loading the reaction and the normal boundary displacement u_N (resp. interface relative displacement $[u_N]$) can define a point on the vertical branch.

iv) In the fig. 1.8d,e certain nonmonotone friction laws are depicted. The first can be applied in geomechanics and rock interface analysis, whereas the second appears between reinforcement and concrete in a concrete structure. Finally the law of fig. 1.8f appears in the tangential direction in adhesively bonded parts and describes the partial cracking and crushing of the adhesive interface material.

v) In fig. 1.9 contact laws with infinite branches are depicted. For instance, the law of fig. 1.9a describes the adhesive contact with a rubber support which presents in compression ideal locking effects (the infinite branch AB). The same happens in two plates in adhesive contact (fig. 1.9b). Here the interface may sustain infinite compressive forces.

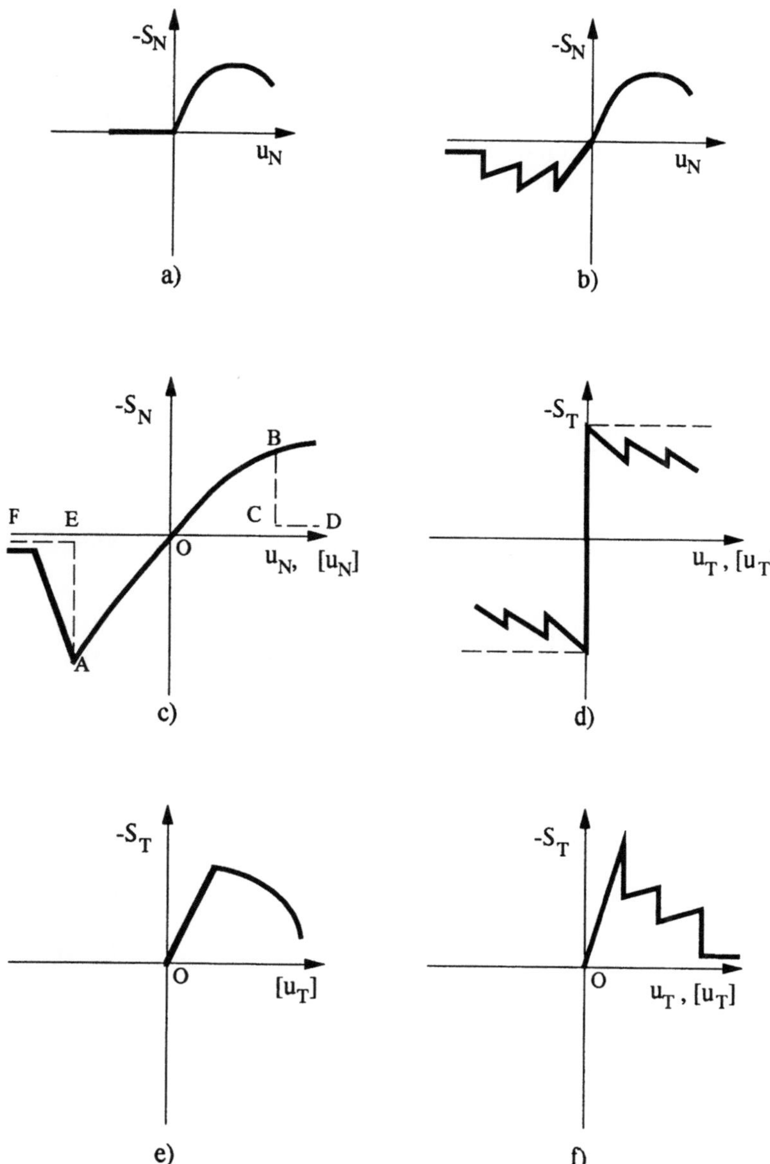

Fig. 1.8: Nonmonotone, possibly multivalued contact laws

All the above contact laws can be put in the general forms

$$-S_N \in \bar{\partial} j_N(u_N) \quad \text{or} \quad -S_T \in \bar{\partial} j_T(u_T)$$

$$(1.141)$$

where u_N and u_T are replaced by the corresponding relative displacements $[u_N]$ and $[u_T]$ in the case of interfaces. Here $\bar{\partial}$ denotes the generalized gradient and j_N, j_T are nonconvex energy functionals called nonconvex superpotentials [Pan85]. They generally are taking values on the extended real line. Then (1.141) introduce by definition the upper subdifferentials $j_N^{\uparrow}(\cdot,\cdot)$ or $j_T^{\uparrow}(\cdot,\cdot)$ (cf. (1.50)) into the variational formulation of the problem. Note that if j_N or j_T are locally Lipschitz (cf. fig. 1.8) then $j_N^{\uparrow}$ or $j_T^{\uparrow}$ are replaced by the more easy expressions j_N^0 and j_T^0 (cf. eq. (1.48)). In the case of unloading (fig. 1.10a) we can again distinguish three types of contact laws and thus we can split the problem into three subproblems. As the loading increases we may encounter the following three types of contact problems. The purely elastic nonmonotone problem (law AA'or ABD), the loading-unloading problem (law BD or BCC') and the unloading-loading problem (law CC' or CBD).

All the aforementioned contact laws are onedimensional; we can obtain their generalizations for threedimensional continua by means of a method described by the second author in [Mor88a]. For instance in order to extend the friction law of fig. 1.8d for a three dimensional body we consider some surfaces defined by the equations $f_j(v_T) = 0$, $j = 1,\ldots,n$, and then we proceed as in [Mor88a p.99]. All these threedimensional extensions can be put in the form

$$-S \in \bar{\partial}j(u \text{ or } [u]). \tag{1.142}$$

Note that instead of (1.141),(1.142) we may have the "inverse" expressions

$$u_N \in \bar{\partial}\tilde{j}_N(-S_N), \quad u_T \in \bar{\partial}\tilde{j}_T(-S_T), \quad u \in \bar{\partial}\tilde{j}(-S) \tag{1.143}$$

where u_N, u_T or u can be replaced by the relative displacements in the case of interface problems. Note that due to the lack of convexity the qualities $\tilde{j}_N$, $\tilde{j}_T$ or $\tilde{j}$ are not the conjugate functions to j_N, j_T or j respectively, in the sense of the notion of conjugacy of Sect. 1.2.1.

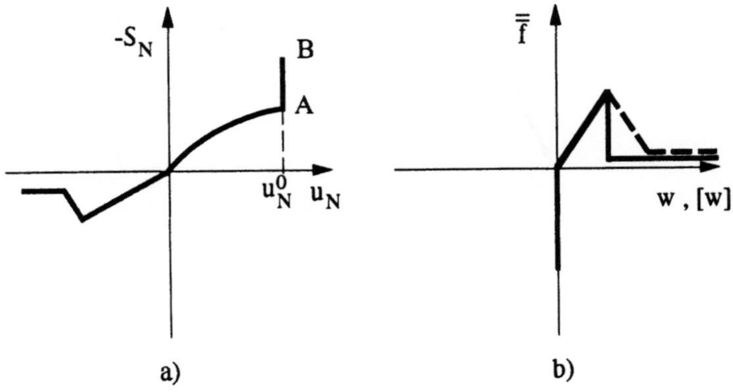

Fig. 1.9: Contact laws with infinite branches

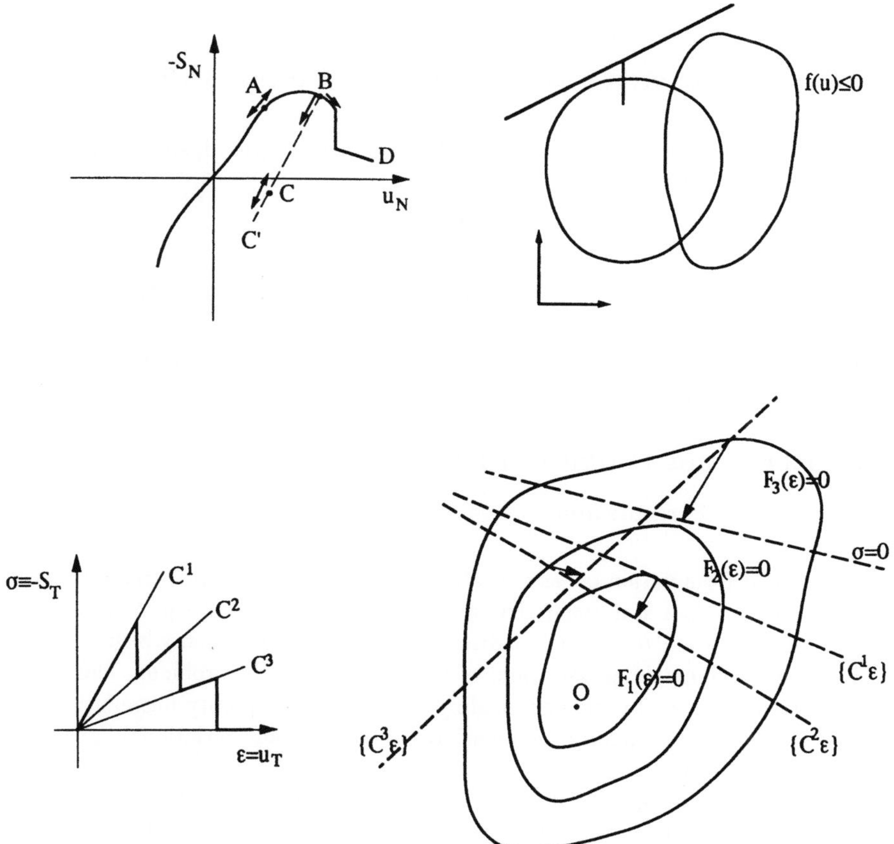

Fig. 1.10: Contact laws with unloading. The threedimensional nonmonotone contact
problem

As in the convex case the previous pointwise boundary conditions can be extended
to function spaces. Then the following proposition holds.

Proposition 1.15 Let H be a separable Hilbert space and $j: \Gamma \times H \to \mathbb{R}$ be a
functional such that (i) $x \to j(x, u)$ is measurable for each u (ii) $u \to j(x, u)$ is a
locally Lipschitz function for each x and (iii) $x \to j(x, 0)$ is finitely integrable. Let
$p \geq 1$ and $c \geq 0$ such that for every (x, u) and for every $f \in \bar{\partial}_u j(x, u)$ the estimate

$$|f| \leq c(1 + |u|^{p-1}) \tag{1.144}$$

holds. Then for $u \in L^p(\Gamma, H)$ the functional $\Phi: L^p(\Gamma, H) \to \mathbb{R}$ defined by

$$\Phi(u) = \int_{\Gamma} j(x, u(x)) d\Gamma \tag{1.145}$$

is Lipschitzian on every bounded subset of $L^p(\Gamma, H)$. Every element $f \in L^p(\Gamma, H)$ of $\bar{\partial}\Phi(u)$ satisfies the relation

$$f(x) \in \bar{\partial}_u j(x, u(x)) \quad \text{a.e on } \Gamma, \tag{1.146}$$

and if $j_u^0(x, u, v) = \tilde{j}_u'(x, u, v)$ a.e. on Γ for every $v \in H$ ($\bar{\partial}$-regularity), then the converse is also true.

For the proof on this proposition, the reader is referred to [Au79] and [Clar83]. This proposition states that

$$\bar{\partial}\Phi(u) \subset \int_\Gamma \bar{\partial}_u j(x, u(x))d\Gamma, \tag{1.147}$$

where the equality holds when j is $\bar{\partial}$-regular. If for instance $H = \mathbb{R}$ then (1.147) permits the definition of $\bar{\partial}\Phi(u_N)$ on $L^2(\Gamma)$ and by considering the restriction $\tilde{\Phi}$ of Φ to $H^{1/2}(\Gamma)$ the definition of $\bar{\partial}\tilde{\Phi}(u_N)$. Note that a proposition of Chang [Cha] implies that $\bar{\partial}\tilde{\Phi}(u_N) \subset \bar{\partial}\Phi(u_N)$ for every $u_N \in H^{1/2}(\Gamma)$.

1.4 Bilateral and Unilateral Problems

1.4.1 Variational Formulations

A variational formulation or, as we call in mechanics for historical reasons, a variational "principle", is a statement that a solution of an operator equation subjected to certain boundary and/or initial conditions satisfies an expression which is equal to zero or nonnegative, involving variations of the quantities of the problem. Thus we distinguish between bilateral problems, leading to variational equality formulations, and unilateral or generally inequality problems. Let us derive some variational "principles" for a deformable body.

Let $\Omega \subset \mathbb{R}^3$ be an open bounded subset occupied by a deformable body in its undeformed state. On the assumption of small strains we can write the relation.

$$\int_\Omega \sigma_{ij}(u)\varepsilon_{ij}(v - u)d\Omega = \int_\Omega \bar{p}_i(v_i - u_i)d\Omega + \langle \sigma_{ij}n_j, (v_i - u_i)\rangle, \quad \forall v \in [H^1(\Omega)]^3 \tag{1.148}$$

for

$$u \in [H^1(\Omega)]^3, \quad \sigma_{ij} \in L^2(\Omega), \quad \bar{p}_i \in L^2(\Omega), \quad i, j = 1, 2, 3.$$

This relation, which is obtained from the operator equations of the problem by applying the Green-Gauss theorem, is the expression of the principle of virtual work for the body when it is considered to be free, i.e., with no constraint on its boundary Γ. Note that for the derivation of (1.148) we have multiplied the equilibrium equation

$$\sigma_{ij,j} + \bar{p}_i = 0, \tag{1.149}$$

where the $\bar{p}_i \in L^2(\Omega)$ is the volume vector, by $v_i - u_i$ and then we have integrated over Ω. On the assumption of "appropriately smooth" functions, we have applied the Green-Gauss theorem by taking into account the strain-displacement relation

$$\varepsilon_{ij} = \frac{1}{2}(u_{i,j} + u_{j,i}). \tag{1.150}$$

An extension by density of the arising functionals leads to the variational equality (1.148).

Let us assume further that the body is linear elastic, i.e. that

$$\sigma_{ij} = C_{ijhk}\varepsilon_{hk} \tag{1.151}$$

where $C = \{C_{ijhk}\}$, $i, j, h, k = 1, 2, 3$, is the elasticity tensor which satisfies the well-known symmetry and ellipticity properties

$$C_{ijhk} = C_{jihk} = C_{khij} \tag{1.151a}$$

$$C_{ijhk}\varepsilon_{ij}\varepsilon_{hk} \geq c\varepsilon_{ij}\varepsilon_{hk} \quad \forall \varepsilon = \{\varepsilon_{ij}\} \in \mathbb{R}^6 \quad c \text{ const} > 0. \tag{1.151b}$$

We denote the bilinear form of linear elasticity by $a(\cdot, \cdot)$, i.e.

$$a(u, v) = \int_\Omega C_{ijhk}\varepsilon_{ij}(u)\varepsilon_{hk}(v)d\Omega. \tag{1.151c}$$

Note also that instead of (1.148) we can write the relation (cf. eq.(1.120))

$$\int_\Omega \sigma_{ij}\varepsilon_{ij}(v - u)d\Omega = \int_\Omega \bar{p}_i(v_i - u_i)d\Omega + \langle S_N, v_N - u_N \rangle_{1/2} + \langle S_T, v_T - u_T \rangle_{H_T} \tag{1.152}$$

$$\forall v \in [H^1(\Omega))]^3.$$

i) Let us assume now that on Γ the classical boundary conditions $S_T = 0$ and $u_N = 0$ hold. Then (1.152) with (1.151c) leads to the following variational equality: Find $u \in V_0 = \{v|v \in [H^1(\Omega)]^3,\ v_N = 0 \text{ on } \Gamma\}$ such that

$$a(u, v) = \int_\Omega \bar{p}_i v_i d\Omega \quad \forall v \in V_0. \tag{1.153}$$

ii) Let us assume that on Γ the Signorini boundary conditions (1.124)(1.125) hold. Then (1.152) with (1.151c) leads to the following variational inequality: Find $u \in K = \{v|v \in [H^1(\Omega)]^3,\ v_N \leq 0 \text{ a.e. on } \Gamma\}$ such as to satisfy the relation

$$a(u, v - u) \geq \int_\Omega \bar{p}_i(v_i - u_i)d\Omega + \langle C_T, v_T - u_T \rangle_{H_T} \quad \forall v \in K. \tag{1.154}$$

iii) Let us assume that on Γ the general monotone possibly multivalued boundary condition (1.115c) holds. Then (1.148) with (1.151c) and with (1.115b) leads to the following variational inequality: Find $u \in [H^1(\Omega)]^3$ such that

$$a(u, v - u) + \bar{\Phi}(\gamma v) - \bar{\Phi}(\gamma u) \geq \int_\Omega \bar{p}_i(v_i - u_i)d\Omega \quad \forall u \in H^1(\Omega)]^3. \tag{1.155}$$

iv) Let us assume that on Γ the general nonmonotone possibly multivalued boundary condition

$$-S_N \in \bar{\partial}\tilde{\Phi}(u_N), \quad S_T = 0 \tag{1.156}$$

holds where $\tilde{\Phi}$ is nonconvex and is defined in Sect. 1.3.3 (see prop. 1.15). Then (1.152) with (1.151c) leads to the following variational expression:Find $u \in [H^1(\Omega)]^3$ such as to satisfy the inequality

$$a(u, v - u) + \tilde{\Phi}^\dagger(u_N, v_N - u_N) \geq \int_\Omega \bar{p}_i(v_i - u_i)d\Omega \quad \forall v \in H^1(\Omega)]^3. \qquad (1.157)$$

This last type of variational expressions have been called by the second author, who introduced them in Mechanics in [Pan81;82;83] and studied them in [Pan85;88, Mor88a], hemivariational inequalities. Their study is based on compactness arguments and not on monotonicity arguments due to the lack of convexity [Pan91, Nan88;89a,b].

v) A combination of the variational inequalities with the hemivariational inequalities is necessary when both monotone and nonmonotone boundary conditions hold, or when the nonmonotone boundary conditions of fig. 1.9 hold. Indeed in fig 1.9a the nonmonotone boundary condition holds together with the inequality $u_N \leq u_N^0$. Thus we have a hemivariational inequality, whose solution must be sought on a convex subset K defined by this inequality. As it is well known (cf. e.g. [Pan85]) we can satisfy this inequality on the whole chosen admissible function space V if the expression $I_K(v) - I_K(u)$ is added to the left hand side of the inequality, where I_K is the indicator function corresponding to K. Thus we are led to the following variational-hemivariational inequality: Find $u \in V$ such that

$$a(u, v-u)+\tilde{\Phi}^\dagger(u_N, v_N-u_N)+\Phi(v_N)-\Phi(u_N) \geq \int_\Omega \bar{p}_i(v_i-u_i)d\Omega \quad \forall v \in V, \qquad (1.158)$$

where now Φ ia a convex l.s.c. and proper functional on V. Variational hemivariational inequalities have been studied in [Pan91].

About the exact relationship of the above variational expressions with the classical formulations of the B.V.Ps we refer to [Duv72, Pan85, Mor88a,b]. Here we briefly note that usually a solution of variational formulations satisfies the operator equations and the boundary conditions of the problem in a generalized sense (e.g. as equalities in the sense of distributions over Ω, or in the sense of $H^{-1/2}(\Gamma)$ etc.). A more elaborate mathematical study may lead to additional more profound regularity results for the variational solution of a B.V.P.

From the standpoint of Mechanics the aforementioned variational formulations express the principle of virtual work in equality or in inequality form. Analogous variational formulations which include the variations of the stresses may be derived . Then we speak about the principle of complementary virtual work (see e.g. [Duv72, Pan85]). A B.V.P. is called bilateral (resp. unilateral) if it leads to variational equality (resp. variational, or hemivariational, or variational-hemivariational inequality) formulations. We call the unilateral problems inequality problems too. Note at this point that the term "unilateral boundary conditions" has been initially used and is until now in use, in order to characterize boundary conditions involving inequalities. But as we have seen, through the introduction of the indicator, the inequality boundary conditions may be put also in the general multivalued forms (1.74a) or (1.142).

As Fourier has noticed [Lan] the inequality form of the principle of virtual or complementary virtual work is due to the fact that the variations of certain variables involved into the problem are "irreversible". For instance, if (1.154) held for $u, v \in V$, where V is a vector space then the substitution $v - u = \pm w$ would lead to a variational equality. But since $u, v \in K$ where K is a closed convex set, we cannot set $v - u = \pm w$, i.e., the variation $v - u$ is irreversible. Irreversible variations are called "unilateral" variations. Unilateral are the variations also in (1.155), unless $\mathrm{grad}\,\bar{\Phi}$ exists everywhere. Indeed in this case (1.155) is equivalent to the variational equality

$$a(u, w) + \langle \mathrm{grad}\,\bar{\Phi}(\gamma u), \gamma w \rangle = \int_{\Omega} \bar{p}_i w_i d\Omega \quad \forall w \in [H^1(\Omega)]^3 \qquad (1.159)$$

as it results easily by setting in (1.155) $v = u \pm \lambda w$, $\lambda \to 0_+$. The converse results easily by setting in (1.159) $w = v - u$ and by applying the inequality ($\bar{\Phi}$ is convex)

$$\bar{\Phi}(\gamma v) - \bar{\Phi}(\gamma u) \geq \langle \mathrm{grad}\,\bar{\Phi}(\gamma u), \gamma(v - u) \rangle \quad \forall v \in [H^1(\Omega)]^3. \qquad (1.159a)$$

Analogously we may argue in the case of hemivariational inequalities. Let us notice also that Prop. 1.2 (resp. relation (1.58)) permits the formulation of minimum (resp. substationary) potential and complementary energy problems which correspond to a variational inequality (resp. to a hemivariational inequality). In most cases also a complete equivalence holds especially in the case of convex energy functionals (i.e. in the case of variational inequalities). See in this context [Lio71, Duv72, Fich72, Pan85, Hl, Mor88a,b]. We refer also to [Pan85] concerning the relation of variational "principles" to the notions of convex and nonconvex superpotentials and to the chosen duality pairing between the "generalized forces" and "generalized displacements" of the problem under consideration. These last ideas constitute generalizations and amelioration of analogous ideas of Tonti concerning the bilateral problems [Tont].

Closing this Section let us note that the variational inequalities, the hemivariational inequalities etc., belong due to the arising nonsmooth energy functionals to the Nonsmooth Mechanics [Mor88a,b], as it has been called this category of problems by the second author in [Pan85 p.374]. For other topics of Nonsmooth Mechanics, especially concerning the introduction into Mechanics of Warga's derivate containers, of Ioffe's, fans, of Demyanov's quasidifferentials we refer respectively to [Pan85;87b, Mor88a, Stav91;92, Pan92]. Concerning finally the numerical treatment of variational and hemivariational inequalities several informations will be given in next Chapters of the book with respect to concrete numerical applications.

1.4.2 Existence Results for Variational and Hemivariational Inequalities

In this Section we shall give certain propositions concerning the existence of the solution of variational and hemivariational inequalities. These propositions will be applied in the sequel to multivalued B.I.Es, which are equivalent to variational or hemivariational inequalities formulated on the boundary.

i) Variational Inequalities I: Let $a(u,v)$ be for $u,v \in V$, where V is a real Hilbert space, a symmetric, continuous bilinear form and let (f,v) be a linear form. Assume that $f \in V'$, and let H be a pivot Hilbert space such that

$$V \subset H \subset V'$$

holds and the injections are continuous and dense. The duality pairing is denoted by $(\cdot,\cdot)$, the norm on H by $|\cdot|$ and on V by $\|\cdot\|$. The following problem is now considered: find $u \in K$ such that

$$a(u, v - u) - (f, v - u) \geq 0, \quad \forall v \in K \tag{1.160}$$

where K is a convex closed subset of V. Then there exists a symmetric, bounded linear operator $A: V \to V'$ such that

$$a(u,v) = (Au,v), \quad \forall u,v \in V.$$

By Prop. 1.2 the solution $u \in K$ of (1.160), if any exists, is a solution of the minimization problem

$$\Pi(u) = \min\{\Pi(v)|v \in K\}, \tag{1.160a}$$

where $\Pi(v) = \frac{1}{2}(Av,v) - (f,v)$, and conversely, on the assumption that $a(v,v) \geq 0$ $\forall v \in V$. Here first, we assume that $a(u,v)$ is coercive on V, i.e.,

$$a(u,u) \geq c\|u\|^2, \quad \forall u \in V, \quad c \text{ const} > 0. \tag{1.161}$$

Proposition 1.16 Suppose that (1.161) holds. Then the variational inequality (1.160) admits a unique solution.

The proof is based on (1.161) which guarantees that (1.6) holds.
Next we study the semicoercive case. Let

$$\ker A = \{v|v \in V, \, Av = 0\}, \tag{1.162}$$

and let Q (resp. $\tilde{Q}$) be the orthogonal projector of V onto $\ker A$ in the topology of H (resp. of V). Let $P = I - Q$ (resp. $\tilde{P} = I - \tilde{Q}$), where I denotes the identity mapping, and assume that

$$a(v,v) \geq c|Pv|^2, \quad \forall v \in V, \quad c \text{ const} > 0. \tag{1.163}$$

Obviously, $\ker A = \{v|v \in V \, a(v,v) = 0\}$. Moreover we assume that

$$\ker A \text{ is finite-dimensional} \tag{1.164}$$

and that

$$c_1\big(a(u,u)^{1/2} + |u|\big) \leq \|u\| \leq c_2\big(a(u,u)^{1/2} + |u|\big) \quad c_1, c_2 \text{ const} > 0. \tag{1.165}$$

Accordingly $a(u,u)^{1/2}$ is a seminorm on V and $|||u||| = a(u,u)^{1/2} + |u|$ is a norm on V equivalent to $\|u\|$. Let U be a subset of V which contains nonzero elements and let for $u \in U \; u \neq 0$

$$p(u,U) = \sup_t \left\{ t \,|\, t \geq 0 \frac{tu}{\|u\|} \in U \right\} \tag{1.166}$$

From (1.163) and (1.165) we obtain that

$$a(v,v) \geq c\|Pv\|^2 \quad \forall v \in V, \quad c \text{ const} > 0. \tag{1.167}$$

and that $\|Pv\|$ is equivalent to $\|\tilde{P}v\|$ for $v \in V$. The kernel of f, $\ker f = \{v \,|\, v \in V, (f,v) = 0\}$, is now considered. We symbolize by L the intersection of $\ker A$ and $\ker f$ and let L_1 be a subspace of $\ker A$ such that

$$\ker A = L \oplus L_1. \tag{1.168}$$

Now we denote by $\bar{Q}$ and $\bar{\bar{Q}}$ the orthogonal projectors of V onto L and L_1 respectively in the topology of V, and by $\bar{P}$ the operator $I - \bar{Q}$. Easily we can show that $\bar{\bar{Q}} = \bar{P} - \tilde{P}$ and $\bar{\bar{Q}} v$ is orthogonal to $\tilde{P}v$. With respect to $u_0 \in K$, we denote by K_{u_0} the set $\{v \,|\, v \in V, v + u_0 \in K\}$. The following propositions hold:

Proposition 1.17 Assume that $(1.163) \div (1.165)$ hold and that a $u_0 \in K$ exists such that

(i) $(f,\rho) < 0$ for $\rho \in \ker A \cap K_{u_0}$, $p(\bar{\bar{Q}} \rho, \bar{\bar{Q}} (\ker A \cap K_{u_0})) = \infty$, and $\tag{1.169}$
(ii) the set $\bar{P}(K_{u_0})$ is closed-in V. $\tag{1.170}$
Then (1.160) has a solution.

Proposition 1.18 Assume that (1.160) has a solution. Then for any $u_0 \in K$ and any $\rho \in \ker A \cap K_{u_0}$ such that $p(\bar{\bar{Q}} \rho, \bar{\bar{Q}} (\ker A \cap K_{u_0})) = \infty$, the condition

$$(f,\rho) \leq 0 \tag{1.171}$$

must hold.

Further we denote by $V/\ker A$ the quotient space and let $[u]$ be an element of it.

Proposition 1.19 If u is a solution of (1.160), then $[u] \in V/\ker A$ is uniquely determined. Every other solution u' of (1.160) can be written as $u' = u + \rho$ where

$$(f,\rho) = 0 \quad \text{and} \quad \rho \in \ker A \cap K_{u_0}. \tag{1.171a}$$

For the proofs of these propositions, see [Pan85] or for the slightly different case in which $V \equiv H$ see [Fich72]. To the same reference we refer for nonsymmetric bilinear forms.

ii) Variational Inequalities II: Further let us study a more general type of variational inequalities. An example of this type is the inequality (1.155). Note that if Φ is the indicator then (1.155) reduces to an inequality like the (1.160) or the (1.154). Let V be a real Hilbert space and V' its dual and let the functional framework of (1.160) hold. For $f \in V'$ we want to find a $u \in V$ such that

$$(T(u), v - u) + \Phi(v) - \Phi(u) \geq (f, v - u) \quad \forall v \in V \tag{1.172}$$

The following proposition holds.

Proposition 1.20 Assume that

(i)   the norm $\|v\|$ on V is equivalent to $p(v) + |v|$, where $p(v)$ and $|v|$ are a seminorm and a norm on V and on H respectively;

(ii) $\mathcal{G} = \{q \mid q \in V, p(q) = 0\}$ is a finite-dimensional subspace of V, Q is the orthogonal project of V onto $\mathcal{G}$ with respect to $|\cdot|$, $P = I - Q$, and

$$|Pv| \leq cp(v), \quad \forall v \in V, \quad c \text{ const} > 0; \tag{1.173}$$

(iii) Φ is a convex, l.s.c., proper functional on V

(iv)   the operator $T: V \to V'$ is pseudomonotone;

(v) $(Tv, q) = 0, \quad \forall v \in V$ and $\forall q \in \mathcal{G};$ \hfill (1.174)

and

(vi)   there exist $b > 1$ such that

$$T(v, v) \geq c\big(p(v)\big)^{b} \quad c \text{ const} > 0 \tag{1.175}$$

holds for every v.

(a) If assumptions (ii) and (v) hold, the relation

$$f|_{\mathcal{G}} \in \overline{R(\partial \Phi_{u_0})}, \quad \forall u_0 \in D(\Phi) \tag{1.176}$$

is a necessary condition for the existence of a solution of (1.172)

(b) If assumptions (i) through (vi) are valid, then a sufficient condition for the existence of a solution of (1.172) is that there exists at least one $u_0 \in D(\Phi)$ such that

$$f|_{\mathcal{G}} \in \text{ relint } R(\partial \Phi_{u_0}). \tag{1.177}$$

For the proof of this proposition which is based on Prop. 1.7 we refer to [Pot, Pan85]. Results of other studies of coercive, semicoercive or noncoercive variational inequalities may be found in [Lio67;71, Fre71, Glo, Ba84;86;88, Gas88a,b, Boi, Kin, Scha].

iii) Hemivariational Inequalities: Here we shall study the following hemivariational inequality: find $u \in V$ as to satisfy

$$a(u, v - u) + \int_{\Omega} j^{0}(u, v - u)d\Omega \geq (f, v - u) \quad \forall v \in V \tag{1.178}$$

where $V, a(\cdot, \cdot), f$ are defined as in (1.160), Ω is an open bounded subset on which V is defined and j is a locally Lipschitz function on $\mathbb{R}$ resulting from $\beta \in L^{\infty}_{\text{loc}} \subset \mathbb{R}$) by means of the procedure described by the relations (1.62)$\div$(1.66). We assume here that $H \equiv L^{2}(\Omega)$ and let

$$V \subset L^{2}(\Omega) \text{ be compact} \tag{1.179}$$

and

$$V \cap L^{\infty}(\Omega) \text{ be dense in } V \text{ for the } V\text{-norm.} \tag{1.180}$$

We assume that $a(\cdot, \cdot)$ is coercive, i.e. that (1.161) holds. The following proposition holds [Pan91].

Proposition 1.21 Suppose that β has the property: there is a $\xi \in \mathbb{R}$ such that

$$\operatorname*{ess\,sup}_{(-\infty,-\xi)} \beta(\xi) \leq \operatorname*{ess\,inf}_{(\xi,\infty)} \beta(\xi). \tag{1.181}$$

Then (1.178) admits at least one solution.

Further we assume that $a(\cdot,\cdot)$ is semicoercive and that (1.163) holds. Moreover let $q \in \ker A$. We denote by q_+ (resp. q_-) the positive (resp. negative) part of q i.e $q_+ = \sup(0,q)$ (resp. $q_- = \sup(0,-q)$) and by $\beta(-\infty)$ (resp. $\beta(+\infty)$) the $\limsup\limits_{\xi\to-\infty} \beta(\xi)$ (resp. $\liminf\limits_{\xi\to-\infty} \beta(\xi)$). Then the following necessary and sufficient conditions can be proved [Pan91]

Proposition 1.22 Let

$$\beta(-\infty) \leq \beta(\xi) \leq \beta(+\infty) \quad \forall \xi \in \mathbb{R} \tag{1.182}$$

Then the inequality

$$\int_\Omega [\beta(-\infty)q_+ - \beta(\infty)q_-]d\Omega \leq (f,q) \leq \int_\Omega [\beta(\infty)q_+ - \beta(-\infty)q_-]d\Omega \quad \forall q \in \ker A \tag{1.183}$$

is a necessary condition for the existence of a solution $u \in V$ of (1.178). If (1.182) holds as a strict inequality the same happens for (1.183).

Proposition 1.23 Suppose that

$$\beta(-\infty) < \beta(+\infty). \tag{1.184}$$

Then if

$$\int_\Omega [\beta(-\infty)q_+ - \beta(\infty)q_-]d\Omega < (f,q) < \int_\Omega [\beta(\infty)q_+ - \beta(-\infty)q_-]d\Omega \quad \forall q \in \ker A,\ q \neq 0, \tag{1.185}$$

the hemivariational inequality (1.178) has at least one solution.

The above types of hemivariational inequalities cover many applications. In this context we refer to [Pan85, Mor88a,b, Pan88;91].

Let us consider here also the case of a plane elastic body which on a part Γ_S of its boundary Γ is subjected to the adhesive contact boundary conditions of fig. 1.8b and of fig. 1.8d in the normal and the tangential directions respectively. Then both the laws (1.141) hold and the problem obeys the following hemivariational inequality for $V = [H^1(\Omega)]^2$: Find $u \in V$ such that

$$a(u,v-u) + \int_{\Gamma_S} j_N^0(u_N, v_N - u_N)d\Gamma + \int_{\Gamma_S} j_T^0(u_T, v_T - u_T)d\Gamma \geq (f,v-u) \quad \forall v \in V \tag{1.186}$$

Then since $H^2(\Omega) \to L^2(\Gamma)$ is compact and since $V \cap \{v|v_N \in L^\infty(\Gamma_S)\}$ is dense in the H^1-norm, propositions similar to 1.22 and 1.23 can be proved. For instance the analogous proposition to Prop. 1.23 reads.

Proposition 1.23a Suppose that (1.184) holds for both β_N and β_T. Then if

$$\int_{\Gamma_S} [\beta_N(-\infty)q_{N+} + \beta_T(-\infty)q_{T+} - \beta_N(\infty)q_{N-} - \beta_T(\infty)q_{T-}]d\Gamma \leq (f,q) \qquad (1.187)$$

$$\leq \int_{\Gamma_S} [\beta_N(\infty)q_{N+} + \beta_T(\infty)q_{T+} - \beta_N(-\infty)q_{N-} - \beta_T(-\infty)q_{T-}]d\Gamma$$

the inequality (1.186) has at least one solution.

Certain other interesting existence results for hemivariational inequalities have been given by Naniewicz [Nan88;89a,b;92], and for variational-hemivariational inequalities in [Pan91].

Chapter 2
The Direct and Indirect B.I.E.M. for Bilateral Problems

2.1 The B.V.P. of Linear Elasticity

It is, in general, impossible to determine solutions of partial differential equations which satisfy all the prescribed boundary conditions of the problem considered. Therefore, one looks for possibly optimal approximations which, by finite procedures produce approximate solutions, either of the partial differential equation in the whole domain Ω considered, or of the corresponding equivalent integral equation along the boundary Γ of this domain.

There are several possibilities for formulating boundary value problems (B.V.Ps) as boundary integral equations (B.I.Es), and, moreover, several methods for deriving these equations. This shall be made apparent in the framework of the classical, linear theory of elasticity.

Let $x_k(k = 1, 2, 3)$ be the co-ordinates of a material point referred to a Cartesian orthogonal coordinate system $Ox_1x_2x_3$ and, $\sigma = \{\sigma_{ij}(x)\}$ and $\varepsilon = \{\varepsilon_{ij}(x)\}$ be the tensor components of the stresses and, respectively, the strains at this point. Then, the following statical and kinematical basic relations have to be satisfied at each point of the domain $\Omega(,_j$ indicates the differentiation with respect to $x_j)$

$$D^\star\sigma = \bar{p} \ \hat{=} \ -\sigma_{ij,j}(x_k) = \bar{p}_i(x_k), \tag{2.1}$$

$$\varepsilon = Du \ \hat{=} \ \varepsilon_{ij}(x_k) = \frac{1}{2}[u_{i,j}(x_k) + u_{j,i}(x_k)]. \tag{2.2}$$

Equation (2.1) describes the equilibrium between the stresses and the given body forces $\bar{p} = \{\bar{p}_i\}$, while relations (2.2) guarantee the kinematic compatibility of the strains and of the displacement components u_i. D is the operator corresponding to (2.2) and $D^\star$ its conjugate.

The general constitutive relations for linear elastic material read

$$\sigma = C\varepsilon \ \hat{=} \ \sigma_{ij} = C_{ijkl}\varepsilon_{kl}, \tag{2.3}$$

where $C = \{C_{ijhk}\}$ is the Hooke's elasticity tensor. In this section we consider the specialized form of (2.3) for a three-dimensional isotropic body which is

$$\sigma_{ij} = \mu[\delta_{ik}\delta_{jl} + \frac{2\nu}{1-2\nu}\delta_{ij}\delta_{kl}]\varepsilon_{kl} = 2\mu[\varepsilon_{ij} + \frac{\nu}{1-2\nu}\delta_{ij}\varepsilon_{kk}]. \tag{2.4}$$

Here δ_{ij} is Kronecker's symbol, i.e. $\delta_{ij} = 0$ if $i \neq j$ and $\delta_{ij} = 1$ if $i = j$. The tensor $C = \{C_{ijkl}\}$, contains in the most general case,(cf. relations $(1.151a,b)$), i.e.

for anisotropic materials, 21 different constants, but for an isotropic material only the shear modulus μ and the Poisson's ratio ν with $\mu > 0$, $0 \leq \nu < 0.5$. These constants are connected with each other via the relation $\mu = E/2(1+\nu)$, where E is the Young's modulus of elasticity for the material.

Using the Eqs. (2.2) and (2.4), the equilibrium conditions (2.1) are transformed into the well-known Cauchy-Navier differential equations:

$$L(u) + \bar{p} = D^\star C D u + \bar{p} = 0 \quad \hat{=} \quad [\mu u_{i,jj} + \frac{\mu}{1-2\nu} u_{j,ji}] + \bar{p}_i = 0. \tag{2.5}$$

The boundary Γ consists of the open usually disjoint parts Γ_1 and Γ_2, i.e. $\Gamma = \bar{\Gamma}_1 + \bar{\Gamma}_2$, where $\bar{\Gamma}_1$ is the closure of Γ_1 etc. Then the boundary condition read:

$$u_i(x_k) = \bar{u}_i(x_k); \qquad x = \{x_k\} \in \Gamma_1, \tag{2.6}$$

$$T_i(x_k) = \sigma_{ij}(x_k) n_j(x_k) = \bar{T}_i(x_k); \qquad x = \{x_k\} \in \Gamma_2, \tag{2.7}$$

i.e. we have prescribed displacements $\bar{u}_i$ on Γ_1 and given boundary tractions $\bar{T}_i$ on Γ_2. There, $n = \{n_j\}$ denotes the outward unit normal vector to Γ.

Instead of (2.6), the displacements may be prescribed along the boundary Γ_1 as a function of the are length s, i.e. $u_i(s)$ may be given. Then the so-called distortion vector $U_i(s) = du_i(s)/ds$, i.e. the arc length derivative of the displacement, can be used as alternative boundary condition [He78a] by prescribing its components. In a complementary way, instead of the prescribed boundary traction vector $T_i(x)$, the so-called boundary-stress vector $t_i(s) = \int T_i(s) ds$ can serve to define the boundary condition on Γ_2, by prescribing its components. The B.V.P. requires that the displacements in the domain Ω and on Γ_2, the stresses in Ω as well as the tractions T_i on Γ_1, be determined.

In the following sub-sections, four different methods will be presented which permit the formulation of the B.V.Ps of elasticity as an integral equation system. The first three methods are direct methods whereas the fourth is an indirect one.

2.2 The Method of Weighted Residuals

Take an approximative solution, i.e a vector $\hat{u}_i$, not satisfying exactly the differential equation (2.5) and the boundary conditions (2.6) and (2.7). Accordingly, we can write the following expressions where R, R_1 and R_2 are residual terms:

$$L(\hat{u}) + \bar{p} \quad \hat{=} \quad \sigma_{ij,j}(\hat{u}) + \bar{p}_i = \mu[\hat{u}_{i,jj} + \frac{1}{1-2\nu} \hat{u}_{j,ji}] + \bar{p}_i =: R(\hat{u}), \tag{2.8}$$

$$G(\hat{u}) - \bar{u} \quad \hat{=} \quad \hat{u}_i - \bar{u}_i =: R_1(\hat{u}) \text{ on } \Gamma_1, \tag{2.9}$$

$$S(\hat{u}) - \bar{T} \quad \hat{=} \quad T_i(\hat{u}) - \bar{T}_i = \mu[\hat{u}_{i,j} + \hat{u}_{j,i} + \frac{2\nu}{1-2\nu} \delta_{ij} \hat{u}_{k,k}] n_j - \bar{T}_i =: R_2(\hat{u}) \text{ on } \Gamma_2. \tag{2.10}$$

The "errors" $R(\hat{u})$, $R_1(\hat{u})$ and $R_2(\hat{u})$ are supposed to be distributed in the domain Ω and on the boundary Γ in such a way that they disappear "in the mean" after a multiplication with certain, "appropriate" weight functions w, $S(w)$ and $G(w)$, respectively.

Thus we may write

$$\int\limits_{\Omega} R(\hat{u}) \cdot w \, d\Omega + \int\limits_{\Gamma_1} R_1(\hat{u}) \cdot S(w) \, d\Gamma - \int\limits_{\Gamma_2} R_2(\hat{u}) \cdot G(w) \, d\Gamma \qquad (2.11)$$

$$= [L(\hat{u}) + \bar{p}, w]_{\Omega} + \{G(\hat{u}) - \bar{u}, S(w)\}_{\Gamma_1} - \{S(\hat{u}) - T, G(w)\}_{\Gamma_2} = 0$$

or, explicitly,

$$\int\limits_{\Omega} (\sigma_{ij,j}(\hat{u}) + \bar{p}_i) w_i \, d\Omega + \int\limits_{\Gamma_1} (\hat{u}_i - \bar{u}_i) T_i(w) \, d\Gamma - \int\limits_{\Gamma_2} (T_i(\hat{u}) - \bar{T}_i) w_i \, d\Gamma = 0. \qquad (2.12)$$

Herein and in the following, it will be supposed that all functions are appropriately regular, i.e. all the integrals exist and all the differentiations are permitted.

By a partial integration of the first term in Eq. (2.12) one obtains that

$$\int\limits_{\Omega} [\sigma_{ij}(\hat{u}) w_{i,j} - \bar{p}_i w_i] d\Omega = \int\limits_{\Gamma_1} [(\hat{u}_i - \bar{u}_i) T_i(w) + T_i(\hat{u}) w_i] d\Gamma + \int\limits_{\Gamma_2} \bar{T}_i w_i \, d\Gamma. \qquad (2.13)$$

Using the symmetry of the stress tensor and the inverse of the constitutive relations (2.4), we obtain that

$$\sigma_{ij}(\hat{u}) w_{i,j} = \sigma_{ij}(\hat{u}) \varepsilon_{ij}(w) = \varepsilon_{ij}(\hat{u}) \sigma_{ij}(w) = \hat{u}_{i,j} \sigma_{ij}(w). \qquad (2.14)$$

Therefore, it is possible again to integrate partially the first term in Eq. (2.13). Thus finally we obtain the relation

$$\int\limits_{\Omega} \hat{u}_i \sigma_{ij,j}(w) \, d\Omega = \int\limits_{\Gamma_1} [\bar{u}_i T_i(w) - T_i(\hat{u}) w_i] d\Gamma \qquad (2.15)$$

$$+ \int\limits_{\Gamma_2} [\hat{u}_i T_i(w) - \bar{T}_i w_i] d\Gamma - \int\limits_{\Omega} \bar{p}_i w_i \, d\Omega.$$

By setting in (2.13) $\hat{u}_i = u_i$ and $w_i = u_i^* - u_i$, where u_i^* satisfies the geometrical boundary conditions on Γ_1, we obtain the expression of the "principle" of virtual work. If further $u_i^* - u_i = \delta u_i$, where δ denotes here the classical variation, (2.13) expresses the fact that at the position of equilibrium the first variation $\delta\Pi$ of the potential energy Π is zero. This is the starting position for finding Ritz-Galerkin approximations for the solution u_i of the B.V.P. by passing to finite dimensions. If, moreover, locally defined shape function are used, we are led to the F.E. approximation u_i^h of the solution [Tör85]. Finally, note that, if in (2.15) the weight functions w_i satisfy the equilibrium condition (2.5), we are led to the basic integral relation of the method of Trefftz.

Now, let us take as special weight function w_i in (2.15) the fundamental solution of the Navier-Cauchy's differential equations (2.5), i.e. the solution of

$$L(w) = \sigma_{ij,j}(w) = \mu[w_{i,jj} + \frac{1}{1 - 2\nu} w_{j,ji}] = -\delta_i^k \delta(x - \xi), \qquad (2.16)$$

Then eq. (2.15) implies, because of the characteristic Dirac functions properties, the relation

$$-\int_\Omega u_i \delta_i^k \delta(x-\xi) d\Omega_x \;=\; -u_k(\xi) = \int_{\Gamma_1} [\bar{u}_i T_i(w^{(k)}) - T_i(u) w_i^{(k)}] d\Gamma_x \qquad (2.17)$$

$$+ \int_{\Gamma_2} [u_i T_i(w^{(k)}) - \bar{T}_i w_i^{(k)}] d\Gamma_x - \int_\Omega \bar{p}_i w_i^{(k)} d\Omega_x.$$

Physically, this fundamental solution describes the displacement $w_i^{(k)}(x,\xi)$ at the observation point x due to a unit-force at source point ξ in the direction x_k. Here $\delta(x-\xi)$ denotes the Dirac measure and δ_i^k is the Kronecker symbol.

In the full space, this displacement field w_i^k can be found for instance in ([Bre84], $r = |x-\xi|$)

$$w_i^{(k)}(x,\xi) = \frac{r^{-1}}{16\pi\mu(1-\nu)}[(3-4\nu)\delta_{ik} + r_{,i}r_{,k}]. \qquad (2.18)$$

From this and the relations (2.9)(2.11)(2.14) the singular boundary traction vector $(r_{,n} = r_{,i}n_i$ at the point x)

$$T_i^{(k)}(x,\xi) = -\frac{r^{-2}}{8\pi(1-\nu)}[(1-2\nu)(r_{,n}\delta_{ik} - r_{,i}n_k + r_{,k}n_i) + 3r_{,n}r_{,i}r_{,k}] \qquad (2.19)$$

is obtained. In the full plain, the displacements produced by a unit-force at ξ in the direction $x_\beta, (\beta = 1, 2)$, are given in the case of plane strain states by (see e.g.[Bre84])

$$w_\alpha^{(\beta)}(x,\xi) = \frac{1}{8\pi\mu}\frac{1}{1-\nu}[-(3-4\nu)\delta_{\alpha\beta}\ln r + r_{,\alpha}r_{,\beta}] \qquad (2.20)$$

while in the case of plane stress states by

$$w_\alpha^{(\beta)}(x,\xi) = \frac{1}{8\pi\mu}[-(3-\nu)\delta_{\alpha\beta}\ln r + (1+\nu)r_{,\alpha}r_{,\beta}]. \qquad (2.21)$$

The corresponding boundary stress vectors can be determined via the displacement-strain relation (2.3), the Hooke's law (2.4) and the equilibrium conditions on the boundary (2.7). They are for plane strain states

$$T_\alpha^{(\beta)}(x,\xi) \;=\; \frac{r^{-1}}{4\pi(1-\nu)}\{-r_{,n}[(1-2\nu)\delta_{\alpha\beta} + 2r_{,\alpha}r_{,\beta}] \qquad (2.22)$$

$$+(1-2\nu)[r_{,\beta}n_\alpha - r_{,\alpha}n_\beta]\}.$$

For plane stress states, instead of (2.4) the following constitutive relation have to be used [Sok]

$$\sigma_{\alpha\beta} = 2\mu[\varepsilon_{\alpha\beta} + \frac{\nu}{1-\nu}\delta_{\alpha\beta}\varepsilon_{\gamma\gamma}]. \qquad (2.23)$$

Then, one obtains

$$T_\alpha^\beta(x,\xi) \;=\; \frac{r^{-1}}{4\pi}\{-r_{,n}[(1-\nu)\delta_{\alpha\beta} + 2(1+\nu)r_{,\alpha}r_{,\beta}] \qquad (2.24)$$

$$+(1-\nu)[r_{,\beta}n_\alpha - r_{,\alpha}n_\beta]\}.$$

Note that the expressions for plane stress states can be obtained from those for plane strain if ν is replaced by $\bar{\nu} = \nu(1 + \nu)$.

Further we give some fundamental solutions for other types of differential equations. These expressions are important for the development of the corresponding B.I.E. approach.

differential equation	**fundamental solution**

one-dimensional $(r = |x - \xi|)$

$$w_{,xx} = \delta(r) \qquad r/2$$

(Laplace equation)

$$w_{,xx} + \lambda^2 w = \delta(r) \qquad \sin(\lambda r)/2\lambda$$

(Helmholtz equation)

two-dimensional $(r = |x - \xi|)$

$$\Delta w = \delta(r) \qquad \frac{1}{4\pi} \ln(r^2)$$

(Laplace equation)

$$k_1 w_{,11} + k_2 w_{,22} = \delta(r) \qquad \frac{1}{4\pi} \frac{1}{\sqrt{k_1 k_2}} \ln \frac{(x_j - \xi_j)^2}{k_j}$$

(Darcy equation)

$$\Delta w + \lambda^2 w = \delta(r) \qquad -H_0^{(2)}(\lambda r)/4i, \ i = \sqrt{-1}$$

(Helmholtz equation)

$$\Delta\Delta w - \lambda^4 w = \delta(r) \qquad \frac{1}{8i}[H_0^{(2)}(\lambda r) - \frac{2i}{\pi} K_0(\lambda r)]/\lambda^2$$

(transformed plate equation)

$$(c_1^2 - c_2^2)u_{i,ij} + c_2^2 u_{j,ii} - s^2 u_j = \delta_j^k \delta(r)$$
$$\frac{-1}{2\pi c_2^2}\left\{[K_2(\frac{sr}{c_2}) - (\frac{c_2}{c_1})^2 K_2(\frac{sr}{c_1})]r_{,j}r_{,k} + \right.$$

(transformed equation of motion)
$$\left. [(\frac{c_2}{c_1})^2 \frac{c_1}{sr} K_1(\frac{sr}{c_1}) - \frac{c_2}{sr} K_1(\frac{sr}{c_2}) - K_0(\frac{sr}{c_2})]\delta_{jk}\right\}$$

three-dimensional $(r = |x - \xi|)$

$$\Delta w = \delta(r) \qquad -\frac{1}{4\pi r}$$

(Laplace equation)

$$\Delta w + \lambda^2 w = \delta(r) \qquad -\frac{1}{4\pi r}e^{-i\lambda r}$$

(Helmholtz equation)

$$k_1 w_{,11} + k_2 w_{,22} + k_3 w_{,33} = \delta(r) \qquad -\frac{1}{4\pi}[k_1 k_2 k_3 \frac{(x_j - \xi_j)^2}{k_j}]^{-1/2}$$

(Darcy equation)

Equation (2.17) is actually an integral equation for the determination of the unknown displacements $u_k(\xi)$ in the interior of the domain Ω. But, it can only be used if either

a)   the fundamental solution $w_i^{(k)}$ corresponds to the homogeneous boundary conditions $w_i^{(k)} = 0$ on Γ_1 and $T_i^{(k)} = 0$ on Γ_2, i.e. it is possible to find the so-called Green's functions of the problem, or

b)   the unknown boundary reactions $T_i(u)$ on Γ_1 and u_i on Γ_2 have been determined before.

Thus the integral equation (2.17) constitutes the starting point of a direct B.I.E. method. In the sequel we shall explain the term "direct" after introducing and describing the so-called "indirect" method. The contrast to the "indirect" method will fully explain the term "direct".

2.3 Generalized Variational Principles

Generalized variational expressions are not only the basis for a direct formulation of the F.E. method, but can also serve as the starting point for the derivation of a direct B.I.E. method.

In the linear theory of elasticity, the most general variational functional is the well-known Hu-Washizu functional [Was]

$$P(u,\sigma,\varepsilon) \;=\; \int_{\Omega} \{U(\varepsilon_{ij}) - \bar{p}_i u_i + \sigma_{ij}[\frac{1}{2}(u_{i,j} + u_{j,i}) - \varepsilon_{ij}]\}d\Omega \tag{2.25}$$

$$- \int_{\Gamma_1} \sigma_{ij}n_j(u_i - \bar{u}_i)d\Gamma - \int_{\Gamma_2} \bar{T}_i u_i d\Gamma.$$

Herein, $U(\varepsilon_{ij})$ denotes the specific deformation energy which gives through $\partial U(\varepsilon_{ij})/\partial \varepsilon_{kl} = \sigma_{kl}$ the constitutive relations (2.4).Obeying to this and, moreover, to equation (2.14), the first variation of the functional in (2.25) yields

$$\delta P(u,\sigma,\varepsilon) \;=\; \partial P/\partial u \cdot \delta u + \partial P/\partial \sigma \cdot \delta \sigma + \partial P/\partial \varepsilon \cdot \delta \varepsilon \tag{2.26}$$

$$= \int_{\Omega} \{-\bar{p}_i \delta u_i + \delta \sigma_{ij}[u_{i,j} - \varepsilon_{ij}] + \sigma_{ij}\delta u_{i,j}\}d\Omega$$

$$- \int_{\Gamma_1} \{\delta \sigma_{ij}(u_i - \bar{u}_i) + \sigma_{ij}n_j\delta u_i\}d\Gamma - \int_{\Gamma_2} \bar{T}_i \delta u_i d\Gamma.$$

Then, integrating partially the second term of the domain integral implies

$$\int_{\Omega} \delta \sigma_{ij}u_{i,j}d\Omega = \int_{\Gamma} \delta \sigma_{ij}n_j u_i d\Gamma - \int_{\Omega} \delta \sigma_{ij,j}u_i d\Omega. \tag{2.27}$$

Then zeroing of the first variation of P yields

$$-\int_{\Omega} \delta\sigma_{ij,j} u_i d\Omega + \int_{\Gamma_1} \delta\sigma_{ij} n_j \bar{u}_i d\Gamma + \int_{\Gamma_2} \delta\sigma_{ij} n_j u_i d\Gamma \qquad (2.28)$$

$$= \int_{\Omega} \{\varepsilon_{ij}\delta\sigma_{ij} - \sigma_{ij}\delta u_{i,j}\} d\Omega + \int_{\Omega} \bar{p}_i \delta u_i d\Omega + \int_{\Gamma_1} \sigma_{ij} n_j \delta u_i d\Gamma + \int_{\Gamma_2} \bar{T}_i \delta u_i d\Gamma.$$

Let us now take into account that the variation of the stresses $\delta\sigma_{ij}$ is dependent on the variation of the displacements δu_i; indeed $\varepsilon_{ij}\delta\sigma_{ij} = \sigma_{ij}\delta u_{i,j}$ holds. Using instead of an arbitrary displacement variation δu_i the special displacements $w_i^{(k)}(\xi, x)$ produced by a unit force at the point ξ in the direction x_k, i.e. the fundamental solution

$$-\delta\sigma_{ij,j}(u) = -\sigma_{ij,j}(\delta u) = \delta_i^k \delta(x - \xi), \qquad (2.29)$$

the equation (2.28) implies the so-called Somigliana identity

$$u_k(\xi) + \int_{\Gamma_1} T_i^{(k)}(x, \xi)\bar{u}_i(x)d\Gamma_x + \int_{\Gamma_2} T_i^{(k)}(x, \xi)u_i(x)d\Gamma_x \qquad (2.30)$$

$$= \int_{\Gamma_1} T_i(x)w_i^{(k)}(x, \xi)d\Gamma_x + \int_{\Gamma_2} \bar{T}_i(x)w_i^{(k)}(x, \xi)d\Gamma_x + \int_{\Omega} \bar{p}_i w_i^{(k)}(x, \xi)d\Omega_x.$$

This is identical to the relation (2.17) which has already been derived by the method of the weighted residuals. Relation (2.30) is called also "reciprocal relation" in the engineering literature for obvious reasons.

2.4 The Use of Reciprocal Theorems

As shown by Tonti et al. [Ton, An86a], in many physical theories the basic differential equations $L(u) + p = 0$ have the property that it is possible to split their differential operator into a "product" D^*CD. Therein, D^* indicates an operator which is formally adjoint to D [Col], and C is assumed to be self-adjoint. Thus, the operator L is formally self-adjoint. In such cases, it is possible to derive not only complementary variational principles [An86a, Ode], but also to formulate "reciprocal relations" of the type

$$[u, D^*CDw]_\Omega + \{Gu, Sw\}_\Gamma = [D^*CDu, w]_\Omega + \{Su, Gw\}_\Gamma. \qquad (2.31)$$

Let us explain it by means of an example: For Poisson's equation $L\Phi = [\varepsilon\Phi_{,i}]_{,i} = -\bar{p}$, defined in the domain Ω with the boundary conditions $G\Phi = \bar{\Phi}$ on Γ_1 and $S\Phi = -\varepsilon\Phi_{,n} = \bar{\psi}$ on Γ_2, the reciprocal relation (2.31) yields

$$\int_{\Omega} \Phi(\varepsilon w_{,i})_{,i} d\Omega + \int_{\Gamma} \Phi(-\varepsilon w_{,n})d\Gamma = \int_{\Omega} (\varepsilon\Phi_{,i})_{,i} w d\Omega + \int_{\Gamma} (-\varepsilon\Phi_{,n})w d\Gamma,$$

which is known as the second Green's theorem.

In the linear theory of elasticity, the reciprocal relation (2.31) yields directly (2.15) and is called Betti-Maxwell [Bet] theorem. In the terminology of structural analysis, where this theorem is of paramount importance, this theorem may be formulated in the following form:

Theorem of BETTI-MAXWELL: In a linearly elastic body and in the case of quasi static deformations, the work A_{12}, performed by a system of forces F_1 for the displacements due to a subsequent loading system F_2, is equal to the work A_{21} which is performed by the system F_2 for the displacements corresponding to the force system F_1.

This can be expressed by the following domain- and boundary integrals:

$$\int_\Omega \overset{2}{u_i}\overset{1}{p_i}\, d\Omega + \int_\Gamma \overset{2}{u_i}\overset{1}{T_i}\, d\Gamma = \int_\Omega \overset{2}{p_i}\overset{1}{u_i}\, d\Omega + \int_\Gamma \overset{2}{T_i}\overset{1}{u_i}\, d\Gamma. \tag{2.32}$$

When the loading system F_2 is chosen to be a unit force at the point ξ of the full space in the direction x_k, i.e. $\overset{2}{p_i}= \delta_i^k \delta(x - \xi)$ and its reactions along the whole boundary Γ are $\overset{2}{T_i}= T_i^{(k)}(x,\xi)$, one obtains with $\overset{2}{u_i}= w_i^{(k)}(x,\xi)$ the aforementioned Somigliana identity

$$u_k(\xi) \;=\; \int_\Gamma [T_i(x)w_i^{(k)}(x,\xi) - u_i(x)T_i^{(k)}(x,\xi)]d\Gamma_x \tag{2.33}$$

$$+ \int_\Omega \bar{p}_i(x)w_i^{(k)}(x,\xi)d\Omega_x \quad \text{for } \xi \in \Omega$$

directly from (2.32) (cf. also (2.30)). Note that in contrast to the use of Betti-Maxwell theorem, in Sect.2.2 and 2.3 the equation (2.33) has been obtained as the final result of several transformation steps.

Since the boundary conditions (2.6) and (2.7) prescribe the displacements along Γ_1 and the tractions along Γ_2, it is sufficient to determine the corresponding boundary reactions $T_i(x)$ on Γ_1 and boundary displacements $u_i(x)$ on Γ_2. The equations for their determination are obtained from (2.33) by shifting the action point ξ of the unit force to a boundary point (in the limit from the interior of the body). This yields the following boundary integral equation:

$$c_i^{(k)}(\xi)u_i(\xi) + \fint_\Gamma T_i^{(k)}(x,\xi)u_i(x)d\Gamma_x \tag{2.34}$$

$$= \int_\Gamma w_i^{(k)}(x,\xi)T_i(x)d\Gamma_x + \int_\Omega \bar{p}_i(x,\xi)w_i^{(k)}(x,\xi)d\Omega_x, \; \xi \in \Gamma.$$

If an integrand is strongly singular – in this case the kernel $T_i^{(k)}(x,\xi)$ – the related integral has to be evaluated as a Cauchy principal value (cf. e.g.[Mus53a],[Mik65b]).

The factors $c_i^{(k)}(\xi)$ of the integral-free terms are dependent on the smoothness of the boundary at the point ξ. In the case of two-dimensional problems, these factors can be given explicitly as functions of the interior angle $\Delta\varphi$ and of the outer normal angles φ_- and φ_+, (see fig. 2.1). For plane strain states, they are [Ha81b]

$$c_i^{(k)}(\xi) = \frac{1}{8\pi(1-\nu)}$$

$$\cdot \begin{bmatrix} 4(1-\nu)\Delta\varphi - \sin 2\varphi_- + \sin 2\varphi_- & 2\sin^2\varphi_- - 2\sin^2\varphi_+ \\ \text{symmetric} & 4(1-\nu)\Delta\varphi + \sin 2\varphi_- - \sin 2\varphi_+ \end{bmatrix}$$

(2.35)

For points ξ on smooth regions of the boundary, $\Delta\varphi = \pi$ and $\varphi_- = \varphi_+$, and thus, we have $c_i^{(k)}(\xi) = 0.5\delta_i^k$ in any case. If ξ is at a corner of the boundary, besides the interior corner angle $\Delta\varphi$, the position of the corner relative to the origin of the co-ordinate system (x_1, x_2) must be taken into account (see also next example).

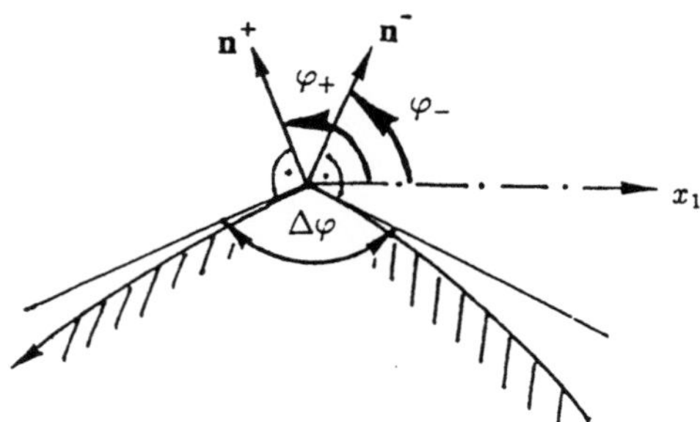

Fig. 2.1: Boundary corner in plane problems. The interior angle $\Delta\varphi$ and the outer normal angles φ_+ and φ_-

In three-dimensional bodies, the factors $c_i^{(k)}(\xi)$ can not be explicitly given for arbitrary corners and edges. In general, they can only be defined by the integral expression (see fig. 2.2)

$$c_i^{(k)}(\xi) = \lim_{\varepsilon \to 0} \int_{\Gamma_\varepsilon(\xi)} T_i^{(k)}(x,\xi)d\Gamma_x = \{\delta_i^k + \lim_{\varepsilon \to 0} \int_{\Gamma_\varepsilon^\varepsilon(\xi)} T_i^{(k)}(x,\xi)d\Gamma_x\}$$

(2.36)

which can be evaluated for some special cases. For points ξ on a smooth surface

$$\{c_i^{(k)}(\xi)\} = \frac{1}{8\pi}\begin{bmatrix} 4\pi & 0 & 0 \\ 0 & 4\pi & 0 \\ 0 & 0 & 4\pi \end{bmatrix},$$

(2.37)

holds. When ξ is a point on an edge parallel to the x_1-axis with an interior angle $\Delta\varphi = 90°$, we find

$$\{c_i^{(k)}(\xi)\} = \frac{1}{8\pi}\begin{bmatrix} 2\pi & 0 & 0 \\ 0 & 2\pi & \frac{2}{1-\nu} \\ 0 & \frac{2}{1-\nu} & 2\pi \end{bmatrix},$$

(2.38)

while for points ξ at a corner with the same interior angle $\Delta\varphi = 90°$ the following factors hold

$$\{c_i^{(k)}(\xi)\} = \frac{1}{8\pi} \left[\begin{array}{ccc} \pi & \frac{1}{1-\nu} & \frac{1}{1-\nu} \\ \frac{1}{1-\nu} & \pi & \frac{1}{1-\nu} \\ \frac{1}{1-\nu} & \frac{1}{1-\nu} & \pi \end{array} \right]. \tag{2.39}$$

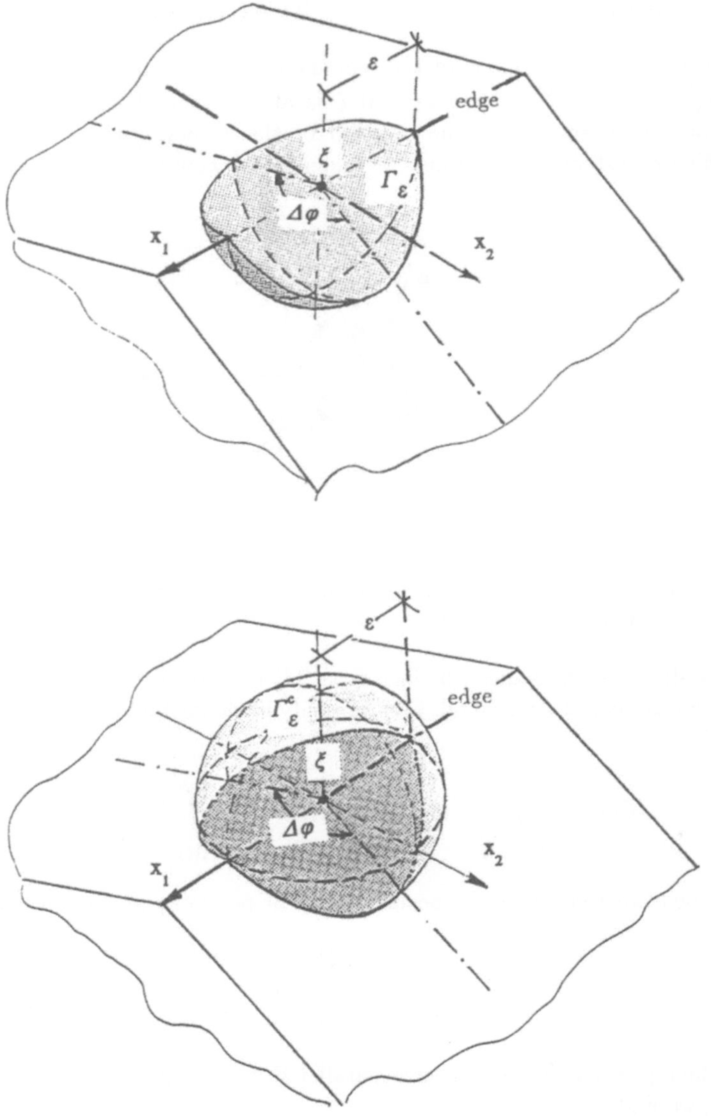

Fig. 2.2: Source point ξ on an edge with an interior angle $\Delta\varphi(\xi)$ and ε-surroundings $\Gamma_\varepsilon(\xi)$ and $\Gamma_\varepsilon^c(\xi)$, respectively

Note that the integral-free terms $c_i^{(k)}(\xi)$ can indirectly be determined, i.e. without evaluating the limit integrals (2.36). Because "rigid-body" displacements $u_i(x) = a_i$ where a_i is a constant, cause no stresses in the body, the expression

$$c_i^{(k)}(\xi) = - \int_\Gamma T_i^{(k)}(x,\xi)d\Gamma_x$$

is valid for any source point ξ, also for corners. In the numerical procedure for an approximative solution of the integral equations, the value of the above integral can easily be found by adding up the respective integrals over all boundary elements. Finally let us illustrate through an example the method for calculating the factors $c_i^{(k)}(\xi)$ for ξ at corners and edges.

a)   In the case of a rectangular slice (plane strain state), assuming $\varphi_+ > \varphi_-$, one obtains from the formula (2.35) for the corners I and III (fig. 2.3a)

$$\{c_\alpha^{(\beta)}\} = \frac{1}{4\pi(1-\nu)} \begin{bmatrix} (1-\nu)\pi & -1 \\ -1 & (1-\nu)\pi \end{bmatrix}, \quad \alpha,\beta = 1,2$$

and for the corners II and IV

$$\{c_\alpha^{(\beta)}\} = \frac{1}{4\pi(1-\nu)} \begin{bmatrix} (1-\nu)\pi & +1 \\ +1 & (1-\nu)\pi \end{bmatrix}, \quad \alpha,\beta = 1,2.$$

b)   In the case of a three-dimensional rod (fig.2.3) one can state that all the corner have an interior 90°-angle, and, therefore, the factors' matrix has the form (2.37) for points ξ on the edges. Depending on the position of the edge, the side-diagonal terms $\frac{1}{8\pi}\frac{2}{1-\nu}$ in the matrix (2.38) are shifted to different positions: from the positions $c_2^{(3)}$ and $c_3^{(2)}$, for a x_1-parallel edge, to the positions $c_1^{(2)}$ and $c_2^{(1)}$, for x_3-parallel edges, and to $c_1^{(3)}$ and $c_3^{(1)}$, for x_2-parallel edges, respectively.

2.5 The Singularity Method (Indirect Method)

Until now we have described three "equivalent" direct methods for the derivation of B.I.Es. This section deals with an indirect method, the singularity method. The requirements and the methodology for solving via the so-called indirect or singularity method are from many respects different from those in the previous sections. Here, it holds that

(i)   The validity, knowledge or derivation of a reciprocal theorem is not necessary.

(ii)   Besides the "usual" fundamental solutions, i.e. the basic fundamental solutions of the considered differential equations, many other singularities or their combinations may be used [He78a, Za85a, Za85b].

(iii)   The singularities must not necessarily act on the boundary of the considered body Ω [He78b].

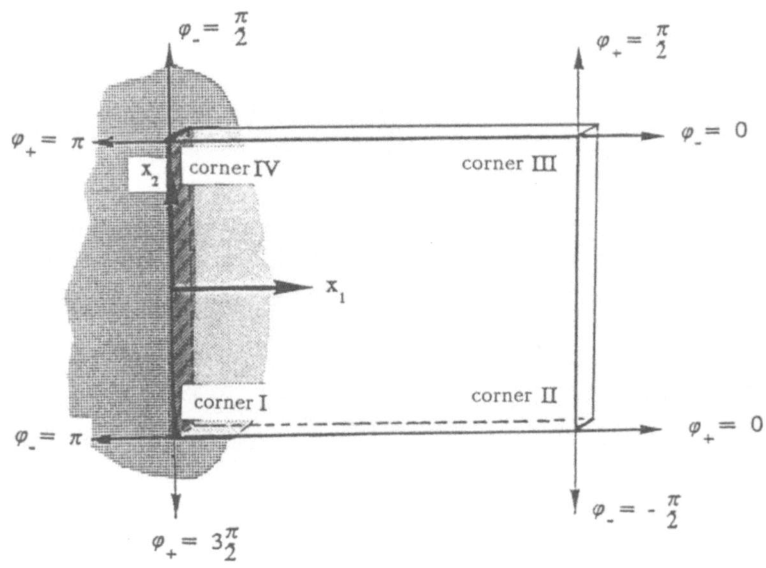

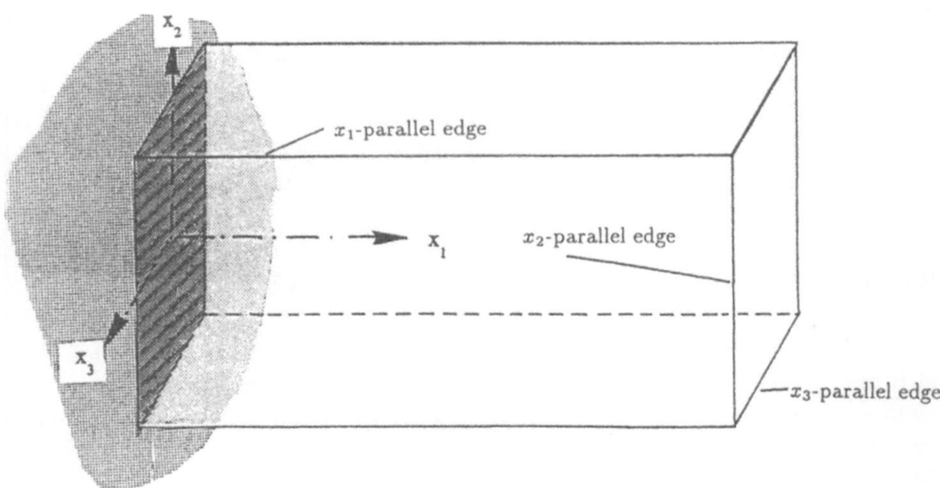

Fig. 2.3: On the calculation of the factors $c_i^{(k)}(\xi)$

These facts give a higher flexibility in formulating B.V.Ps as integral equations. But, in the application of this "indirect" method it has to be assumed that a particular solution equilibrating the body forces $\bar{p}_i$, if the latter are taken into account, is known. Then, the solution of a homogeneous differential equation system can be taken into account under the condition that only the boundary conditions will appropriately change due to the "split-off" of the particular solution.

Let us illustrate the singularity method by considering certain B.V.Ps of two-dimensional elastic bodies. At the points ξ on a surface Γ^+ – either coinciding with the real boundary Γ of the body Ω or a certain, in general, constant distance of it – (see Figure 2.4) – basic singularities, e.g. unit forces in the direction x_k, are acting. The corresponding displacements, caused by them, are known at each observation point x from the influence functions of the unit forces.

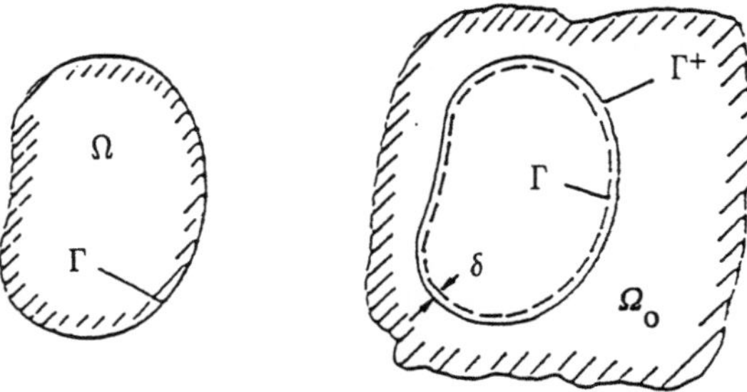

Fig. 2.4: Body Ω^+ embedded in the full space $\Omega_0 = \mathbb{R}^3$ or $\mathbb{R}^n$ with its boundary Γ and the fictitious boundary Γ^+ at a distance δ

These influence functions are the connection links between the variables, describing the state of an elastic body, and the singularities, causing this state. Obviously fundamental solutions are special influence functions. Indeed, the fundamental solution $u_i^{(k)}(x,\xi)$ is the influence function $(uF)_{i.j}(x,\xi)$ which gives the displacement u_i at the point x due to an unit force $F_j = \delta(x - \xi)\delta_j^k$ located at the point ξ and acting in the direction x_k, i.e.

$$u_i(x) = (uF)_{i.j}(x,\xi)F_j(\xi). \tag{2.40}$$

Using such influence functions, a boundary value problem will be solved according to the following idea:

Solution principle: Determine layers of basic singularities, e.g. of forces $F_j(\xi)d\Gamma(\xi)$, on a fictitious boundary Γ^+ embedded in the full space (or in the full plane) Ω_o such that the states caused at the boundary Γ of the real body Ω, take the prescribed boundary values of the problem. Note that the case $\Gamma^+ \to \Gamma$ can be also considered.

Different types of basic singularities can be used. Additionally to the well-known "statical" singularities, so-called "geometrical" singularities are notable. The most important basic forms of these two types shall be presented in the following.

The above mentioned unit single forces $F_j = \delta_j^k\delta(x - \xi)$ acting at the source points ξ into the direction x_k are the most known of the statical singularities. But besides, singularities of higher order exist, e.g. single moments (dipoles $m_{\alpha\beta}$) which can be produced by force couples (see fig. 2.5) when the distance 2ε tends to zero. Analytically, the corresponding influence functions are obtained by differentiating with respect to the source point co-ordinates ξ_j [Gir,Nit].

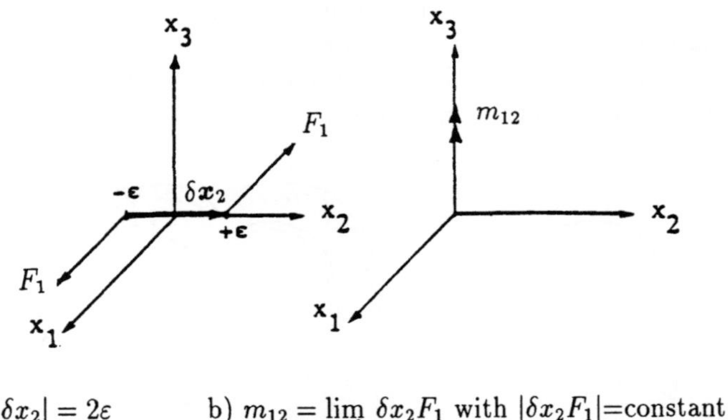

a) $|\delta x_2| = 2\varepsilon$ b) $m_{12} = \lim_{\varepsilon \to 0} \delta x_2 F_1$ with $|\delta x_2 F_1|$=constant

Fig. 2.5: a) Force couple F_1 with distance vector δx_2 b) force dipole m_{12}

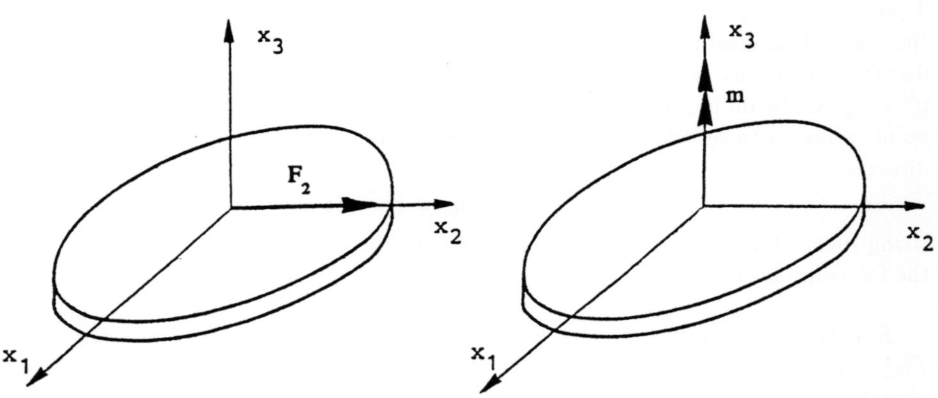

Fig. 2.6: Static singularities in slice problems

For plane problems, besides the two single forces F_β acting in both the directions $x_\beta, (\beta = 1, 2)$, the dipole $m_{\alpha\beta}$ or better the resulting single moment $m = e_{\alpha\beta} m_{\alpha\beta}$ ([He78a]), around the axis x_3 which is orthogonal to the plane $x_1 - x_2$, are the main statical singularities (fig. 2.6)

On the surface (the boundaries) of the body under consideration, only certain components of tensorial state terms, the so-called "torso" state variables [He78a] are essential. The vector of the boundary stresses T_i is the best known of such states; it is obtained via the equation (2.7) from the stress tensor σ_{ij} and the outward unit normal vector n_j. Analogous multiplications etc with the unit normal vector, but also with the unit tangential vector are applied to other tensorial states, also: for example, along the boundary Γ the force-dipol vector m_β [He78a] will be deduced from the tensor of the force-dipoles $m_{\alpha\beta}$ by multiplication with the unit tangential vector s_α and addition with respect to the index α, i.e $m_\beta = -s_\alpha m_{\alpha\beta}$. Thus, plane elastic states can be investigated by using either the influence functions of single forces $F_\beta(\xi), \beta = 1, 2$, or the ones produced by force-dipoles, by the single moment $m(\xi)$, and by the force-dipole vector $m_\beta(\xi)$, respectively.

In the neighborhood of the source points, these influence functions show a singular behavior which is dependent on the type of the singularity and, moreover, on the actually considered state variable. Hence, the state variable "displacement field" u_α, caused by the singularity "single force" at the point ξ acting in the direction x_β

$$u_\alpha(x) = (uF)_{\alpha\cdot\beta}(x,\xi)F_\beta(\xi), \tag{2.41}$$

has only a logarithmically singular influence function. In the case of plane stress states, it is explicitly (cf. (2.21))

$$(uF)_{\alpha\cdot\beta}(x,\xi) = \frac{1}{8\pi\mu}[-(3-\nu)\delta_{\alpha\beta}\ln r + (1+\nu)r_{,\alpha}\,r_{,\beta} + \frac{1}{2}\delta_{\alpha\beta}]. \tag{2.42}$$

On the other hand, the influence functions of the corresponding boundary traction $T_\alpha(x) = \sigma_{\alpha\rho}(x)n_\rho(x)$ (see 2.24) is given by the expression

$$(TF)_{\alpha\cdot\beta}(x,\xi) = \frac{1}{4\pi}\frac{1}{r}\{-r_{,\rho}n_\rho(x)[(1-\nu)\delta_{\alpha\beta} + 2(1+\nu)r_{,\alpha}\,r_{,\beta}] \tag{2.43}$$
$$+ (1-\nu)[r_{,\beta}\,n_\alpha(x) - r_{,\alpha}\,n_\beta(x)]\}$$

and those of the displacements $u_\alpha(x)$, due to a single moment m or a force dipole m_β at the point ξ [He78a] are respectively given by the expressions

$$(um)_{\alpha\cdot}(x,\xi) = \frac{1}{4\pi\mu}\frac{1}{r}e_{\beta\alpha}r_{,\beta} \tag{2.44}$$

$$(um)_{\alpha\cdot\beta}(x,\xi) = \frac{1}{8\pi\mu}\frac{1}{r}\{(1+\nu)r_{,\rho}[e_{\rho\alpha}r_{,\beta} + e_{\rho\beta}r_{,\alpha}]r_{,\gamma}n_\gamma(\xi) - (3-\nu)\delta_{\alpha\beta}r_{,\gamma}s_\gamma(\xi)\}. \tag{2.45}$$

Relations (2.43)÷(2.45) are singular like r^{-1}, i.e. they are strongly singular. This means that integrals with such kernels have a meaning only as Cauchy's principal values [Had].

The influence functions of the boundary tractions $T_\alpha(x)$ due to a single moment m and to a force dipole m_β [He78a] are respectively,

$$(Tm)_{\alpha\cdot}(x,\xi) = \frac{1}{2\pi}\frac{r_{,\gamma}}{r^2}\{r_{,\alpha}s_\gamma(x) - e_{\rho\alpha}r_{,\rho}n_\gamma(x)\}, \tag{2.46}$$

$$(Tm)_{\alpha,\beta}(x,\xi) \;=\; \frac{r_{,\gamma}r_{,\rho}}{4\pi r^2}\{[n_\rho(x)s_\gamma(\xi) + s_\rho(x)n_\gamma(\xi)][(1-\nu)\delta_{\alpha\beta}$$
$$+2(1+\nu)r_{,\alpha}r_{,\beta}] + n_\rho(x)n_\gamma(\xi)[(3+\nu)e_{\alpha\beta} - 4(1+\nu)e_{\lambda\beta}r_{,\lambda}r_{,\alpha}]$$
$$-s_\rho(x)s_\gamma(\xi)(1-\nu)e_{\alpha\beta}\} \qquad (2.47)$$

and thus they are hypersingular. For such kernels, Cauchy principal values exist only under additional assumptions concerning the singularity densities [Had]. In this case, the singularities single moments $m(\xi)$ and force dipoles $m_\beta(\xi)$, must fullfil these assumptions. Often (see e.g. [Ioa82]), such integrals are evaluated by certain regularized versions, the so-called "integrals in the sense of Hadamard" [Had, Hac].

Besides the statical singularities, the so-called geometrical singularities can be used to formulate integral equations. These singularities are discontinuities of the displacement field, producing, however, unique and continuous strain and stress fields. In correspondence to the possible translations and rotations in the three-dimensional space, there exist three different translation discontinuities, or dislocations, b_i and three different rotational jumps, or disclinations, β_i (see the types I-1, 2, 3 and II-1, 2, 3, respectively, in Figure 2.7). In the case of plane problems, only two different dislocations $D_\alpha = b_\alpha$ and one rotational jump $d = \beta_3$ can be defined (see fig. 2.8).

It is obvious that these singularities (see their visualizations with respect to a thick cylindrical shell, fig. 2.7) can be produced only first by a "cutting" of the elastic medium, then by a shift or a rotation, and by a subsequent "re-sticking" of the two sides of the section.

The following influence functions represent the discontinous displacement fields of such geometrical singularities in the case of plane stress problems [He78a, Za85b]

$$(uD)_{\alpha,\beta}(x,\xi) = \frac{1}{4\pi}\{(1-\nu)e_{\alpha\beta}\ln r + (1+\nu)e_{\beta\rho}r_{,\rho}r_{,\alpha} + \frac{3-\nu}{2}e_{\alpha\beta} + 2\delta_{\alpha\beta}\varphi\} \qquad (2.48)$$

and [He78a]

$$(ud)_{\alpha.}(x,\xi) = \frac{r}{4\pi}\{(1-\nu)r_{,\alpha}\ln r - \frac{1-\nu}{2}r_{,\alpha} - 2e_{\alpha\rho}r_{,\rho}\varphi\}, \qquad (2.49)$$

respectively. Here, the discontinuity is given by the angle function

$$\varphi = \arctan\frac{x_2 - \xi_2}{x_1 - \xi_1} = \arctan\frac{r_{,2}}{r_{,1}}(\text{ or in intrinsic coordinates } \varphi = \arctan\frac{r_{,s}}{r_{,n}})$$

which jumps along the line $x_1 = \xi_1$ (or along the tangent, i.e. for $r_{,n} = 0$) by the value π. The boundary tractions produced by the dislocations D_β and the disclination d, respectively, are described by the following continuous but singular influence functions:

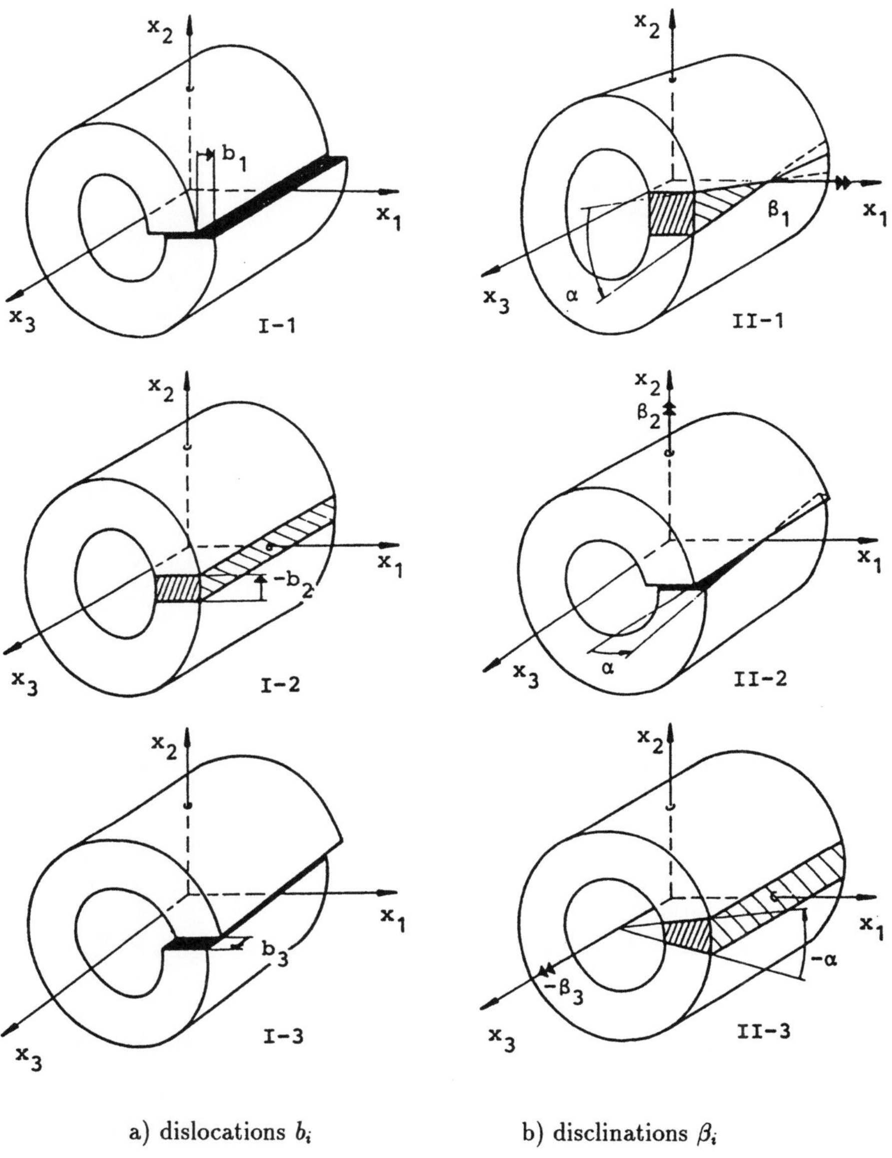

a) dislocations b_i b) disclinations β_i

Fig. 2.7: On the derivation of the geometric singularities in a 3-D elastic medium

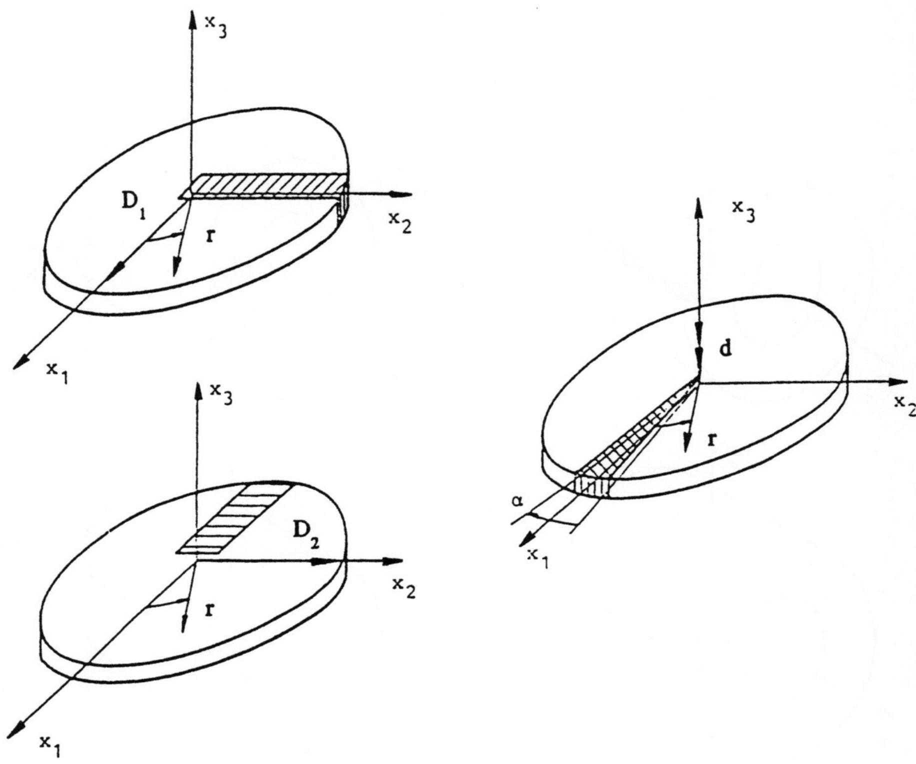

a) edge dislocations D_α b) wedge disclination d

Fig. 2.8: On the derivation of the geometric singularities in a slice

$$(TD)_{\alpha.\beta}(x,\xi) = -\frac{\mu(1+\nu)}{2\pi r}\{s_\lambda(x)r_{,\lambda}\delta_{\alpha\beta} + n_\lambda(x)r_{,\lambda}r_{,\rho}(e_{\alpha\rho}r_{,\beta} - e_{\rho\beta}r_{,\alpha})\} \tag{2.50}$$

and

$$(Td)_{\alpha.}(x,\xi) = -\frac{\mu(1+\nu)}{2\pi}\{n_\alpha(x)[\ln r + \frac{3}{2}] - r_{,\alpha}r_{,\beta}n_\beta(x)\}. \tag{2.51}$$

Using these few influence functions given herein, it is possible via the ideas of the indirect method to formulate already a multitude of different integral equations for the same B.V.P.

For example, if the displacements $\bar{u}_\alpha$ are prescribed along the whole boundary Γ, four different systems of integral equations can be formulated, as follows:

$$\int_{\Gamma+} (uF)_{\alpha.\beta}(x,\xi)F_\beta(\xi)d\Gamma_\xi = \bar{u}_\alpha(x) \tag{2.52}$$

$$[um]_{\alpha.}(x)m(x) + \int_{\Gamma^+} (um)_{\alpha.}(x,\xi)m(\xi)d\Gamma_\xi + \int_{\Gamma} (ud)_{\alpha}(x,\xi)d(\xi)d\Gamma_\xi = \bar{u}_\alpha(x) \qquad (2.53)$$

$$\int_{\Gamma^+} (um)_{\alpha.\beta}(x,\xi)m_\beta(\xi)d\Gamma_\xi = \bar{u}_\alpha(x) \qquad (2.54)$$

$$\int_{\Gamma^+} (uD)_{\alpha.\beta}(x,\xi)D_\beta(\xi)d\Gamma_\xi = \bar{u}_\alpha(x). \qquad (2.55)$$

The equations (2.52) and (2.55) have weakly singular kernels and, therefore, no "integral -free" terms exist, i.e. they are integral equations of the first kind [Kre, Tri]. In the integral equations (2.53) and (2.54), the influence functions of the statical singularities are strongly singular. Nevertheless, only one integral-free term appears when the observation point x on the boundary Γ coincides with a source point ξ on the boundary Γ^+. This term arises only in the case of a layer of moments $m(\xi)$, because in the limit $\Gamma^+ \to \Gamma$ the integration over $\Gamma_\varepsilon(\varepsilon \to 0$, see fig. 2.2) yields on a smooth boundary curve Γ the result

$$[um]_{\alpha.}(x) = \frac{1}{4\pi\mu}e_{\beta\alpha}[-\pi n_\beta(x)] = \frac{1}{4\mu}t_\alpha(x). \qquad (2.56)$$

On the contrary we obtain that in the case of a force dipole vector

$$[um]_{\alpha.\beta}(x) = 0. \qquad (2.57)$$

Note that it is possible to find these results without any explicit integration because it is known [He78a, Hei76] that the following limits hold as $\varepsilon \to 0$.

$$r_{,\beta}/r \to -\pi n_\beta$$

$$r_{,\alpha}r_{,\beta}r_{,\gamma}/r \to -0.5\pi(n_\alpha\delta_{\beta\gamma} + n_\beta\delta_{\alpha\gamma} + n_\gamma\delta_{\alpha\beta} - 2n_\alpha n_\beta n_\gamma).$$

Thus, when these aforementioned influence functions are used, only the integral equation (2.53) is of second kind.

But, if instead of the given boundary displacements $\bar{u}_\alpha$ the respective distortion vector $U_\alpha(x) = \partial\bar{u}_\alpha(x)/\partial s$ is assumed to be prescribed, layers of singularities $D_\beta(\xi)$ give rise to a second kind integral equation. The relevant influence function is obtained from (2.41) by differentiating with respect to the tangential direction

$$(UD)_{\alpha.\beta}(x,\xi) = s_\rho(x)\frac{\partial}{\partial x_\rho}(uD)_{\alpha.\beta}(x,\xi) \qquad (2.58)$$

$$= \frac{1}{4\pi r}\{n_\rho(x)r_{,\rho}[(1-\nu)\delta_{\alpha\beta} + 2(1+\nu)r_{,\alpha}r_{,\beta}] + s_\rho(x)r_{,\rho}(1-\nu)e_{\alpha\beta}\}.$$

This influence function related to dislocation layers $D_\beta(\xi)$ is strongly singular, and leads to the integral equation of the second kind

$$-\frac{1}{2}\delta_{\alpha\beta}D_\beta(x) + \int_{\Gamma} (UD)_{\alpha.\beta}(x,\xi)D_\beta(\xi)d\Gamma_\xi = \bar{U}_\alpha(x),\, x \in \Gamma. \qquad (2.59)$$

The corresponding transformation cannot be successfully applied to force layers $F_\beta(\xi)$ because the relevant influence functions $(UF)_{\alpha.\beta}(x,\xi)$ have a strong singularity [He78a]. However, their integral-free terms $[UF]_{\alpha.\beta}(x)$ are disappearing.

When statical boundary values, for instance, the boundary tractions $\bar{T}_\alpha$, are prescribed along the whole boundary Γ, again, the above given influence functions, i.e. (2.43), (2.46), (2.47) and (2.50)(2.51) can be used to formulate several systems of integral equations. Thus, layers of forces $F_\beta(\xi)$ and layers of dislocations $D_\beta(\xi)$ produce respectively the following integral equations for which in the singular case, i.e. for $\Gamma \equiv \Gamma^+$, the integrals have to be determined as Cauchy's principal values:

$$[TF]_{\alpha.\beta}(x)F_\beta(x) + \fint_\Gamma (TF)_{\alpha.\beta}(x,\xi)F_\beta(\xi)d\Gamma_\xi = \bar{T}_\alpha(x), x \in \Gamma. \tag{2.60}$$

$$[TD]_{\alpha.\beta}(x)D_\beta(x) + \fint_\Gamma (TD)_{\alpha.\beta}(x,\xi)D_\beta(\xi)d\Gamma_\xi = \bar{T}_\alpha(x), x \in \Gamma. \tag{2.61}$$

But, only the equation (2.60) is of the second kind because it has a non-zero integral-free term; it reads in the case of smooth boundary contours

$$[TF]_{\alpha.\beta}(x) = \frac{1}{2}\delta_{\alpha\beta}. \tag{2.62}$$

At least in the case of smooth contours Γ, the integral-free term in the equation (2.61) is identically zero, i.e. we have an integral equation of the first kind.

The hypersingular influence functions (2.46) and (2.47), connecting the boundary tractions $T_\alpha(x)$ with the single moment $m(\xi)$ and the force-dipole $m_\beta(\xi)$, respectively, can only be used to formulate integral equations for the second boundary value problem, i.e. for determining the reactions due to prescribed boundary tractions $T_\alpha(x)$ along Γ, if either a small, but finite distance remains between Γ and Γ^+, i.e. between the boundary of the real structure, and the fictitious boundary Γ^+ containing the layer of singularities [He78b], or the boundary integrals are evaluated in the sense of the so-called Hadamard-integrals [Kay, Riz, Ioa].

Thus, the same influence functions can be used either to formulate regular integral equations of the first kind, i.e. with force-dipole layers $m_\beta(\xi)$

$$\int_{\Gamma^+} (Tm)_{\alpha.\beta}(x,\xi)m_\beta(\xi)d\Gamma_\xi = \bar{T}_\alpha(x), \qquad x \in \Gamma \tag{2.63}$$

or with mixed layers of single moments $m(\beta)$ and disclinations $d(\xi)$

$$\int_{\Gamma^+} [(Tm)_{\alpha.}(x,\xi)m(\xi) + (Td)_{\alpha.}(x,\xi)d(\xi)]d\Gamma_\xi = \bar{T}_\alpha(x), x \in \Gamma, \tag{2.64}$$

or interval equations of the hyper-singular type with the "regularized Hadamard integrals".

Until now, hypersingular integral equations have been used only to solve a few special problems, such as the Laplace equation ($\Delta u = 0$ in Ω) with prescribed boundary flux ($\partial u / \partial n = \varphi$ on Γ) via double layer potentials [Had], also in crack problems [Ioa, Kay] as well as in the study of the propagation and refraction of acoustic and elastic waves [Riz].

Analogously to the geometric B.V.Ps. in which, instead of the displacements $\bar{u}_\alpha(x)$ the distortion vector $\bar{U}_\alpha(x)$ can be used as the prescribed "right-hand side" (see Eq. (2.59)), in the case of statical B.V.Ps the boundary stress function vector $t_\alpha(s) = \int T_\alpha(s)ds$, i.e the state resulting from the boundary tractions by an integration along the boundary, can be considered as given. Because of this integration the respective influence functions are one level less singular, i.e the before hyper-singular functions will now be only strongly singular etc. Thus, now, a force-dipole layer $m_\beta(\xi)$ yields with

$$(tm)_{\alpha.\beta}(x,\xi) = -\frac{\nu}{4\pi}\frac{r_{,\rho}}{r}\{n_\rho(\xi)[(1-\nu)\delta_{\alpha\beta} + 2(1+\nu)r_{,\alpha}\,r_{,\beta}] + s_\rho(\xi)(1-\nu)e_{\alpha\beta}\} \quad (2.65)$$

the singular integral equation of the second kind

$$\frac{1}{2}m_\alpha(x) + \int_\Gamma (tm)_{\alpha.\beta}(x,\xi)m_\beta(\xi)d\Gamma_\xi = t_\alpha(x), x \in \Gamma, \quad (2.66)$$

whose integrals are no longer hypersingular (as in the case of the influence function (2.47)), but can be evaluated as a usual Cauchy principal value.

For mixed B.V.Ps one must choose an appropriate type of singularities. Then, the related integral equations have to be formulated correspondingly to the boundary values prescribed along the relevant boundary parts. When, for example, displacements $\bar{u}_\alpha(x)$ and tractions $\bar{T}_\alpha(x)$ are prescribed along Γ_1 and Γ_2, respectively, the following coupled system of integral equations can be formulated:

$$\int_\Gamma (uF)_{\alpha.\beta}(x,\xi)F_\beta(\xi)d\Gamma_\xi = \bar{u}_\alpha(x), x \in \Gamma_1, \quad (2.67)$$

$$[TF]_{\alpha.\beta}(x)F_\beta(x) + \int_\Gamma (TF)_{\alpha.\beta}(x,\xi)F_\beta(\xi)d\Gamma_\xi = \bar{T}_\alpha(x), x \in \Gamma_2. \quad (2.68)$$

Chapter 3
Boundary Integral Formulations for Some Special Elastostatic B.V.Ps

3.1 Bending of Beams and Stretching of Bars

The study of the static and dynamic behaviour of structures made up of bars and beams, i.e. in trusses and frames, is one of the main tasks in applied mechanics and engineering. Although, in this field, the F.E.M is already very well developed and almost exclusively sovereighing, there are certain problems, e.g. shape optimization, where the boundary indeed plays the predominant role and, because of that, boundary methodologies had better been prefered. In this chapter we deal with certain boundary integral formulations for some of the basic beam, bar and plate problems. To elucidate the derivation possibilities of Ch.2, all the previously introduced methods are applied in this Chapter in order to obtain the integral equations. Let us consider first the bending of beams. Bending is assumed to be independent of the stretching in the framework of a geometrically linear theory (see Fig. 3.1). The longitudinal displacement $u(x) = u_1(x)$ and the deflection $w(x) = u_3(x)$ have to satisfy the two independent basic differential equations

$$\frac{d^2}{dx^2}\left[EI\frac{d^2}{dx^2}w(x)\right] = q(x) \qquad \text{on } 0 \le x \le 1, \tag{3.1}$$

$$-\frac{d}{dx}\left[EA\frac{d}{dx}u(x)\right] = p(x) \qquad \text{on } 0 \le x \le 1. \tag{3.2}$$

Eqs. (3.1),(3.2) describe the equilibrium between the exterior forces, the transversal and longitudinal loading, respectively, $q(x)$ and $p(x)$, and the internal "forces", i.e. the bending moment $M(x)$, the shear force $Q(x)$, and the normal force $N(x)$. The internal forces are connected with the displacements via the relations

$$M(x) = -EI\frac{d^2}{dx^2}w(x) = -EIw_{,xx} \tag{3.3}$$

$$Q(x) = -EI\frac{d^3}{dx^3}w(x) = -EIw_{,xxx} \tag{3.4}$$

$$N(x) = EA\frac{d}{dx}u(x) = EAu_{,x}\,. \tag{3.5}$$

Therein, besides the Young's modulus E, the term $I = I_y$ denotes the moment of inertia of the cross-section area A with respect to the y-axis (see fig. 3.1). Both I and A as well as E are taken to remain constant along the beam.

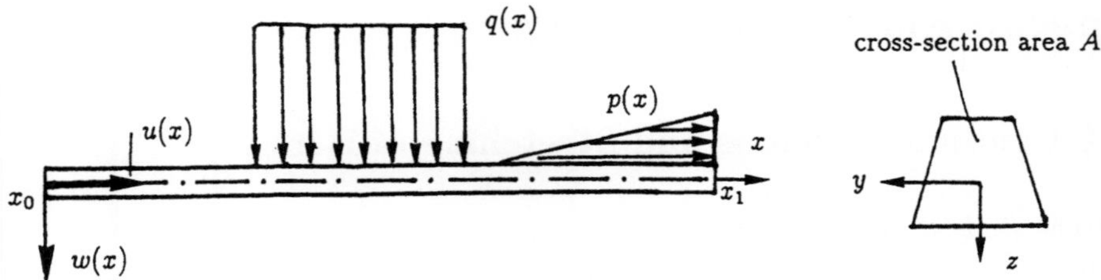

Fig. 3.1: Geometry and Exterior Loading of a slender bar/beam structure

The boundary conditions, corresponding to the differential equations (3.1) and (3.2), are simply pointwise conditions for one of the following conjugate quantities:

$$u(x) \text{ or } N(x) \qquad \text{for the stretching or the so-called bar problem}$$

$$\left.\begin{array}{ll} w(x) & \text{or} \quad Q(x) \\ w_{,x}(x) & \text{or} \quad M(x) \end{array}\right\} \qquad \text{bending or the so-called beam problem}$$

Thus, e.g. a fixed hinge at x_0 implies the boundary condition

$$u(x_0) = \bar{u} = 0 \qquad \text{for the bar,} \tag{3.6}$$

$$\left.\begin{array}{l} w(x_0) = \bar{w} = 0 \\ M(x_0) = \bar{M} = 0 \end{array}\right\} \qquad \text{for the beam,} \tag{3.7}$$

while in the case of a hinge at x_1, the relevant boundary conditions read

$$N(x_1) = \bar{N} = 0 \qquad \text{for the bar,} \tag{3.8}$$

$$\left.\begin{array}{l} w(x_1) = \bar{w} = 0 \\ M(x_1) = \bar{M} = 0 \end{array}\right\} \qquad \text{for the beam.} \tag{3.9}$$

Now, the method of weighted residuals will be used to derive the B.I. formulation of the problem. Assume for simplicity that $x_0 = 0$ and $x_1 = l$. Then the actual boundary conditions are written in the form

$$\left.\begin{array}{l} u(x = 0) = \bar{u}_0 \\ N(x = l) = \bar{N}_l \end{array}\right\} \qquad \text{for the stretching problem,} \tag{3.10}$$

$$\left.\begin{array}{l} w(x = 0) = \bar{w}_0 \\ w(x = l) = \bar{w}_l \\ M(x = 0) = \bar{M}_0 \\ M(x = l) = \bar{M}_l \end{array}\right\} \qquad \text{for the bending problem,} \tag{3.11}$$

and the respective residual terms (cf. Sect.2.2) become

$$\int_0^l [-EAu_{,xx}(x) - \bar{p}(x)] \overset{*}{u}(x,\xi)dx + [N(l) - \bar{N}_l] \overset{*}{u}(l,\xi) + [u(0) - \bar{u}_0] \overset{*}{N}(0,\xi) = 0 \tag{3.12}$$

and

$$\int_0^l [EI w_{,xxxx}(x) \; - \; \bar{q}(x)] \; \overset{\star}{w} \, (x,\xi) dx \tag{3.13}$$

$$+ \; [M(0) - \bar{M}_0] \, \overset{\star}{w}_{,x} (0,\xi) - [M(l) - \bar{M}_l] \, \overset{\star}{w}_{,x} (l,\xi)$$

$$+ \; [w(0) - \bar{w}_0] \, \overset{\star}{Q} (0,\xi) - [w(l) - \bar{w}_l] \, \overset{\star}{Q} (l,\xi) = 0.$$

Here the quantities with the star are the weight functions for the problem considered. After partial integrations, and the use of the relations (3.3)-(3.5) we obtain the relations

$$\int_0^l EA u_{,xx}(x) \, \overset{\star}{u} \, (x,\xi) dx = [N(x) \, \overset{\star}{u} \, (x,\xi)]_{x=0}^l - \int_0^l u_{,x}(x) \, \overset{\star}{N} \, (x,\xi) dx \tag{3.14}$$

and

$$\int_0^l EI w_{,xxxx}(x) \, \overset{\star}{w} \, (x,\xi) dx \; = \; [-Q(x) \, \overset{\star}{w} \, (x,\xi) + M(x) \, \overset{\star}{w}_{,x} \, (x,\xi)]_{x=0}^l \tag{3.15}$$

$$- \int_0^l w_{,xx}(x) \, \overset{\star}{M} \, (x,\xi) dx.$$

Thus, the residual terms (3.12) and (3.13) can be transformed respectively into

$$\int_0^l u_{,x}(x) \, \overset{\star}{N} \, (x,\xi) dx \; = \; \int_0^l \bar{p}(x) \, \overset{\star}{u} \, (x,\xi) dx \tag{3.16}$$

$$-[u(0) - \bar{u}_0] \, \overset{\star}{N} \, (0,\xi) - N(0) \, \overset{\star}{u} \, (0,\xi) + \bar{N}_l \, \overset{\star}{u} \, (l,\xi)$$

and

$$\int_0^l w_{,xx}(x) \, \overset{\star}{M} \, (x,\xi) dx \; = \; - \int \bar{q}(x) \, \overset{\star}{w} \, (x,\xi) dx + [w(0) - \bar{w}_0] \, \overset{\star}{Q} (0,\xi) \tag{3.17}$$

$$-[w(l) - \bar{w}_l] \, \overset{\star}{Q} (l,\xi) - Q(l) \, \overset{\star}{w} \, (l,\xi) + Q(0) \, \overset{\star}{w} \, (0,\xi)$$
$$-\bar{M}_0 \, \overset{\star}{w}_{,x} (0,\xi) + \bar{M}_l \, \overset{\star}{w}_{,x} (l,\xi).$$

Again, in the left-hand sides of these equations an integration by parts is performed. Finally, considering the basic differential equations (3.1) and (3.2), it results that

$$\int_0^l u(x) \, \overset{\star}{p} \, (x,\xi) dx + u(l) \, \overset{\star}{N} \, (l,\xi) - \bar{u}_0 \, \overset{\star}{N} \, (0,\xi) \tag{3.18}$$

$$= \int_0^l \overset{\star}{u}(x,\xi)\bar{p}(x)dx + \overset{\star}{u}(l,\xi)\bar{N}_l - \overset{\star}{u}(0,\xi)N(0)$$

and

$$\int_0^l w(x)\overset{\star}{q}(x,\xi)dx + \bar{w}_l\overset{\star}{Q}(l,\xi) - \bar{w}_0\overset{\star}{Q}(0,\xi) \tag{3.19}$$

$$- w_{,x}(l)\overset{\star}{M}(l,\xi) + w_{,x}(0)\overset{\star}{M}(0,\xi)$$

$$= \int_0^l \overset{\star}{w}(x,\xi)\bar{q}(x)dx + \overset{\star}{w}(l,\xi)Q(l) - \overset{\star}{w}(0,\xi)Q(0)$$

$$- \overset{\star}{w}_{,x}(l,\xi)\bar{M}_l + \overset{\star}{w}_{,x}(0,\xi)\bar{M}_0.$$

These equations represent the statements of Betti's reciprocal theorem for the considered bar and the beam, respectively. Then, if the fundamental solutions of single unit forces, i.e. the solutions of the equations

$$\overset{\star}{N}_{,x}(x,\xi) = -\overset{\star}{p}(x,\xi) = -\delta(x-\xi), \tag{3.20}$$

$$\overset{\star}{M}_{,xx}(x,\xi) = -\overset{\star}{q}(x,\xi) = -\delta(x-\xi), \tag{3.21}$$

are taken as special weight functions, the eqs. (3.17) (3.18) imply the so-called Somigliana identities

$$u(\xi) = u(l)\overset{\star}{N}(l,\xi) - \overset{\star}{u}(l,\xi)\bar{N}_l - \bar{u}_0\overset{\star}{N}(0,\xi) + \overset{\star}{u}(0,\xi)N(0) \tag{3.22}$$

$$- \int_0^l \overset{\star}{u}(x,\xi)\bar{p}(x)dx$$

$$= [u(x)\overset{\star}{N}(x,\xi) - \overset{\star}{u}(x,\xi)N(x)]_{x=0}^l - \int_0^l \overset{\star}{u}(x,\xi)\bar{p}(x)dx$$

and

$$w(\xi) = -[w(x)\overset{\star}{Q}(x,\xi) - w_{,x}(x)\overset{\star}{M}(x,\xi) - \overset{\star}{w}(x,\xi)Q(x) \tag{3.23}$$

$$+ \overset{\star}{w}_{,x}(x,\xi)M(x)]_{x=0}^l + \int_0^l \overset{\star}{w}(x,\xi)\bar{q}(x)dx.$$

where no distinction is made between prescribed boundary values and unknown reactions.

Explicitly, these fundamental solutions can be given by $(a, b_1, b_2, b_3 \in R)$

$$\overset{\star}{u}(x,\xi) = \frac{1}{EA}\{-[x-\xi]_+^1 + a\} \tag{3.24}$$

$$\overset{\star}{w}(x,\xi) = \frac{1}{EI}\{\frac{1}{6}[x-\xi]_+^3 - \frac{1}{2}b_1 x^2 - b_2 x - b_3\} \tag{3.25}$$

$$\text{where} \quad [y]_+^n := \begin{cases} y^n & \text{for } y > 0, \\ 0 & \text{for } y \le 0 \end{cases} \quad n = 0, 1, \ldots .$$

The conditions which must be satisfied by the functions (3.24) and (3.25) when the latter are fundamental solutions, are

$$\lim_{\varepsilon \to 0}\{\overset{\star}{N}(\xi + \varepsilon, \xi) - \overset{\star}{N}(\xi - \varepsilon, \xi)\} = -1,$$

$$\lim_{\varepsilon \to 0}\{\overset{\star}{Q}(\xi + \varepsilon, \xi) - \overset{\star}{Q}(\xi - \varepsilon, \xi)\} = 1,$$

and these conditions are satisfied for rather arbitrary constants a and b_1, b_2, b_3, respectively. Thus, these constants can be used to fulfill some of the conditions for the respective Green's function [Ha81a].

The eqs. (3.22) and (3.23) can be used to determine the longitudinal displacement and the deflection at any point ξ in the intervall $[0, l]$. However, first one must determine the values for the unknown boundary data. In these one-dimensional problems, this is achieved simply by means of collocations at the boundary points, approaching them from the interior, i.e. at $\xi = 0 + \varepsilon$ and $\xi = l - \varepsilon(\varepsilon \to 0)$. For the stretching problem, this yields the following system of algebraic equations

$$\begin{bmatrix} \overset{\star}{u}(0,0) & \overset{\star}{N}(l,\varepsilon) \\ \overset{\star}{u}(0,l) & 1 + \overset{\star}{N}(l,l-\varepsilon) \end{bmatrix} \cdot \begin{bmatrix} N(0) \\ u(1) \end{bmatrix} = \tag{3.26}$$

$$= \begin{bmatrix} \overset{\star}{N}(0,\varepsilon) - 1 & \overset{\star}{u}(l,0) \\ \overset{\star}{N}(0,l-\varepsilon) & \overset{\star}{u}(l,l) \end{bmatrix} \cdot \begin{bmatrix} \bar{u}_0 \\ \bar{N}_l \end{bmatrix} + \int_0^l \begin{bmatrix} \overset{\star}{u}(x,0) \\ \overset{\star}{u}(x,l) \end{bmatrix} \bar{p}(x)dx$$

or using (3.24), the system

$$\begin{bmatrix} a & -l \\ a & 0 \end{bmatrix} \cdot \begin{bmatrix} N(0)/EA \\ u(l) \end{bmatrix} = \begin{bmatrix} -1 & a-l \\ 0 & a \end{bmatrix} \cdot \begin{bmatrix} u(0) \\ N(l)/EA \end{bmatrix} \tag{3.27}$$

$$+ \int_0^l \begin{bmatrix} -x+a \\ l-x+a \end{bmatrix} \frac{\bar{p}(x)}{EA}dx.$$

Obviously, the constant a can have any arbitrary value except zero, and the prescribed boundary values must fullfil (3.27). Let us give an example to illustrate this method: We consider a bar of length l with constant stiffness EA which is fixed at $x = 0$. If at its free end $x = l$, a force $\bar{N}(l)$ is positive (noncompressive), the above equations yield immediately

$$N(0) = N(l) \quad \text{and} \quad u(l) = lN(l)/EA.$$

Since the eq. (3.23) delivers by collocation only two algebraic equations, although four boundary values are unknown, a second "integral" equation has to be deduced.

When as a second fundamental solution, a single unit moment is considered, i.e. the solution of

$$\overset{\star\star}{M}_{,x}(x,\xi) = \overset{\star\star}{Q}(x,\xi) = -\delta(x-\xi) \tag{3.28}$$

is taken as the weight function, the partial integration of the left-hand side of (3.17) implies that

$$\int_0^l w_{,xx}(x)\,\overset{\star\star}{M}(x,\xi)dx = [w_{,x}(x)\,\overset{\star\star}{M}(x,\xi)]_{x=0}^l - \int_0^l w_{,x}(x)\,\overset{\star\star}{M}_{,x}(x,\xi)dx \tag{3.29}$$

$$= [w_{,x}(x)\,\overset{\star\star}{M}(x,\xi)]_{x=0}^l + w_{,x}(\xi).$$

Thus, one obtains as a second "integral" equation the relation

$$w_{,x}(\xi) = [-w_{,x}(x)\,\overset{\star\star}{M}(x,\xi) + \overset{\star\star}{w}_{,x}(x,\xi)M(x) - \overset{\star\star}{w}(x,\xi)Q(x)]_{x=0}^l \tag{3.30}$$

$$- \int_{x=0}^l \overset{\star\star}{w}(x,\xi)\bar{q}(x)dx,$$

where the respective fundamental solutions are given by the expression

$$\overset{\star\star}{w}(x,\xi) = \frac{1}{EI}\{\frac{1}{2}[x-\xi]_+^2 + c_1 x + c_2\}. \tag{3.31}$$

Then, as in the stretching problem, collocations at the boundary points yield an appropriate system of algebraic equations, which is, if $\bar{w}_0, \bar{w}_0', \bar{w}_l$ and $\bar{M}_l$, are for example, prescribed.

$$\begin{bmatrix} -\overset{\star}{w}(0,0) & \overset{\star}{w}_{,x}(0,0) & \overset{\star}{w}(l,0) & \overset{\star}{M}(l,\varepsilon) \\ -\overset{\star}{w}(0,l) & \overset{\star}{w}_{,x}(0,l) & \overset{\star}{w}(l,l) & \overset{\star}{M}(l,l-\varepsilon) \\ -\overset{\star\star}{w}(0,0) & \overset{\star\star}{w}_{,x}(0,0) & \overset{\star\star}{w}(l,0) & \overset{\star\star}{M}(l,\varepsilon) \\ -\overset{\star\star}{w}(0,l) & \overset{\star\star}{w}_{,x}(0,l) & \overset{\star\star}{w}(l,l) & 1+\overset{\star\star}{M}(l,l-\varepsilon) \end{bmatrix} \cdot \begin{bmatrix} Q(0) \\ M(0) \\ Q(l) \\ w'(l) \end{bmatrix}$$

$$= \begin{bmatrix} +1-\overset{\star}{Q}(0,\varepsilon) & \overset{\star}{M}(0,\varepsilon) & \overset{\star}{Q}(l,\varepsilon) & \overset{\star}{w}_{,x}(l,0) \\ -\overset{\star}{Q}(0,l-\varepsilon) & \overset{\star}{M}(0,l-\varepsilon) & 1+\overset{\star}{Q}(l,l-\varepsilon) & \overset{\star}{w}_{,x}(l,l) \\ 0 & -1+\overset{\star\star}{M}(0,\varepsilon) & 0 & \overset{\star\star}{w}_{,x}(l,0) \\ 0 & \overset{\star\star}{M}(0,l-\varepsilon) & 0 & \overset{\star\star}{w}_{,x}(l,l) \end{bmatrix} \cdot \begin{bmatrix} \bar{w}_0 \\ \bar{w}_0' \\ \bar{w}_l \\ \bar{M}_l \end{bmatrix}$$

$$- \int_0^l \begin{bmatrix} \overset{\star}{w}(x,0) \\ \overset{\star}{w}(x,l) \\ \overset{\star\star}{w}(x,0) \\ \overset{\star\star}{w}(x,l) \end{bmatrix} \bar{q}(x)dx . \tag{3.32}$$

In the above relations we denote by a prime the derivation $\frac{d}{dx}$. From of these equations, it can be easily found that the main condition for the constants b_1, b_2, b_3 and c_1, c_2 of the fundamental solutions (3.25) and (3.31), respectively, is given by $b_2 \cdot c_2 \neq b_3 \cdot c_1 \neq 0$. Therefore, the simple choice $b_1 = b_2 = c_2 = 0, b_3 = b > 0$ and $c_1 = c > 0$ is possible in any case, if the beam is free of rigid body motions. Let us illustrate the method by an example: Consider a beam of length l and with constant stiffness EI, where the deflection w and the slope w' are known at $x = 0$, while at $x = l$ the deflection and an exterior single moment are prescribed. Then the unknown boundary reactions can be determined from the system

$$
\begin{bmatrix}
b & 0 & -b + \frac{l^3}{6} & -l \\
b & 0 & -b & 0 \\
0 & c & \frac{l^2}{2} + cl & -l \\
0 & c & cl & 0
\end{bmatrix}
\cdot
\begin{bmatrix}
Q(0)/EI \\
M(0)/EI \\
Q(l)/EI \\
w_{,x}(l)
\end{bmatrix}
$$

$$
=
\begin{bmatrix}
1 & 0 & -1 & \frac{l^2}{2} \\
0 & 0 & 0 & 0 \\
0 & -1 & 0 & l + c \\
0 & 0 & 0 & c
\end{bmatrix}
\cdot
\begin{bmatrix}
\bar{w}_0 \\
\bar{w}_0' \\
\bar{w}_l \\
\bar{M}_l/EI
\end{bmatrix}
-
\int\limits_0^l
\begin{bmatrix}
-b + \frac{x^3}{6} \\
-b \\
\frac{x^2}{2} + cx \\
c
\end{bmatrix}
\bar{q}(x)dx \; .
$$

(a)   In the special case of a beam which is clamped at $x = 0$, i.e $w_0 = w_0' = 0$, and simply supported by a hinge at $x = l$, i.e $w_l = 0$ and $M_l = 0$, the boundary reactions due to a constant loading $\bar{q}(x) = q_0, 0 \leq x \leq l$, are

$$
Q(0) = \frac{5}{8}q_0 l \quad , \quad Q(l) = -\frac{3}{8}q_0 l
$$

$$
M(0) = -\frac{1}{8}q_0 l^2 \quad , \quad w_{,x}(l) = -\frac{1}{48}\frac{q_0 l^3}{EI}.
$$

(b)   If the beam is clamped at $x = 0$ and the hinge at $x = l$ is pulled down a certain value $\bar{w}_l \neq 0$, there are boundary reactions also for an unloaded beam, i.e. for $\bar{q}(x) \equiv 0$ and they have the values

$$
Q(0) = Q(l) = 3\bar{w}_l EI/l^3,
$$

$$
M(0) = -3\bar{w}_l EI/l^2 \text{ and } w_{,x}(l) = \frac{3}{2}\bar{w}_l/l.
$$

3.2 A Direct B.I.E.M. for Kirchhoff Plates

A plate is a plane body whose thickness, h, is considerably smaller than its in-plane dimensions. The plane $x_1 - x_2$ is taken to coincide with the mean surface of the plate. As long as the deflection of a thin plate is small in comparison with its thickness, the bending of this plate can be described by the classical linear Kirchhoff plate theory. In this theory, the bending is considered to be independent from the in-plane stretching of the plate.

In order to treat the bending of thick plates, many approximate theories have been proposed (cf. e.g [Iga]). Following Reissner [Re45], the effect of shear deformation has to be taken into account, at least. For both, Kirchhoff and Reissner plate theory, direct boundary element formulations will be derived in this Section and in the next one: first, for Kirchhoff plates the relevant reciprocal theorem will be applied in this Section for the formulation of a direct B.I.E.M. while, for the more refined Reissner plate model the corresponding Hu-Washizu functional is taken as the starting-point in the next Section.

Within the context of Kirchhoff plate model, many publications have appeared concerning the applications of boundary integrals for solving thin plate bending problems. Among these, several indirect formulations have been presented (see e.g. M.A. Jaswon & A. Maiti [Jas], E. Hansen [Han] and H. Glahn [Gla], and the literature given there) but there, certain transformations are necessary to determine all the interesting boundary quantities. The first direct integral equation methods seem to be the ones by Bezine [Bez] and Stern [Ste79], where the aforementioned disadvantages do not occur.

Consider a thin plate subjected to transverse forces $q(x)$ (per unit area) inside the plate domain Ω. Moreover, the bending moment M_n and the so-called Kirchhoff shear force K_n given by

$$K_n = Q_n + \frac{\partial}{\partial s} M_{nt}$$

act (per unit of length) along the boundary Γ. There, Q_n and M_{nt} denote the real shearing force and the twisting moment on Γ, respectively, and s is the arc-length in the direction of the unit tangent t on Γ. Besides, if the boundary has N corners A_i, at these points A_i with curvilinear abscissa s_i, the twisting moment M_{nt} is subjected to a jump thus giving rise to the so-called corner forces $T_c = [(M_{nt})_{si-} - (M_{nt})_{si+}]$. The boundary loadings K_n, T_c and M_n correspond respectively to the following kinematic quantities: the deflection w along Γ, the deflections w at the corners A_i, and the rotation $\varphi_n = -w_{,n}$, respectively ("$_{,n}$" indicates the derivation with respect to the outward unit normal on Γ).

Thus, for two systems of forces acting on a plate, the reciprocal theorem (2.32) simply yields the relation

$$\int_\Omega \overset{1}{q}\overset{2}{w}\, d\Omega \;+\; \int_\Gamma \{\overset{1}{K}_n\overset{2}{w} + \overset{1}{M}_n\overset{2}{\varphi}_n\} d\Gamma + \sum_{j=1}^{N} \overset{1}{T}_c\,(x_{A_j})\, \overset{2}{w}\,(x_{A_j}) \tag{3.33}$$

$$= \int_\Omega \overset{2}{q}\overset{1}{w}\, d\Omega + \int_\Gamma \{\overset{2}{K}_n\overset{1}{w} + \overset{2}{M}_n\overset{1}{\varphi}_n\} d\Gamma + \sum_{j=1}^{N} \overset{2}{T}_c\,(x_{A_j})\, \overset{1}{w}\,(x_{A_j}).$$

Suppose now that the loading system "2" is chosen to be a unit transversal force at the point ξ of an infinite plate with stiffness D, i.e. [Ha81a]

$$\overset{2}{q}= \delta(x - \xi). \tag{3.34}$$

Its reactions are given by (κ denotes the curvature of the plate boundary)

$$\overset{2}{K}_n=\overset{\star}{K}_n\,(x,\xi) = -\frac{1}{4\pi}\frac{1}{r}\{2r_{,n} + (1 - \nu)(r_{,n} - \kappa r)(r_{,n}^2 - r_{,s}^2)\}, \tag{3.35}$$

$$\overset{2}{M}_n = \overset{*}{M}_n(x,\xi) = -\frac{1}{8\pi}\{(1+\nu)(\ln r^2 + 1) + 2r_{,n}^2 + 2\nu r_{,s}^2\}, \tag{3.36}$$

$$\overset{2}{T}_c = \overset{*}{T}_c(x_{Aj},\xi) = [\overset{*}{M}_{nt}(s)]_{s=s_{j+}}^{s_{j-}} = -\frac{1-\nu}{4\pi}[r_{,n}r_{,s}]_{s=s_{j+}}^{s_{j-}}. \tag{3.37}$$

Then the corresponding deformation $\overset{2}{w}=\overset{*}{w}$ fullfil the equation:

$$D\Delta\Delta\,\overset{*}{w}(x,\xi) = \delta(x,\xi), \tag{3.38}$$

from which we obtain that

$$\overset{*}{w}(x,\xi) = \frac{1}{16\pi D}r^2\ln r^2.$$

Moreover $\overset{2}{\varphi}_n = \overset{*}{\varphi}$ is given by

$$\overset{*}{\varphi}_n(x,\xi) = -\frac{1}{8\pi D}rr_{,n}(1+\ln r^2). \tag{3.39}$$

The relations $(3.34)\div(3.39)$ combined with (3.33) yields a directly formulated integral equation for determining the deflection at a point ξ in the interior of the plate Ω. Indeed since

$$\int_\Omega \overset{2}{q}\overset{1}{w}\,d\Omega_x = \int_\Omega \delta(x-\xi)w(x)d\Omega_x = w(\xi) \text{ for } \xi \in \Omega, \tag{3.40}$$

the equation (3.33) yields

$$\begin{aligned}
c(\xi)w(\xi) &= \int_\Omega q(x)\,\overset{*}{w}(x,\xi)d\Omega_x + \sum_j[T_c(x_{Aj})\,\overset{*}{w}(x_{Aj},\xi) \\
&\quad - \overset{*}{T}_c(x_{Aj},\xi)w(x_{Aj})] + \int_\Gamma \{K_n(x)\,\overset{*}{w}(x,\xi) + M_n(x)\,\overset{*}{\varphi}_n(x,\xi) \\
&\quad - \overset{*}{K}_n(x,\xi)w(x) - \overset{*}{M}_n(x,\xi)\varphi_n(x)\}d\Gamma_x
\end{aligned} \tag{3.41}$$

with $c(\xi) = 1$ for $\xi \in \Omega$ and $c(\xi) = 0$ for $\xi \notin \Omega$.

By shifting the point ξ on the boundary Γ one obtains a B.I.E. for the determination of the unknown boundary reactions. Then the equation (3.41) remains formally unchanged, but the coefficient $c(\xi)$ depends on the angle $\Delta\varphi$ (cf. Fig. 3.2), i.e.

$$c(\xi) = \begin{cases} \Delta\varphi/2\pi & \text{at corners,} \\ & \text{for } \xi \in \Gamma \\ 0.5 & \text{on smooth boundary parts.} \end{cases}$$

Since in Kirchhoff plate theory at each boundary point two quantities are unknown, a second integral equation has to be derived. Therefore, analogously to the case of beam theory, a single unit moment is taken as a second singular loading. As shown by Bezine [Bez], the new integral equation can be obtained by forming simply the directional derivative of eq. (3.41) in a fixed direction n_0, with respect to the coordinates of the source point ξ ($\xi \in \Omega$ with $c(\xi) = 1$). We obtain the relation

$$
\begin{aligned}
w_{,n_0}(\xi) \;=\; & \int_\Omega q(x)\, \overset{*}{w}_{,n_0}(x,\xi)\,d\Omega_x \\[2mm]
& + \sum_j \{ T_c(x_{A_j})\, \overset{*}{w}_{,n_0}(x_{A_j},\xi) - \overset{*}{T}_{c,n_0}(x_{A_j},\xi) w(x_{A_j}) \} \\[2mm]
& + \int_\Gamma \{ K_n(x)\, \overset{*}{w}_{,n_0}(x,\xi) + M_n(x)\, \overset{*}{\varphi}_{n,n_0}(x,\xi) \\[2mm]
& \quad - \overset{*}{K}_{n,n_0}(x,\xi) w(x) - \overset{*}{M}_{n,n_0}(x,\xi)\varphi_n(x) \} d\Gamma_x .
\end{aligned}
\tag{3.42}
$$

Then, shifting the point ξ to a boundary point, this fixed direction $n_0(\xi)$ is chosen to be the direction of the outward normal on the boundary Γ. This yields the necessary second boundary integral equation which reads:

$$
\begin{aligned}
c_\alpha(\xi) w_{,\alpha}(\xi) \;=\; & \int_\Omega q(x)\, \overset{**}{w}(x,\xi)\,d\Omega_x \\[2mm]
& + \sum_{j=1}^{N} [T_c(x_{A_j})\, \overset{**}{w}(x_{A_j},\xi) - \overset{**}{T}_c(x_{A_j},\xi) w(x_{A_j})] \\[2mm]
& + \int_\Gamma \{ K_n(x)\, \overset{**}{w}(x,\xi) + M_n(x)\, \overset{**}{\varphi}_n(x,\xi) \\[2mm]
& \quad - \overset{**}{K}_n(x,\xi) w(x) - \overset{**}{M}_n(x,\xi)\varphi_n(x) \} d\Gamma_x ,
\end{aligned}
\tag{3.43}
$$

where "**" denotes the new fundamental solution quantities, i.e. the reactions caused by a single unit moment at the point ξ (n_0 indicates the outward normal $n(\xi)$), which are supplied by the formulas

$$
\overset{**}{w}(x,\xi) = \frac{\partial}{\partial n_0}\, \overset{*}{w}(x,\xi) = -\frac{r}{8\pi D} r_{,n_0}(1 + \ln r^2)
\tag{3.44}
$$

$$
\overset{**}{\varphi}_n(x,\xi) = \frac{1}{8\pi D}[(1 + \ln r^2)(r_{,n}r_{,n_0} + r_{,s}r_{,s_0}) + 2 r_{,n}r_{,n_0}]
\tag{3.45}
$$

$$
\overset{**}{M}(x,\xi) = \frac{1}{4\pi}\frac{1}{r}[(1 + \nu)r_{,n} + 2(1 - \nu)r_{,n_0}r_{,s}r_{,s_0}]
\tag{3.46}
$$

$$
\begin{aligned}
\overset{**}{K}(x,\xi) \;=\; & \frac{1}{4\pi}\frac{1}{r^2}[(3 - \nu - 2(1 - \nu)r_{,s_0}^2)(r_{,s_0}r_{,s} - r_{,n_0}r_{,n}) \\
& + 4(1 - \nu)(r_{,n_0} - \kappa(x)r)r_{,n_0}r_{,s}r_{,n_0}].
\end{aligned}
\tag{3.47}
$$

Again, the factor $c_\alpha(\xi)$ depends on the interior angles φ_1 and φ_2 ($\varphi_1 > \varphi_2$) ahead and behind a boundary point ξ (see Fig. 3.2, [Ha81a]) and has the form:

$$
c_1(\xi) = \frac{-1}{2\pi}\{ n_{1_0}\Delta\varphi + \nu[(n_{1_0}\cos\varphi + n_{2_0}\sin\varphi)\sin\varphi]_{\varphi_2}^{\varphi_1} \}
\tag{3.48}
$$

$$c_2(\xi) = \frac{-1}{2\pi}\{n_{2_0}\Delta\varphi + \nu[(n_{1_0}\sin\varphi - n_{2_0}\cos\varphi)\sin\varphi]_{\varphi_2}^{\varphi_1}\}. \tag{3.49}$$

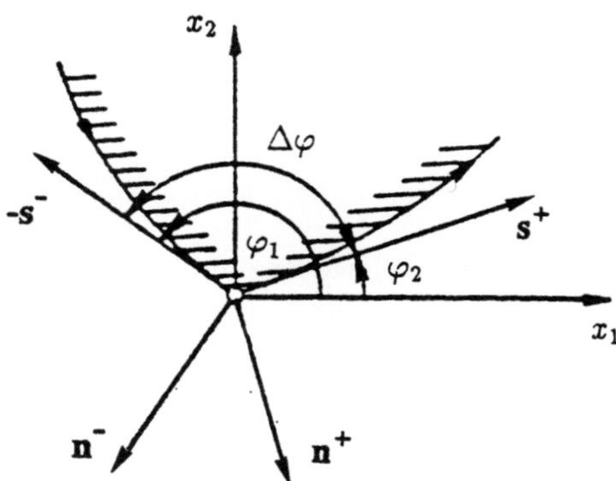

Fig. 3.2: Point ξ at a boundary corner in the plate; definition of φ_1 and φ_2

For every point ξ on a smooth boundary part, the interior angle is $\Delta\varphi = \varphi_1 - \varphi_2 = \pi$ which gives simply that

$$c_\alpha(\xi)w_{,\alpha}(\xi) = -\frac{1}{2}w_{,n}(\xi) = \frac{1}{2}\varphi_n(\xi). \tag{3.50}$$

Note at this point that the kernel $\overset{\star\star}{K}_n(x,\xi)$ in (3.47) is singular like r^{-2}. Therefore the related integral in eq. (3.43) cannot be evaluated directly. But, since it can be proved [Ha81a], that

$$\int_\Gamma \overset{\star\star}{K}_n(x,\xi)d\Gamma_x = 0\,,$$

we can use the extended integral term

$$\int_\Gamma \overset{\star\star}{K}_n(x,\xi)[w(x) - w(\xi)]d\Gamma_x \tag{3.51}$$

in (3.43). Then this integral can be determined as a Cauchy principal value. It should be mentioned here that other regularization techniques are also possible (see e.g. [Kat, Bal]) to avoid such strongly singular integrals.

Using these eqs. (3.41) and (3.42)(3.43) it is possible to solve all plate bending problems with any combination of the following boundary conditions:

(i) $K_n = \bar{K}_n$ and $M_n = \bar{M}_n$ on a free boundary part
      $T_c = \bar{T}_c$ at a free corner

(ii) $w = \bar{w}$ and $M_n = \bar{M}_n$ on a simply supported boundary part including corners

(iii) $w = \bar{w}$ and $\varphi_n = \bar{\varphi}_n$ on a clamped boundary part including corners.

In general, the values with the overbar are prescribed to be zero. But, if a particular solution $w_p(x)$ of the inhomogeneous equilibrium equation

$$D\Delta\Delta w_p(x) = q(x)$$

is used to avoid in the integral equations the domain integrals which contain the transverse force distribution $q(x)$, then, the boundary conditions have to be changed, e.g. along a clamped boundary

$$\bar{w} = -\bar{w}_p \text{ and } \bar{\varphi}_n = -\varphi_n(\bar{w}_p).$$

Let us close this Section by the following example: We consider a rectangular plate which is simply supported along its whole boundary Γ. Then the boundary reactions due to a transversal loading $\bar{q}(x)$ can be determined by the following system of integral equations $(w_{,n} = -\varphi_n)$

$$\int_\Gamma \{\overset{*}{M}_n (x,\xi)\varphi_n(x) - \overset{*}{w}(x,\xi)K_n(x)\}d\Gamma_x - \sum_j T_c(x_{A_j}) \overset{*}{w}(x_{A_j},\xi)$$

$$= \int_\Omega \bar{q}(x) \overset{*}{w}(x,\xi)d\Omega_x$$

$$\frac{1}{2}\varphi_n(\xi) + \int_\Gamma \{\overset{**}{M}_n (x,\xi)\varphi_n(x) - \overset{**}{w}(x,\xi)K_n(x)\}d\Gamma_x - \sum_j T_c(x_{A_j}) \overset{**}{w}(x_{A_j},\xi)$$

$$= \int_\Omega \bar{w}(x) \overset{**}{w}(x,\xi)d\Omega_x$$

where the point ξ is assumed to be placed as a smooth boundary Γ, not at a corner x_{A_j}. For the numerical, approximate solution of these equations, the reader is referred to the next Chapter.

3.3 A Direct B.I.E.M. for Reissner Plates

The research on the integral equations formulations for the more refined plate model of Reissner (a 6th order plate model) started very late. The reason is that only in 1982 van der Ween [We82b] found the basic fundamental solutions. Therefore, the number of publications on this topic is rather small. Besides some work of the first author on both the direct [An84a] and the indirect method [An86b], and on some new fundamental solutions [An85b, An84b], there are only a few papers reporting on the applications and the special numerical treatment (Brebbia et al [Bre87], and Telles et al. [Kar]).

Let us consider a plate with constant thickness h where the plane $x_1 - x_2$ is assumed to coincide with the mean surface. This mean surface is subjected to distributed transverse loads $q(x_1, x_2)$, as well as to body forces $F_\alpha(x_1, x_2), (\alpha = 1, 2)$ and $F_3(x_1, x_2)$

per unit area, which cause generalized displacements u_i: u_α indicates the rotations φ_α, and u_3 the deflection w in the direction of the thickness x_3. The related strains, the bending strains $\kappa_{\alpha\beta}$ and the transverse shear strains γ_α can be obtained from the generalized displacements through the relations

$$\kappa_{\alpha\beta} = \frac{1}{2}(\varphi_{\alpha,\beta} + \varphi_{\beta,\alpha}), \tag{3.52}$$

$$\gamma_\alpha = \varphi_\alpha + w_{,\alpha}. \tag{3.53}$$

The corresponding "generalized internal forces", (called also "generalized stresses"), i.e. the bending stress-couples $M_{\alpha\beta}$ and the shear-stress-resultants Q_α are related to these strains through the constitutive equations

$$
\begin{aligned}
M_{\alpha\beta} &= D(1-\nu)[\kappa_{\alpha\beta} + \frac{\nu}{1-\nu}\delta_{\alpha\beta}\kappa_{\lambda\lambda}] + \frac{\nu}{1-\nu}\delta_{\alpha\beta}\lambda^{-2}q \\
&= \hat{M}_{\alpha\beta} + \frac{\nu}{1-\nu}\delta_{\alpha\beta}\lambda^{-2}q
\end{aligned}
\tag{3.54}
$$

$$Q_\alpha = \lambda^2 D\frac{1-\nu}{2}\gamma_\alpha, \tag{3.55}$$

where $D = Eh^3/12(1-\nu^2)$ is the bending rigidity and $\lambda^2 = 10/h^2$ is a characteristic quantity related to Reissner's model.

The equilibrium equations within the domain Ω of the plate are given by

$$\hat{M}_{\alpha\beta,\beta} - Q_\alpha = -p_\alpha = -(F_\alpha + \frac{\nu}{1-\nu}\lambda^{-2}q_{,\alpha}), \tag{3.56}$$

$$Q_{\alpha,\alpha} = -p_3 = -(F_3 + q). \tag{3.57}$$

In this theory, the most general variational functional, the Hu-Washizu functional is expressed by

$$P(\varphi_\alpha, w; \hat{M}_{\alpha\beta}, Q_\alpha; \kappa_{\alpha\beta}, \gamma_\alpha) \tag{3.58}$$

$$
\begin{aligned}
:= &\int_\Omega \{U(\kappa_{\alpha\beta}, \gamma_\alpha) - (\bar{p}_\alpha\varphi_\alpha + \bar{p}_3 w) + \hat{M}_{\alpha\beta}[\frac{1}{2}(\varphi_{\alpha,\beta} + \varphi_{\beta,\alpha}) - \kappa_{\alpha\beta}] \\
&\qquad + Q_\alpha[(\varphi_\alpha + w_{,\alpha}) - \gamma_\alpha]\}d\Omega \\
&- \int_{\Gamma_1} \{\hat{M}_{nn}(\varphi_n - \bar{\varphi}_n) + \hat{M}_{nt}(\varphi_t - \bar{\varphi}_t) + Q_n(w - \bar{w})\}d\Gamma \\
&- \int_{\Gamma_2} \{\bar{\hat{M}}_{nn}\varphi_n + \bar{\hat{M}}_{nt}\varphi_t + \bar{Q}_n w d\Gamma,
\end{aligned}
$$

where the specific deformation energy $U(\kappa_{\alpha\beta}, \gamma_\alpha)$ is given by

$$U(\kappa_{\alpha\beta}, \gamma_\alpha) = D\frac{1-\nu}{2}\{[\kappa_{\alpha\beta} + \frac{\nu}{1-\nu}\delta_{\alpha\beta}\kappa_{\lambda\lambda}]\kappa_{\alpha\beta} + \frac{\lambda^2}{2}\gamma_\alpha\gamma_\alpha\}. \tag{3.59}$$

Γ_1 and Γ_2 denote those parts of the plate boundary Γ which have prescribed "displacements" $(\varphi_n, \varphi_t, w)$ and "forces" $(\hat{M}_{nn}, \hat{M}_{nt}, Q_n)$, respectively. They are given by the expressions

$$\varphi_n = \varphi_\alpha n_\alpha \text{ and } \varphi_t = \varphi_\alpha s_\alpha \tag{3.60}$$

$$\hat{M}_{nn} = \hat{M}_{\alpha\beta}n_\alpha n_\beta \quad \text{and} \quad \hat{M}_{nt} = \hat{M}_{\alpha\beta}n_\alpha s_\beta, \tag{3.61}$$

$$Q_n = Q_\alpha n_\alpha. \tag{3.62}$$

Taking into account that

$$\frac{\partial U(\kappa_{\alpha\beta}, \gamma_\alpha)}{\partial \kappa_{\rho\lambda}} = \hat{M}_{\rho\lambda} \quad \text{and} \quad \frac{\partial U(\kappa_{\alpha\beta}, \gamma_\alpha)}{\partial \gamma_\rho} = Q_\rho, \tag{3.63}$$

$$\hat{M}_{\alpha\beta}\frac{1}{2}(\varphi_{\alpha,\beta} + \varphi_{\beta,\alpha}) = \hat{M}_{\alpha\beta}\varphi_{\alpha,\beta}, \tag{3.64}$$

the first variation of the functional of (3.58) becomes

$$\delta P(\varphi_\alpha, w; \hat{M}_{\alpha\beta}, Q_\alpha; \kappa_{\alpha\beta}, \gamma_\alpha) \tag{3.65}$$

$$= \int_\Omega \left\{ -(\bar{p}_\alpha \delta\varphi_\alpha + \bar{p}_3 \delta w) + \hat{M}_{\alpha\beta}\delta\varphi_{\alpha,\beta} + Q_\alpha(\delta\varphi_\alpha + \delta w_{,\alpha}) \right.$$

$$\left. + \delta\hat{M}_{\alpha\beta}(\varphi_{\alpha,\beta} - \kappa_{\alpha\beta}) + \underline{\delta Q_\alpha}[(\varphi_\alpha + \underline{w_{,\alpha}}) - \gamma_\alpha] \right\} d\Omega$$

$$- \int_{\Gamma_1} \left\{ \hat{M}_{nn}\delta\varphi_n + \hat{M}_{nt}\delta\varphi_t + Q_n\delta w \right.$$

$$\left. + \delta\hat{M}_{nn}(\varphi_n - \bar{\varphi}_n) + \delta\hat{M}_{nt}(\varphi_t - \bar{\varphi}_t) + \delta Q_n(w - \bar{w}) \right\} d\Gamma$$

$$- \int_{\Gamma_2} \left\{ \bar{\hat{M}}_{nn}\delta\varphi_n + \bar{\hat{M}}_{nt}\delta\varphi_t + \bar{Q}_n\delta w \right\} d\Gamma.$$

When the underlined terms in eq. (3.65) are integrated by parts, the resulting relation is

$$\int_\Omega \left\{ \varphi_{\alpha,\beta}\delta\hat{M}_{\alpha\beta} + w_{,\alpha}\delta Q_\alpha \right\} d\Omega = \int_{\Gamma_1 + \Gamma_2} \left\{ \varphi_\alpha\delta\hat{M}_{\alpha\beta}n_\beta + w\delta Q_\alpha n_\alpha \right\} d\Gamma \tag{3.66}$$

$$- \int_\Omega \left\{ \varphi_\alpha\delta\hat{M}_{\alpha\beta,\beta} + w\delta Q_{\alpha,\alpha} \right\} d\Omega.$$

By introducing (3.66) into (3.65) and by setting δP equal to zero we obtain that

$$- \int_\Omega \left\{ \varphi_\alpha[\delta\hat{M}_{\alpha\beta,\beta} - \delta Q_\alpha] + w\delta Q_{\alpha,\alpha} \right\} d\Omega \tag{3.67}$$

$$+ \int_{\Gamma_1} \left\{ \bar{\varphi}_n\delta\hat{M}_{nn} + \bar{\varphi}_t\delta\hat{M}_{nt} + \bar{w}\delta Q_n \right\} d\Gamma + \int_{\Gamma_2} \left\{ \varphi_n\delta\hat{M}_{nn} + \varphi_t\delta\hat{M}_{nt} + w\delta Q_n \right\} d\Gamma$$

$$= \int_\Omega \left\{ \bar{p}_\alpha\delta\varphi_\alpha + \bar{p}_3\delta w \right\} d\Omega$$

$$+ \int_\Omega \left\{ [\delta\hat{M}_{\alpha\beta}\kappa_{\alpha\beta} + \delta Q_\alpha\gamma_\alpha] - [\hat{M}_{\alpha\beta}\delta\varphi_{\alpha,\beta} + Q_\alpha(\delta\varphi_\alpha + \delta w_{,\alpha})] \right\} d\Omega$$

$$+ \int_{\Gamma_1} \hat{M}_{nn}\delta\varphi_n + \hat{M}_{nt}\delta\varphi_t + Q_n\delta w \right\} d\Gamma + \int_{\Gamma_2} \left\{ \bar{\hat{M}}_{nn}\delta\varphi_n + \bar{\hat{M}}_{nt}\delta\varphi_t + \bar{Q}_n\delta w \right\} d\Gamma.$$

Then, from (3.54)(3.52), and from (3.55) (3.53) we obtain easily that

$$\kappa_{\alpha\beta}\delta\hat{M}_{\alpha\beta} = \hat{M}_{\alpha\beta}\delta\varphi_{\alpha,\beta}, \tag{3.68}$$

$$\gamma_\alpha\delta Q_\alpha = Q_\alpha(\delta\varphi_\alpha + \delta w_{,\alpha}), \tag{3.69}$$

and taking instead of arbitrary displacement variations the special displacements $\varphi_\alpha^{(k)}(x,\xi)$, $w^{(k)}(x,\xi)$ due to unit loadings at the point ξ of an infinite plate, i.e. the solutions of

$$\hat{M}_{\alpha\beta,\beta}(\varphi_\rho^{(k)}, w^{(k)}) - Q_\alpha(\varphi_\rho^{(k)}, w^{(k)}) = -\delta(x - \xi)\delta_\alpha^k \tag{3.70}$$

$$k = 1, 2, 3$$

$$Q_{\alpha,\alpha}(\varphi_\rho^{(k)}, w^{(k)}) = -\delta(x - \xi)\delta_3^k, \tag{3.71}$$

the following integral equations result from $((k = 1, 2, 3))$:

$$\varphi_\alpha(\xi)\delta_\alpha^k + w(\xi)\delta_3^k = \int\limits_\Omega \left\{\bar{p}_\alpha(x)\varphi_\alpha^{(k)}(x,\xi) + \bar{p}_3(x)w^{(k)}(x,\xi)\right\}d\Omega_x \tag{3.72}$$

$$+ \int\limits_\Gamma \left\{\hat{M}_{nn}(x)\varphi_n^{(k)}(x,\xi) + \hat{M}_{nt}(x)\varphi_t^{(k)}(x,\xi) + Q_n(x)w^{(k)}(x,\xi)\right\}d\Gamma_x$$

$$- \int\limits_\Gamma \left\{\varphi_n(x)\hat{M}_{nn}^{(k)}(x,\xi) + \varphi_t(x)\hat{M}_{nt}^{(k)}(x,\xi) + w(x)Q_n^{(k)}(x,\xi)\right\}d\Gamma_x.$$

Since the boundary conditions for stress couples prescribe values for M_{nn} and M_{nt}, but not for $\hat{M}_{nn}$ and $\hat{M}_{nt}$, the respective boundary integral has to be transformed:

$$\int\limits_\Gamma \hat{M}_{nn}(x)\varphi_n^{(k)}(x,\xi)d\Gamma_x \tag{3.73}$$

$$= \int\limits_\Gamma M_{nn}(x)\varphi_n^{(k)}(x,\xi)d\Gamma_x - \frac{\nu}{1-\nu}\frac{1}{\lambda^2}\int\limits_\Gamma q(x)\varphi_\alpha^{(k)}(x,\xi)n_\alpha(x)d\Gamma_x$$

$$= \int\limits_\Gamma M_{nn}(x)\varphi_n^{(k)}(x,\xi)d\Gamma_x - \int\limits_\Omega \frac{\nu}{1-\nu}\frac{1}{\lambda^2}[q(x)\varphi_\alpha^{(k)}(x,\xi)]_{,\alpha}d\Gamma_x.$$

Finally, after combining the domain integrals, one obtains

$$\varphi_\alpha(\xi)\delta_\alpha^k + w(\xi)\delta_3^k = \int\limits_\Omega \left\{\bar{F}_\alpha(x)\varphi_\alpha^{(k)}(x,\xi) + \bar{F}_3(x)w^{(k)}(x,\xi)\right. \tag{3.74}$$

$$\left. + \bar{q}(x)[w^{(k)}(x,\xi) - \frac{\nu}{1-\nu}\frac{1}{\lambda^2}\varphi_{\alpha,\alpha}^{(k)}(x,\xi)]\right\}d\Omega_x$$

$$+ \int\limits_\Gamma \left\{M_{nn}(x)\varphi_n^{(k)}(x,\xi) + M_{nt}(x)\varphi_t^{(k)}(x,\xi) + Q_n(x)w^{(k)}(x,\xi)\right\}d\Gamma_x$$

$$- \int\limits_\Gamma \left\{\varphi_n(x)\hat{M}_{nn}^{(k)}(x,\xi) + \varphi_t(x)\hat{M}_{nt}^{(k)}(x,\xi) + w(x)Q_n^{(k)}(x,\xi)\right\}d\Gamma_x.$$

By using the equations (3.74) for $k = 1, 2, 3$, the rotations $\varphi_\alpha(\xi)$ and the deflection $w(\xi)$ can be determined at any arbitrary point ξ in the interior of the plate Ω, when all unknown boundary reactions have been determined before. The boundary integral equations, which are necessary for their determination, are gained by locating the point ξ on the boundary Γ. This yields the relations (body forces are neglected)

$$c_{\beta\alpha}(\xi)\varphi_\alpha(\xi) = \int_\Omega \bar{q}(x)[w^{(\beta)}(x,\xi) - \frac{\nu}{1-\nu}\frac{1}{\lambda^2}\varphi_{\gamma,\gamma}^{(\beta)}(x,\xi)]d\Omega_x \tag{3.75}$$

$$+ \int_\Gamma \{M_{nn}(x)\varphi_n^{(\beta)}(x,\xi) + M_{nt}(x)\varphi_t^{(\beta)}(x,\xi) + Q_n(x)w^{(\beta)}(x,\xi)\}d\Gamma_x$$

$$- \int_\Gamma \{\varphi_n(x)\hat{M}_{nn}^{(\beta)}(x,\xi) + \varphi_t(x)\hat{M}_{nt}^{(\beta)}(x,\xi) + w(x)Q_n^{(\beta)}(x,\xi)\}d\Gamma_x, \tag{3.76}$$

$$c(\xi)w(\xi) = \int_\Omega \bar{q}(x)[w^{(3)}(x,\xi) - \frac{\nu}{1-\nu}\frac{1}{\lambda^2}\varphi_{\gamma,\gamma}^{(3)}(x,\xi)]d\Omega_x \tag{3.76}$$

$$+ \int_\Gamma \{M_{nn}(x)\varphi_n^{(3)}(x,\xi) + M_{nt}(x)\varphi_t^{(3)}(x,\xi) + Q_n(x)w^{(3)}(x,\xi)\}d\Gamma_x$$

$$- \int_\Gamma \{\varphi_n(x)\hat{M}_{nn}^{(3)}(x,\xi) + \varphi_t(x)\hat{M}_{nt}^{(3)}(x,\xi) + w(x)Q_n^{(3)}(x,\xi)\}d\Gamma_x,$$

where $c_{\beta\alpha}(\xi) = 0.5\delta_{\beta\alpha}$ and $c(\xi) = 0.5$ on smooth curves (cf. also ref. [We82a] in this context). Moreover, in order to have a complete intrinsic component formulation (i.e. $\varphi_n(\xi)$ and $\varphi_t(\xi)$ also on the left-hand side) both sides of eq. (3.75) have to be multiplied with $n_\beta(\xi)$ and $s_\beta(\xi)$, respectively.

The singular solutions in the eqs. (3.75)(3.76) are [We82a, An84a, Kar]:

$$\varphi_\alpha^{(\beta)} = \frac{1}{8\pi D}\{[\frac{8}{1-\nu}B(z) - \ln z^2 + 1]\delta_{\alpha\beta} - [\frac{8}{1-\nu}A(z) + 2]r_{,\alpha}r_{,\beta}\} \tag{3.77}$$

$$w^{(\beta)} = -\varphi_\beta^{(3)} = \frac{1}{8\pi D}rr_{,\alpha}(\ln z^2 - 1) \tag{3.78}$$

$$w^{(3)} = \frac{1}{16\pi D}\frac{1}{\lambda^2}\{[z^2 - \frac{8}{1-\nu}]\ln z^2 - 2z^2\} \tag{3.79}$$

where

$$A(z) = K_0(z) + \frac{2}{z}(K_1(z) - \frac{1}{z}) \tag{3.80}$$

$$B(z) = K_0(z) + \frac{1}{z}(K_1(z) - \frac{1}{z}) \tag{3.81}$$

contain the modified Bessel functions $K_o(z)$ and $K_1(z)$ with the argument $z = \lambda r$ where $\lambda^2 = 10/h^2$. $A(z)$ is continous while $B(z)$ has the singularity $\ln(z)$ [We82a].

The moments $\hat{M}_{nn}^{(k)}, \hat{M}_{nt}^{(k)}$ and the shear stress resultants $Q_n^{(k)}$ can be obtained using the relations $(3.52)\div(3.55)$ [14] ($r_{,n} = r_{,\alpha}n_\alpha(x)$ and $r_{,s} = r_{,\alpha}s_\alpha(x)$):

$$Q_n^{(\beta)} = Q_\alpha^{(\beta)}n_\alpha(x) = \frac{\lambda^2}{2\pi}[B(z)n_\beta(x) - A(z)r_{,\beta}r_{,n}] \tag{3.82}$$

$$Q_n^{(3)} = Q_\alpha^{(3)}n_\alpha(x) = -\frac{1}{2\pi}\frac{1}{r}r_{,n} \tag{3.83}$$

$$\hat{M}_{nn}^{(\beta)} = \frac{1}{2\pi}\frac{1}{r}\{[zA'(z) - \frac{1-\nu}{2}]2r_{,s}r_{,n}e_{\beta\rho}r_{,\rho} + [2A(z)(r_{,n}^2 - r_{,s}^2) - \frac{1+\nu}{2}]r_{,\beta}\} \tag{3.84}$$

$$\hat{M}_{nn}^{(3)} = -\frac{1}{8\pi}\{(1+\nu)\ln z^2 + (1-\nu)(r_{,n}^2 - r_{,s}^2)\} \tag{3.85}$$

$$\hat{M}_{nt}^{(\beta)} = \frac{1}{\pi}\frac{1}{r}\{r_{,s}^2(zA'(z) - \frac{1-\nu}{2})e_{\beta\rho}r_{,\rho} + 2A(z)r_{,n}r_{,s}r_{,\beta}\} \tag{3.86}$$

$$\hat{M}_{nt}^{(3)} = -\frac{(1-\nu)}{4\pi}r_{,n}r_{,s}, \tag{3.87}$$

where $e_{\alpha\beta}$ is the permutation tensor, and

$$A'(z) = \partial A(z)/\partial z = -\frac{2}{z}K_o(z) - K_1(z) - \frac{4}{z^2}(K_1(z) - \frac{1}{z}). \tag{3.88}$$

Finally, it should be mentioned that in the case of uniform load $q(x) = q_0$, the domain integrals in eqs. (3.75),(3.76) are convertible to boundary integrals. With the functions [We82a, An84a]

$$v^{(\alpha)} = \frac{r^3}{128\pi D}r_{,\alpha}(2\ln z^2 - 5) \tag{3.89}$$

$$v^{(3)} = \frac{r^2}{256\pi D}\{[r^2 - \frac{32}{\lambda^2(1-\nu)}]\ln z^2 + \frac{64}{\lambda^2(1-\nu)} - 3r^2\}, \tag{3.90}$$

which satisfy the Poisson equation $\Delta v^{(k)} = w^{(k)}$, the divergence theorem yields:

$$q_0\int_\Omega[w^{(k)} - \frac{\nu}{1-\nu}\lambda^{-2}\varphi_{\alpha,\alpha}^{(k)}]d\Omega = q_0\int_\Gamma[v_{,\alpha}^{(k)} - \frac{\nu}{1-\nu}\lambda^{-2}\varphi_\alpha^{(k)}]n_\alpha d\Gamma. \tag{3.91}$$

Besides these directly derived integral equations (3.75)(3.76), the indirect method which use static as well as geometric fundamental solutions shall be employed in the case of Reissner's plates. Let us describe this method. The basic types of statical singularities are the single forces. The rotations $\varphi_\alpha(x)$ and the deflection $w(x)$ due to the unit single forces $p_k = \delta(x - \xi)\delta_k^i$ acting at a point ξ of the infinite plate Ω^0 in a direction i are

$$\varphi_\alpha(x) = (\varphi p)_{\alpha.k}(x,\xi)p_k(\xi) \quad \text{and} \quad w(x) = (wp)_{.k}(x,\xi)p_k(\xi) \tag{3.92}$$

where the influence functions are given by (3.77),(3.78),(3.79), i.e.

$$(\varphi p)_{\alpha.k}\hat{=}\varphi_\alpha^{(k)} \quad \text{and} \quad (wp)_{.k}\hat{=}w^{(k)} \quad . \tag{3.93}$$

The influence functions describing the behaviour of the stress couples $\hat{M}_{nn}$ and $\hat{M}_{nt}$ as well as the stress resultant Q_n due to single forces have also been given before by eqs. (3.82),(3.83),(3.84) and we set

$$(\hat{M}p)_{nn.k} \hateq \hat{M}_{nn}^{(k)}, \quad (\hat{M}p)_{nt.k} \hateq \hat{M}_{nt}^{(k)} \text{ and } (Qp)_{n.k} \hateq Q_n^{(k)}. \tag{3.94}$$

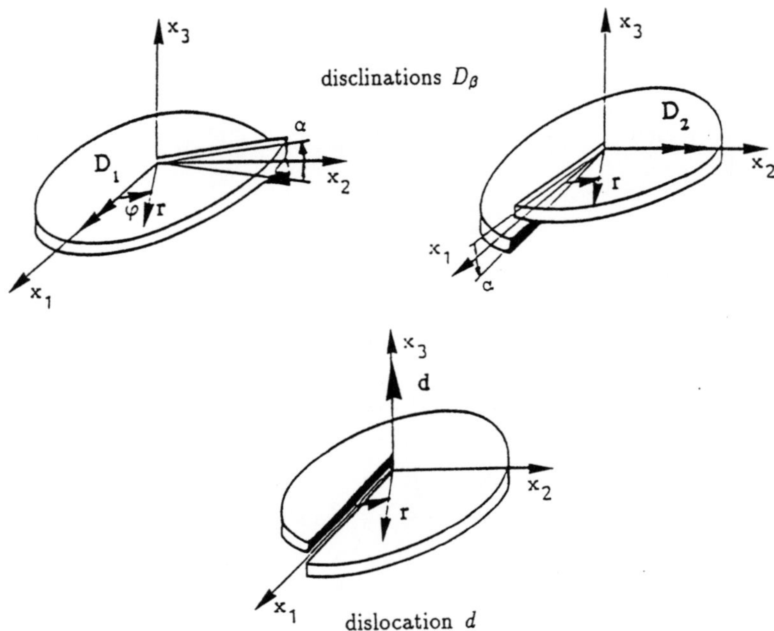

Fig. 3.3: Disclinations and dislocations in a Reissner plate

The basic geometrical singularities of Reissner plates are the so-called twist disclination D_α and the screw dislocation d (see Fig. 3.3). Mathematically, these states are singular inhomogeneties, i.e. fundamental solutions of the compatibility conditions [An84a]

$$\frac{1}{2}e_{\beta\rho}\{\kappa_{\alpha\beta,\rho} - \kappa_{\alpha\rho,\beta} + \gamma_{\rho,\alpha\beta}\} = D_\alpha, \tag{3.95}$$

where $e_{\beta\rho}$ is the permutation tensor. They represent discontinuities of the rotations φ_α and of the deflection w, respectively. The influence functions $(\varphi D)_{\alpha.\beta}$ and $(wD)_{,\beta}$, describe the "displacements" at the observation point x due to a unit rotation discontinuity D_β at ξ. They have been found [An85a, An87a] to be

$$(\varphi D)_{\alpha.\beta} = e_{\beta\rho}\{e_{\alpha\rho}[2B(z) - 2A(z)r_{,\alpha}^2 + \frac{1-\nu}{2}r_{,\rho}^2 + \frac{1+\nu}{4}\ln(z^2)]\delta_{\alpha\beta} \tag{3.96}$$

$$+ [e_{\rho\mu}r_{,\rho}r_{,\mu}(2A(z) + \frac{1-\nu}{2}) - \varphi]\}/2\pi$$

$$(wD)_{,\beta} = re_{\rho\beta}\{e_{\mu\rho}r_{,\mu}(1+\nu)[\ln(z^2)-2] - 4r_{,\rho}\varphi\}/8\pi \ . \tag{3.97}$$

Due to the behaviour of the functions A(z) and $B(z)$, the influence functions $(\varphi D)_{\alpha.\beta}$, $\alpha \neq \beta$, are weakly singular while the others are regular, but, due to the inverse circular tangent $\varphi = \arctan(r_{,1}/r_{,2})$, they are discontinuous.

Since the screw dislocation d may also be interpreted as a dipole of two opposite disclinations D_β [An85a], its influence function can be obtained by differentiation of (3.96) and (3.97) with respect to the source point co-ordinates ξ. The result is simply

$$(\varphi d)_{\beta.} = e_{\beta\rho}r_{,\rho}r^{-1}[1 - zK_1(z)]/2\pi \quad \text{and} \quad (wd)_. = \varphi/2\pi \ . \tag{3.98}$$

These influence functions are strongly singular and discontinuous, respectively.

Since influence functions are transformed via the same relations as the corresponding state variables, the knowledge of one influence function of a certain type, for a statical or a geometrical singularity, respectively, is sufficient for the calculation of the other influence functions. Hence, we obtain the influence functions of the corresponding strains and stresses by using the adequate relations $(3.52)\div(3.55)$. In this Section only the expressions for the "internal forces" caused by singular twist disclinations D_ρ are given, because some influence functions of the screw dislocation d have a r^{-2} singularity and need for this reason a special treatment, i.e. a regularization.

$$(\hat{M}D)_{\alpha\beta.\rho} = N(1-\nu)r^{-1}\left\{[r_{,\alpha}\delta_{\beta\rho} + r_{,\beta}\delta_{\alpha\rho}][zA'(z) + \frac{1+\nu}{2}]\right. \tag{3.99}$$

$$\left.-r_{,\rho}\delta_{\alpha\beta}[2A(z) + \frac{3+\nu}{2}] + 2r_{,\rho}r_{,\alpha}r_{,\beta}[2A(z) - zA'(z) + \frac{1-\nu}{2}]\right\}/2\pi$$

$$(QD)_{\alpha.\rho} = N(1-\nu)\lambda^2\{B(z)\delta_{\alpha\rho} - A(z)r_{,\alpha}r_{,\rho}\}/2\pi \ . \tag{3.100}$$

For an actual B.V.P. any of these singularities, either the static or the geometric ones, or a combination of both, can be used. The only condition to observe is that one has to employ the same three different singularity layers with unknown intensity, for example, the three orthogonal components of the singular "forces", in formulating the three integral equations for the three prescribed boundary values. When, for example, the bending stress-couple $\hat{M}_{nn}(s)$ is prescribed along some part of the plate boundary, the three layers can be those of the mentioned three components of the singular forces $p_k(\xi)$. Then, introducing intrinsic co-ordinates both for the source point $\bar{s}(\hat{=}\xi)$ and for the observation point $s(\hat{=}x)$, one gets, if the fictitious boundary Γ^+ coincides with the real boundary Γ, the expression

$$\hat{M}_{nn}(s) \ = \ \frac{1}{2}p_n(s) + \int_\Gamma \{(\hat{M}p)_{nn.\bar{n}}(s,\bar{s})p_{\bar{n}}(\bar{s}) \tag{3.101}$$

$$+(\hat{M}p)_{nn.\bar{t}}(s,\bar{s})p_{\bar{t}}(\bar{s}) + (\hat{M}p)_{nn.3}(s,\bar{s})p_3(\bar{s})\}d\bar{s} \ ,$$

where $(\hat{M}p)_{nn.\bar{n}}$, for example, can easily be obtained from (3.84), i.e.

$$(\hat{M}p)_{nn.\bar{n}} = r^{-1}\left\{2r_{,\bar{s}}r_{,s}r_{,n}[\frac{1-\nu}{2} - zA'(z)]\right. \tag{3.102}$$

$$\left.+ r_{,\bar{n}}[2A(z)(r_{,n}^2 - r_{,s}^2) - \frac{1+\nu}{2}]\right\}/2\pi \ .$$

It is not an easy task to decide about, which combination of singularity layers may be optimal, for a problem under consideration. Based on investigations of the spectra of integral operators, some criteria for the choice of the singularities, and thus, of the adequate integral equations, are known in the case of plane elastostatics [He85], but not, until now, for the present Reissner equations. Therefore, herein, the only fact which is generally correct will be used as a criterium: We avoid integral equations of the first kind, because their numerical treatment may be problematic.

Chapter 4
On the Numerical Implementation of Boundary Element Equations

4.1 General Methods

It is not generally possible to compute the boundary integrals and solve the boundary integral equations analytically. Therefore, a numerical approach has to be used. Recent advances, such as the use of isoparametric boundary elements and the careful analytical treatment of singular integrals, have had a major impact on the competitiveness of the B.E.M. in routine two-dimensional as well as in three-dimensional static and dynamic analyses. For more on these developments, aimed at establishing the BEM as a general engineering tool, the reader is referred, for instance to [Ba81a] and [Ba81b].

There are two major approaches concerning the numerical treatment procedure:

i) the nodal point collocation approach

ii) the Galerkin approach.

The first step of the numerical implementation is the same in both B.E.M. approaches: the boundary has to be discretized into elements using piecewise polynomial approximations for the boundary geometry as well as for the prescribed and for the unknown boundary values. First, the boundary is decomposed into smooth subsets between the existing corners. If a subset is a straight line or an arc of a circle, then it is easily specified by the end points and, if necessary, by a centre of curvature. In any case, it is desirable to define each subset of the boundary to be smooth enough, that the normal vector and the curvature be continuous. Each subset is then partitioned into elements. Thus, the integrals over the whole boundary are decomposed into a sum of $e \in [1, N_e]$ integrals over single elements Γ_e:

$$\int_{\Gamma} [\ldots] d\Gamma_x = \sum_{e=1}^{N_e} \int_{\Gamma_e} [\ldots] d\Gamma_x. \tag{4.1}$$

For example, in the case of the elastostatic boundary integral equations (2.34) when the body force term is neglected, we have the discretized equation

$$c_i^{(k)}(\xi) u_i(\xi) \; + \; \sum_{e=1}^{N_e} \!\!\!\!\int_{\Gamma_e} T_i^{(k)}(x, \xi) u_i(x) d\Gamma_x \tag{4.2}$$

$$- \; \sum_{e=1}^{N_e} \int_{\Gamma_e} w_i^{(k)}(x, \xi) T_i(x) d\Gamma_x = 0, \qquad \xi \in \Gamma.$$

Now, on each element Γ_e, nodal points have to be chosen; their number and location depends on the shape functions approximating the boundary quantities. Thus, for eq. (4.2), the boundary functions u_i and T_i have to be approximated in every boundary element Γ_e by means of the nodal point values u_i^{me} and T_i^{me}, and the shape functions $\Theta^m(\zeta)$ which depend on the local coordinate $\zeta \in [-1, +1]$. The approximate forms of the above functions read

$$u_i(\zeta) = \sum_{m=1}^{p} \Theta^m(\zeta) u_i^{me}, \tag{4.3}$$

$$T_i(\zeta) = \sum_{m=1}^{p} \Theta^m(\zeta) T_i^{me}, \tag{4.4}$$

where p depends on the grade of the approximation. It is important to note that the variables in this formulation do not involve the boundary geometry explicitly, and therefore, if necessary, the shape of the boundary may be approximated independently by linear or by curved elements. Then, in order to express the integrals in local coordinates, transformation relations with the Jacobian $J(\zeta)$ have been used. Thus if the transformation $x = x(\zeta)$ is induced by the curved boundary we shall have the relation

$$d\Gamma_x = |J(\zeta)| d\Gamma(\zeta) \tag{4.5}$$

Finally, this yields the following approximate formulation of the eq. (2.34) for any $\xi \in \Gamma_l$ with the local coordinate ζ^l

$$RuT_k(\xi) \doteq c_i^{(k)}(\xi) u_i^{ml} \Theta^m(\zeta^l) \tag{4.6}$$

$$+ \sum_{e=1}^{N_e} \sum_{m=1}^{p} u_i^{me} \int_{\Gamma_e} T_i^{(k)}(x(\zeta), \xi) \Theta^m(\zeta) |J(\zeta)| d\Gamma(\zeta)$$

$$- \sum_{e=1}^{N_e} \sum_{m=1}^{p} T_i^{me} \int_{\Gamma_e} w_i^{(k)}(x, \xi) \Theta^m(\zeta) |J(\zeta)| d\Gamma(\zeta) = 0.$$

Now, as in Sect. 2.2, it is possible to use in eq. (4.6) again the "method of weighted residuals" (see [Kuh]); i.e one obtains a sufficient set of algebraic equations for the unknown nodal point values, through the relation

$$\int_{\Gamma} RuT_k(\xi) w(\xi) d\Gamma(\xi) = 0. \tag{4.7}$$

Using (4.7) implies that both the already mentioned approaches, the nodal point collocation approach and the Galerkin approach, are made dependent on the applied weight function w.

The remaining characteristics of both approaches are described in the two next Sections for Kirchhoff's (see [Ste83]) and Reissner's plate integral equations using the point collocation and the Galerkin approach respectively .

4.2 Kirchhoff Plate Boundary Element Equations by the Point Collocation Method

The primary variables of the Kirchhoff plate are the deflection w, the normal slope φ_n, the bending moment M_n, the Kirchhoff shear force K_n at each point of the boundary and the corner forces T_c at each boundary corner. Note that at each corner there are two distinct limiting values of φ_n, M_n and K_n. Thus, one has to approximate the boundary functions $u_i = (w, \varphi_n)$ and $S_i = (K_n, M_n)$ by means of their values u_i^{me} and S_i^{me} at the nodal points in the n elements Γ_e, and by adequate shape functions. One obtains the following discretized, approximating form (i $= 1,2$)

$$RuS_i(\xi) \hat{=} c_{ik}(\xi)u_k(\xi) \tag{4.8}$$

$$+ \sum_{e=1}^{N_e} \sum_{m=1}^{p} u_k^{em} \int_{\Gamma_e} T_{ik}[\xi, x(\zeta)]\Theta^m(\zeta)|J(\zeta)|d\Gamma(\zeta) + \sum_{j=1}^{C} T_{ic}[\xi, x_{A_j}]u_1(x_{A_j})$$

$$- \sum_{e=1}^{N_e} \sum_{m=1}^{p} S_k^{em} \int_{\Gamma_e} U_{ik}[\xi, x(\zeta)]\Theta^m(\zeta)|J(\zeta)|d\Gamma(\zeta) + \sum_{j=1}^{C} U_{i1}[\xi, x_{A_j}]T_c(x_{A_j}).$$

Here linear shape functions $\Theta^m(\zeta)$ may be taken, for instance,

$$\Theta^1(\zeta) = \frac{1}{2}(1 - \zeta), \tag{4.9}$$

$$\Theta^2(\zeta) = \frac{1}{2}(1 + \zeta), \tag{4.10}$$

where $-1 \leq \zeta \leq 1$. The coefficients $c_{ik}(\xi)$ are simply $\frac{1}{2}\delta_{ik}$ for all points ξ on smooth boundary parts, and the kernels T_{ik} and $U_{ik}(i, k = 1, 2)$ represent the singular "generalized tractions and displacements" of the unit loadings:

$$[T_{ik}] \hat{=} \begin{bmatrix} \overset{*}{K}_n(x,\xi) & \overset{*}{M}_n(x,\xi) \\ \overset{**}{K}_n(x,\xi) & \overset{**}{M}_n(x,\xi) \end{bmatrix} \text{ and } [U_{ik}] \hat{=} \begin{bmatrix} \overset{*}{w}_n(x,\xi) & \overset{*}{\varphi}_n(x,\xi) \\ \overset{**}{w}_n(x,\xi) & \overset{**}{\varphi}_n(x,\xi) \end{bmatrix}. \tag{4.11}$$

If ξ is a corner point of the boundary, the integral-free terms of eq. (4.8) must be changed as in (3.48), i.e.

$$c_{11}(\xi)u_1(\xi) = \frac{\Delta\alpha}{2\pi}w(\xi) \tag{4.12}$$

$$c_{12}(\xi)u_2(\xi) = c_{21}(\xi)u_1(\xi) = 0. \tag{4.13}$$

In the general case the relations (3.48)(3.49) hold, which here have the form

$$c_{22}(\xi)u_2(\xi) \hat{=} \frac{1}{2\pi}\left\{\Delta\alpha n_{1o} + \nu\left[\sin\alpha(n_{1o}\cos\alpha + n_{2o}\sin\alpha)\right]_{\alpha_2}^{\alpha_1}\right\}\varphi_1(\xi) \tag{4.14}$$

$$\frac{1}{2\pi}\left\{\Delta\alpha n_{2o} + \nu\left[\sin\alpha(n_{1o}\sin\alpha + n_{2o}\cos\alpha)\right]_{\alpha_2}^{\alpha_1}\right\}\varphi_2(\xi).$$

Here α_1 and α_2 are the interior angles ahead and behind a boundary point ξ and (n_{1o}, n_{2o}) denote the components of the direction vector which determines the directional derivations which form the singular kernels of the integral equation (3.43). Thus, since the integral terms of eq. (4.8) are depending on this direction, as well, two independent B.I.Es can be obtained for two different directions, e.g. for n^- and n^+, i.e. the outward normal vectors ahead and behind the corner. These two equations constitute a regular linear system for the two values $\varphi_\alpha(\xi), \alpha = 1, 2$, and via

$$\varphi_\alpha(\xi) = n_\alpha^+ \varphi_n^+ = n_\alpha^- \varphi_n^-$$

for the two distinct limiting values of the normal derivative, φ_n^- and φ_n^+.

In order to get a system of algebraic equations for the unknown nodal point values via the point collocation approach, one has to take in eq. (4.7) a linearly independent combination of N_e times p Dirac- functions $\delta(\xi, x^{me})$ as the weight function

$$w(\xi) := \sum_{e=1}^{N_e} \sum_{m=1}^{p} \beta_{em} \delta(\xi, x^{me}). \tag{4.15}$$

Since the coefficients β_{em} are arbitrary constants, and the integration of eq. (4.7) leads, due to the Dirac functions, simply to the pointwise evaluation of its integrand, one obtains the conditions

$$RuS_i(x^{me}) = 0, \qquad \text{for} \quad m = 1, \dots, p \quad \text{and} \quad e = 1, \dots, N_e, \tag{4.16}$$

i.e. the pointwise evaluation of eq. (4.8) at all nodes. Then, the elements of the matrix of the final system of algebraic equations are accomplished by an analytical or numerical computation of the boundary integrals over the elements Γ_e. To achieve it, eight boundary integrals per element have to be calculated, if all prescribed boundary values are non-zero:

i) for the regular and the weakly singular element integrals the usual or the weighted Gaussian quadrature formula are sufficient,

ii) for integrals with strongly singular integrands - here those with the kernels $\overset{*}{K}_n\left(x(\zeta), \xi\right) \Theta^m(\zeta)$, $\overset{**}{M}_n\left(x(\zeta), \xi\right)\Theta^m(\zeta)$ and $\overset{**}{K}_n\left(x(\zeta), \xi\right)[\Theta^m(\zeta) - \Theta^m(\xi)]$ - the Cauchy principal values should be calculated either analytically or using special numerical integration techniques [Ha89, Ioa77, The76,77a,b,d].

Finally, one obtains a system of $2[N_e(p-1) + CN_e] + C$ (N_e indicates the number of elements, p the order of the shape functions and C the number of the corners) linear algebraic equations which can be written in matrix form as

$$(C + T) \cdot u = \hat{T} \cdot u = U \cdot t. \tag{4.17}$$

The diagonal elements of the matrix $\hat{T}$, which contain the term $c_{ik}(\xi)$ and are found as principal values of singular integrals, can be determined implicitly by considering the B.E. formulation of a rigid body motion [Du] (for more details, see also next Section). Combined with the prescribed boundary conditions, eq. (4.17) is sufficient to determine the unknown nodal values of the boundary states $u_i(x)$ and $S_i(x)$. For the present B.V.P., eq. (4.17) has to be rearranged in the form

$$[\hat{T}_q - U_q] \cdot \left\{ \begin{array}{c} u \\ t \end{array} \right\} = [U_p - \hat{T}_p] \cdot \left\{ \begin{array}{c} \bar{t} \\ \bar{u} \end{array} \right\}, \tag{4.18}$$

where the indices q and p indicate the unknown and the prescribed boundary values, respectively. The matrices in eq. (4.18) are fully occupied and neither symmetric nor positive definite. Only these properties constitute a substantial disadvantage of the BEM and may bring to naught all its other advantages, like the significant reduction of the set of equations due to the discretization only at the boundary.

4.3 The Galerkin B.E.M. in Reissner Plate Theory

In the plate model of Reissner, the three boundary "generalized forces" S_i, ($i = 1, 2, 3$), i.e. the bending moment M_{nn}, the twisting moment M_{nt} and the real shearing force Q_n, and the three corresponding boundary "generalized displacements" u_i, the rotations around the unit tangent vector and around the outward unit normal vector, φ_n and φ_t, respectively, and the deflection w are considered. Therefore, in this model, six boundary value functions have to be approximated in each element by shape functions the corresponding nodal point values u_i^m and S_i^m. Substituting the expressions (4.3) and (4.4) into the discretized integral equations (3.75) and (3.76), we obtain the following approximate boundary integral formulation, (when $q(x)$ is neglected) ($i, k = 1, 2, 3$):

$$(RuS)_i(\xi) := c_{ik}(\xi)u_k(\xi) \tag{4.19}$$

$$+ \sum_{e=1}^{N_e} \sum_{m=1}^{p} u_k^{em} \int_{\Gamma_e} T_{ik}[\xi, x(\zeta)]\Theta^m(\zeta)|J(\zeta)|d\Gamma(\zeta)$$

$$- \sum_{e=1}^{N_e} \sum_{m=1}^{p} S_k^{em} \int_{\Gamma_e} U_{ik}[\xi, x(\zeta)]\Theta^m(\zeta)|J(\zeta)|d\Gamma(\zeta).$$

Therein, the kernels T_{ik} and $U_{ik}(i, k = 1, 2, 3)$ mean the singular Reissner "tractions" and "displacements" of the unit loadings

$$[T_{ik}] \;\hat{=}\; [M_{nn}^{(i)}(x, \xi), M_{nt}^{(i)}(x, \xi), Q_n^{(i)}(x, \xi)], \tag{4.20}$$

$$[U_{ik}] \;\hat{=}\; [\varphi_n^{(i)}(x, \xi), \varphi_t^{(i)}(x, \xi), w^{(i)}(x, \xi)]. \tag{4.21}$$

The special Kirchhoff corner terms hold in both the Reissner and the Kirchhoff plate theory. The boundary element equations are in both plate theories formally the same, but, in Reissner's more refined theory there is obviously one more unknown quantity at each boundary point.

In this section the computation of the influence matrices will be performed by following the Galerkin approach. In this approach (4.7) will be weighted through a linearly independent combination of the above introduced shape functions $\Theta^m(\eta)$ on the N_e boundary elements Γ_e:

$$w(\xi) = w(\eta(\xi)) = \sum_{e'=1}^{N_e} \sum_{m'=1}^{p} \beta^{e'm'}\Theta^{m'}(\eta(\xi)). \tag{4.22}$$

Since the coefficients $\beta^{e'm'}$ are arbitrary constants, this leads to the following N_e times

p conditions ($e' = 1, 2, \ldots N_e; m' = 1, 2, \ldots p$)

$$\int\limits_{\Gamma_{e'}} RuS_i(\xi)\Theta^{m'}(\xi)d\Gamma(\xi) = 0, \qquad i = 1, 2, 3. \tag{4.23}$$

Using the notation of eq. (4.17), relations (4.23) give rise formally to the same matrix equation as the collocation approach in the Kirchhoff plate B.I.Es. They read again

$$(\boldsymbol{C} + \boldsymbol{T}) \cdot \boldsymbol{u} = \hat{\boldsymbol{T}} \cdot \boldsymbol{u} = \boldsymbol{U} \cdot \boldsymbol{t}. \tag{4.24}$$

But, in the Galerkin approach, the coefficients of the matrices are determined by one integration more along the boundary elements than in the collocation approach, i.e,

$$\boldsymbol{C} \doteq \left\{ \sum_{m=1}^{p} \int\limits_{\Gamma_{e'}} c_{ik}(\xi(\eta))\Theta^m(\xi(\eta))\Theta^{m'}(\xi(\eta)|J(\eta)|d\Gamma(\eta) \right\} \tag{4.25}$$

$$\boldsymbol{T} \doteq \left\{ \sum_{e=1}^{N_e} \sum_{m=1}^{p} \int\limits_{\Gamma_{e'}} \int\limits_{\Gamma_e} \Theta^{m'}(\xi(\eta))T_{ik}[\xi(\eta), x(\zeta)]\Theta^m(\zeta)|J(\zeta)|d\Gamma(\zeta)|J(\eta)|d\Gamma(\eta) \right\} \tag{4.26}$$

$$\boldsymbol{U} \doteq \left\{ \sum_{e=1}^{N_e} \sum_{m=1}^{p} \int\limits_{\Gamma_{e'}} \int\limits_{\Gamma_e} \Theta^{m'}(\xi(\eta))U_{ik}[\xi(\eta), x(\zeta)]\Theta^m(\zeta)|J(\zeta)|d\Gamma(\zeta)|J(\eta)|d\Gamma(\eta) \right\}. \tag{4.27}$$

As already mentioned, the integrals containing the strongly singular kernels T_{ik} can be computed together with the corresponding C_{ik} coefficients indirectly by considering in the equation (4.23) the three rigid body displacements of the form [We82a, We82b, Kar]

(i) $u_1 = \varphi_n = n_1; \quad u_2 = \varphi_t = s_1; \quad u_3 = w = x_1(\xi) - x_1(x),$

(ii) $u_1 = \varphi_n = n_2; \quad u_2 = \varphi_t = s_2; \quad u_3 = w = x_2(\xi) - x_2(x),$

(iii) $u_1 = \varphi_n = 0; \quad u_2 = \varphi_t = 0; \quad u_3 = w = 1.$

Then, using again the notation of eq. (4.17), the leading diagonal submatrices of $\hat{\boldsymbol{T}}$ are calculated by the formula [Kar]

$$\hat{\boldsymbol{T}}_{ll} = -\sum_{\substack{e=1 \\ e \neq l}}^{N} \hat{\boldsymbol{T}}_{le}\boldsymbol{D}_{el}, \quad l = 1, 2, \ldots, N, \tag{4.22}$$

where N is the total number of nodal points, $\hat{\boldsymbol{T}}_{le}$ are 3×3 submatrices of $\hat{\boldsymbol{T}}$ and the matrices $\boldsymbol{D}_{el}$ contain the rigid body displacements defined above, and they have the form

$$\boldsymbol{D}_{el} = \begin{bmatrix} n_1 & 0 & 0 \\ 0 & n_2 & 0 \\ x_1(l) - x_1(e) & x_2(l) - x_2(e) & 1 \end{bmatrix}. \tag{4.23}$$

Chapter 5
Extension to Dynamic Problems

5.1 Generalities

The phenomenon of wave propagation is encountered in several engineering problems. In particular the dynamic interaction between elastic structures and adjacent media (soil or fluids) has in recent years, turned out to be very substantial in dynamic analyses and design of structures. The catastrophic consequences of a dam failure, for example, make it important that dams can safely withstand strong earthquakes. The same holds for nuclear power plants, tanks in chemical industries etc. Thus, in aseismic design the knowledge of the details of wave propagation, e.g. of wave focussing due to special bedrock geometries [An88a], is essential.

The propagation of small-amplitude waves through a homogeneous, elastic and isotropic medium obeys the linear equations of motion

$$(c_1^2 - c_2^2)u_{j,ji} + c_2^2 u_{i,jj} - \ddot{u}_i = -b_i/\rho. \tag{5.1}$$

In eq. (5.1), the commas and the overdots denote the partial space ($u_{j,i} = \partial u_j/\partial x_i$) and time ($\dot{u}_i = \partial u_i/\partial t$) derivatives of the displacement components $u_i(x, t)$ respectively. The propagation velocities of the dilatational (or pressure) and distortional (or shear) waves are given by c_1 and c_2, respectively; they are defined e.g. for the case of plane strain, by the formulas

$$c_1^2 = \frac{E(1 - \nu)}{\rho(1 + \nu)(1 - 2\nu)}; \quad c_2^2 = \frac{E}{2\rho(1 + \nu)}, \tag{5.2}$$

where E is Young's modulus and ν is Poisson's ratio. Also, ρ denotes the mass density of the elastic medium and b_i is the body force distribution.

Further, the motion of the structure satisfies boundary as well as initial conditions which can be summarized as follows:
i) conditions along the boundary $\Gamma = \Gamma_1 \cup \Gamma_2$:

$$u_i(x, t) = \bar{u}_i(x, t) \qquad \text{for } t > 0 \text{ on } \Gamma_1 \tag{5.3a}$$

$$T_i(x, t) = \sigma_{ij}(x, t)n_j(x) = \bar{T}_i(x, t) \qquad \text{for } t > 0 \text{ on } \Gamma_2 \tag{5.3b}$$

ii) conditions at the initial time $t = 0$ on $\Omega \cup \Gamma$

$$u_i(x, 0) = \bar{u}_{i0}(x) \tag{5.4a}$$

$$\dot{u}_i(x, 0) = \bar{v}_{i0}(x) \tag{5.4b}$$

Here the prescribed values are indicated by overbars. As before, T_i denotes the traction vector along the boundary with the outward normal n_j. Its components can be expressed as functions of the displacement derivatives according to

$$T_i = \rho[\delta_{ij}(c_1^2 - 2c_2^2)u_{k,k} + c_2^2(u_{i,j} + u_{j,i})]n_j. \tag{5.5}$$

In a compressible fluid which is initially at rest, the propagation of small acoustic disturbances causes a small amplitude velocity field v_i^f. Moreover when assuming a barotropic fluid or an ideal gas, which behaves as inviscid and as non-heat-conducting and where the specific entropy s remains constant, the flow induced by the disturbances is irrotational and

$$p = p(\rho_f) \quad \text{and} \quad c^2 \hat{=} \frac{\partial p}{\partial \rho_f}$$

hold. $p(x,t)$ stands for the hydrodynamic pressure (in excess of the hydrostatic pressure p_{stat}), ρ_f is the fluid density and c is the velocity of sound. The influence of gravity will be ignored and a time-independent hydrostatic pressure p_{stat} is assumed. With these assumptions, i.e. in a lineralized theory, one can derive the hydrodynamic equations of motion to be

$$\rho_f \dot{v}_i^f = -(p - p_{\text{stat}})_{,i} \tag{5.6}$$

and the local mass balance equation, the continuity equation as

$$\frac{1}{c^2}\dot{p} = -\rho_f(v_{i,i}^f - \gamma). \tag{5.7}$$

In this equation, $\gamma(x,t)$ is the given space and time dependence of the source density in the fluid. By elimination of the velocity field v_i^f, we obtain the so-called scalar wave equation

$$p_{,ii} - \frac{1}{c^2}\ddot{p} = -\rho_f \dot{\gamma} + (p_{\text{stat}})_{,ii}. \tag{5.8}$$

Besides this formulation (5.8) using the hydrodynamic pressure $p(x,t)$, it is also possible to describe the wave propagation by the velocity potential Φ, i.e. applying the relations

$$v_i^f = -\Phi_{,i} \quad \text{and} \quad p - p_{\text{stat}} = \rho_f \frac{\partial \Phi}{\partial t}. \tag{5.9}$$

Then, equation (5.7) becomes

$$\Phi_{,ii} - \frac{1}{c^2}\frac{\partial^2 \Phi}{\partial t^2} = -\gamma. \tag{5.10}$$

The appropriate initial and boundary conditions for equation (5.8) can be defined as follows:

i) conditions along the boundary $\Gamma = \Gamma_{1f} \cup \Gamma_{2f}$ for $t > 0$

$$p(x,t) = \bar{p}(x,t) \quad \text{or} \quad \Phi(x,t) = \bar{\Phi}(x,t) \text{ on } \Gamma_{1f} \tag{5.11a}$$

$$q(x,t) = p_{,j}n_j^f = \bar{q}(x,t) \quad \text{or} \quad \Psi(x,t) = \Phi_{,j}n_j^f = \bar{\Psi}(x,t) \text{ on } \Gamma_{2f} \tag{5.11b}$$

ii) conditions at the initial time $t = 0$ on Γ_f and in Ω_f

$$p(x,0) = \bar{p}_0(x) \qquad or \qquad \Phi(x,0) = \Phi_0(x) \qquad (5.12a)$$

$$\dot{p}(x,0) = b_0(x) \qquad or \qquad \dot{\Phi}(x,0) = \bar{p}_0(x)/\rho_f. \qquad (5.12b)$$

For example, homogeneous conditions have to be prescribed for p at the free-surface boundary Γ_{1f} of a fluid domain Ω_f and for q along a hard reflecting boundary Γ_{2f}. Absorbing surfaces are characterized by the condition of Robin

$$q(x,t) = -\frac{1}{c}\left[\frac{1 - \alpha_r(x)}{1 + \alpha_r(x)}\right]\frac{\partial}{\partial t}p(x,t), \qquad (5.13)$$

where $\alpha_r(x)$ represents the reflection factor of the surface.

Obviously, the initial-boundary value problem of lineralized wave propagation in compressible non-viscous fluids may be regarded, roughly speaking, as a special case of the more general wave propagation problem in elastic media. This is due to the fact that eq. (5.8) is obtained, together with the appropriate boundary and initial conditions, from eq. (5.1) and the corresponding boundary and initial conditions by formally setting $c_1 = c_2 = c$, and by considering scalar instead of vector-valued states.

Whenever dynamic problems involving infinite domains must be studied, the most appropriate method for solving the corresponding initial-boundary value problems seems to be the B.E.M., because it takes into account automatically the very important dynamic effect of radiation damping.

In any case, there are two possibilities for the treatment of dynamic problems:

(i)   Use of Fourier-Laplace transformations in the frequency domain. Thus the solution of a transient dynamic problem reduces to the solution of a "static-like" B.V.P. Finally, the inverse transformationm must be performed numerically. Only in the case of harmonic excitations the steady-state solution can be obtained directly, i.e. without inversion.

(ii)   Formulation of the problem in the time domain. Then the solution can be obtained directly by using a numerical step-by-step time integration.

In the next Sections we show the application of the B.E.M. to wave propagation problems in elastic media as well as in compressible fluids.

5.2. Steady State and Harmonic Problems

In many applications it is important to predict the dynamic behavior under harmonic, i.e. sinusoidally varying excitations. The response is then a function of the angular frequency ω; the initial conditions (5.4) and (5.10), in solids and fluids respectively, can be neglected by assuming that a sufficiently long time has elapsed so that a steady state is reached. In this case, the time dependence can be removed by transforming the governing equations via the Laplace transform. This transforms the equations of motion (5.1) into the equation

$$(c_1^2 - c_2^2)\hat{u}_{j,ji} + c_2^2\hat{u}_{i,jj} - s^2\hat{u}_i = -(\frac{1}{\rho}\hat{b}_i + s\bar{u}_{i0} + \bar{v}_{i0}) = -\frac{1}{\rho}\hat{q}_i \qquad (5.14)$$

and the scalar wave equations (5.8a,b) into

$$c^2\hat{p}_{i,jj} - s^2\hat{p} = -(\rho_f c^2 s\hat{a} + s\bar{p}_0 + \bar{b}_0) = -\rho_f c^2 s\hat{\gamma} \qquad (5.15a)$$

$$\hat{p}_{,jj} - \left(\frac{s}{c}\right)^2\hat{p} = -(\rho_f s\hat{a} + \frac{s}{c^2}\bar{p}_0 + \frac{1}{c^2}\bar{b}_0) = -\rho_f s\hat{\gamma}, \qquad (5.15b)$$

where quantities marked now by " ˆ " indicate the transformed variables, and s is the complex transformation parameter of the Laplace transform. The initial values are grouped in 5.15a together with $\hat{b}_i$ and $\hat{a}$ in a modified body force term $\hat{q}_i$ and modified sound source density $\hat{\gamma}$, respectively. The transformed equations must be solved together with the transformed pertinent boundary conditions on Γ_1 and Γ_2.

The mentioned special case of harmonic excitations is contained in the above equations (5.14)(5.15) when s is replaced by $i\omega$, and the initial conditions are neglected. It is worth noting that since the above eqs. (5.14), (5.15) are identical, up to the different coefficients, to the Cauchy-Navier equations (2.5), all the possibilities described in Ch. 2 in order to reformulate this differential equation as a boundary integral equation problem can be applied also here. Hence this reformulation can be achieved by using the method of weighted residuals (Sect. 2.2), or, because the Laplace-transformed differential equations are of elliptic type and self-adjoint, by using the usual reciprocal theorems (cf. Sect. 2.4). Here, in order to avoid the repetition of Ch. 2, the derivation of the integral equation will be given only for eq. (5.15), the so-called Helmholtz equation. The respective formulation for elastic media problems may be found in [An86c]. By using the reciprocal theorem of Green and adequate fundamental solutions, one may obtain the following B.I.E. which is equivalent to the differential equation (5.15).

$$d(\xi)\hat{p}(\xi) = \int_{\Gamma} \left[\hat{q}(x)\hat{p}^\star(x,\xi;\kappa) - \hat{p}(x)\hat{q}^\star(x,\xi;\kappa)\right]d\Gamma_x \qquad (5.16)$$

$$-\rho i\omega \int_{\Omega} \hat{\gamma}(x)\hat{p}^\star(x,\xi;\kappa)d\Omega_x.$$

Here $\hat{p}^\star$ (resp. $\hat{q}^\star$) denotes the fundamental solution (resp. its normal derivative $p^\star_{,n}$. The "jump" factor $d(\xi)$ is 1 or 0.5 for points ξ in the interior of Ω or on a smooth part of the boundary Γ, respectively. At the corners of the boundary Γ, we must distinguish between two dimensional (2-D) problems and three dimensional (3-D) problems. It holds that

$$d(\xi) = \frac{\Delta\varphi}{2\pi} \qquad \text{in 2-D}, \qquad (5.17a)$$

$$d(\xi) = \frac{\Delta\Omega}{4\pi} \qquad \text{in 3-D}, \qquad (5.17b)$$

where $\Delta\varphi$ is the interior angle and $\Delta\Omega$ is the inner solid angle at corners and edges in 2-D and in 3-D problems respectively. If, as a special case, L different sources $\hat{\gamma}$ with a prescribed strength $g_l(\omega)$ are concentrated at interior points x_l [An90a], i.e.

$$\hat{\gamma}(x) = \sum_{l=1}^{L} g_l(\omega)\delta(x - x_l), \qquad (5.18)$$

the source term in eq. (5.16) becomes

$$-\rho i\omega \int_{\Omega} \hat{\gamma}(x)\hat{p}^*(x,\xi;\omega)d\Omega_x = -\rho i\omega \sum_{l=1}^{L} \hat{g}_l(\omega)\hat{p}^*(x_l,\xi;\omega).$$ (5.19)

The adequate fundamental solution $\hat{p}^*(x,\xi;\omega)$ of eq. (5.15) and its normal derivative $\hat{q}^*(x,\xi;\omega)$ are given by

$$p^* = \frac{1}{2\pi}K_0(i\kappa r)$$ (5.20a)

in 2-D,

$$q^* = -\frac{i\kappa}{2\pi}K_1(i\kappa r)\frac{\partial r}{\partial n}$$ (5.21a)

$$p^* = \frac{1}{4\pi}\frac{1}{r}e^{-i\kappa r}$$ (5.20b)

in 3-D,

$$q^* = -\frac{1}{4\pi}\frac{1+i\kappa r}{r^2}e^{-i\kappa r}\frac{\partial r}{\partial n}$$ (5.21b)

where K_0 and K_1 are the modified zero- and first-order Bessel functions of the second kind, respectively, and $\kappa = \omega/c$ denotes the wavenumber.

A treatment analogous to that described in Sect. 4.2, i.e. through point collocations and using interpolation functions as simple as possible (e.g. assuming in 2-D problems along each boundary line element Γ^e and in 3-D problems for every plane triangular surface segment Γ^e, that both the pressure $\hat{p}$ and the flux $\hat{q}$ have constant values), implies the following system of algebraic equations:

$$(\frac{1}{2}I + \hat{Q})\cdot\hat{p} = \hat{P}\cdot\hat{q} + \hat{r},$$ (5.22)

where the elements of the matrices $\hat{Q}$ and $\hat{P}$ are obtained by using Gaussian quadrature formula. They have the form ($r_j = |x - \xi_j|$)

$$\hat{Q} \hat{=} -\{\frac{i\kappa}{2\pi}\int_{\Gamma^e} K_1(i\kappa r_j)\frac{\partial r_j}{\partial x_k}n_k(x)d\Gamma_x\}$$ (5.23a)

$$\hat{P} \hat{=} \{\frac{1}{2\pi}\int_{\Gamma^e} K_0(i\kappa r_j)d\Gamma_x\}$$ (5.23b)

in 2-D problems, and the form

$$\hat{Q} \hat{=} -\{\frac{1}{4\pi}\int_{\Gamma^e} \frac{1+i\kappa r_j}{r_j^2}e^{-i\kappa r_j}\frac{\partial r_j}{\partial x_k}n_k(x)d\Gamma_x\}$$ (5.24a)

$$\hat{P} \hat{=} \{\frac{1}{4\pi}\int_{\Gamma^e} \frac{1}{r_j}e^{-i\kappa r_j}d\Gamma_x\}.$$ (5.24b)

in 3-D problems. The source term, vector $\hat{r}$, is derived from (5.19) during the application of the collocation at all the nodes ξ_j. The result for interior point sources at x_l reads:

in two-dimensional problems

$$\hat{r} \hat{=} - \left\{ \rho \frac{ic\kappa}{2\pi} \sum_{l=1}^{L} \hat{g}_l(\kappa) K_0(i\kappa|x_l - \xi_j|) \right\}, \qquad (5.25a)$$

and in three-dimensional problems

$$\hat{r} \hat{=} - \left\{ \rho \frac{ic\kappa}{4\pi} \sum_{l=1}^{L} \hat{g}_l(\kappa) \frac{1}{|x_l - \xi_j|} e^{-i\kappa|x_l - \xi_j|} \right\}. \qquad (5.25b)$$

For exterior problems, the Helmholtz integral formulation fails to provide a unique solution at certain frequencies corresponding to all eigenvalues of the enclosed interior regions. We can deal with this difficulty by using an over-determined system of equations including the system generated from eq. (5.16) through surface collocations at ξ_j (with $d(\xi_j) = 0.5$ for ξ_j on a smooth surface) and additional equations generated from collocations at points ξ_j of the enclosed region (i.e. $d(\xi_j) = 0$). In this context we refer to [Sche].

The systems of linear equations generated by the B.E.M. are considerably smaller than those of an equivalent finite-difference or finite-element approximation. However, in many applications the dimension of the system may still be large enough, so that iterative techniques are needed for their solution. But, for such systems the fast equation solvers developed for finite element equations are not applicable, because the system of linear equations obtained in the B.E.M. contain fully populated and generally non-symetric matrices. Besides the algorithms applied to this type of linear systems (cf. e.g[Go][Str]), special iterative schemes, the multigrid methods [McCo][Hac][Br] are applicable to the systems of equations obtained by the B.E.M. They have already shown their efficiency in B.E. solutions of acoustic problems [An88b].

The applicabilty and the efficiency of the method developed in this Section is illustrated by some numerical studies presented in the following Section.

5.3 Numerical Applications

5.3.1 Convergence Studies and Cut-off Errors

In order to show and to analyze some of the numerical phenomena appearing in the case of frequency dependent problems, a very simple example will be examined [An89a, An90b]: In front of an infinitely extended, hard reflecting wall Γ_1 a small noise source (loud speaker) of $l = 1$ m height is placed at a distance $d = 2$ m. (Fig. 5.1a) This excitation source produces a noise with a frequency $f = 10$ Hz and constant intensity, which propagates with a wavespeed of $c = 340$ m/s. The boundary conditions of the problem are

$$q = 0 \text{ along } \Gamma_1 \qquad \text{and} \qquad q = \bar{q} = \text{const along } \Gamma_2.$$

Now, we want to determine the pressure at point A lying at a distance $D = 4$ m behind the hardreflecting wall. We shall use the B.E.M. of the previous Section which leads to the algebraic equation (5.22). Obviously, the exact value of the pressure at any point behind the wall must be zero; thus $p_A = 0$. But, due to the discretization error of the boundary element approximation and, moreover, due to the cut-off of the infinite boundary Γ_1 (only a finite part of length L is taken for modeling the infinite wall) there will be certainly obtained some sound pressure $p_A \neq 0$ at this point. In fig.5.1b, this pressure p_A is depicted related to its value $\bar{p}_A$ in the undisturbed case, i.e. in the case in which the reflecting wall does not exist. Thus influence of both the mentioned error causes becomes apparent. The following results are obtained:

- If the discretized boundary of length L is shorter than the wavelength $\lambda = c/f = 34$ m, the cut-off error exceeds considerably the error due to the discretization. We consider the cases of $E = 10,\ 20,\ 40,\ 80$ elements per wavelength along the boundary Γ_1.

- On the other hand, if the infinite wall is discretized along a sufficiently long part of it, e.g., if it is approximated by elements along a length $L > 3\lambda = 102$ m, the cut-off error becomes negligible in comparison with the discretization error. But, also this error decreases, if the discretization is enough refined and converges to zero. For instance, if more than 40 elements per wave length are used, this error almost disappears (for 40 elements: $0,25\%$).

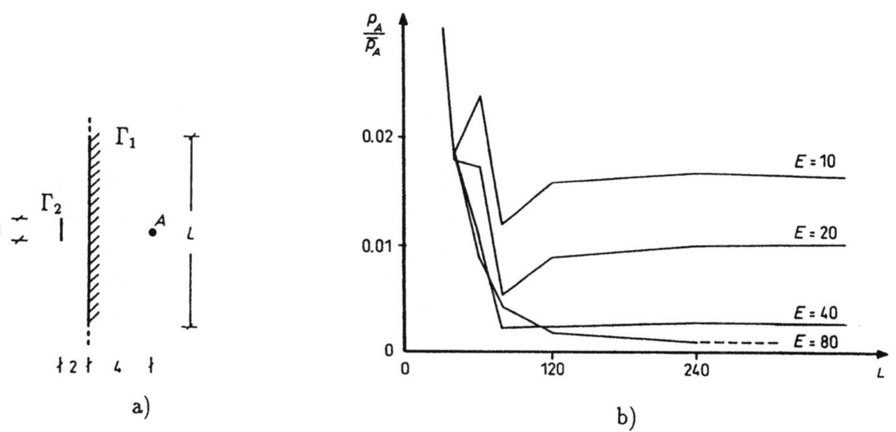

Fig. 5.1: a) Geometry of the sytem "loudspeaker-wall", b) Relative sound pressure at point A behind the wall

As another application we consider now the effect of adding a second wall Γ_3 between the point A and the wall Γ_1 (fig. 5.2). Without approximations, i.e. if at least one of the walls has infinite length, the analytical, exact solution at point A remains $p_A = 0$. The addition of the second wall should not cause any difference. Moreover, in using a refined discretization (40 elements per wave length) along the finite wall models of equal length L, one should expect numerical results which are at least as "exact" as in the first example. But, as the comparison of both results shows (fig. 5.2b) this "two-walls model" produces much worse numerical results. Although both walls are approximated along $L = 120$ m using a discretization of 40 elements per wavelength, the relative pressure at point A is 5%, i.e. 20 times higher than before in the case of the single wall obstacle. This behavior makes us to doubt about the applicability of the B.E.M. to such problems. But again, the increasing the discretized length of the wall leads to correct results if enough elements are used to approximate the boundary. This fact

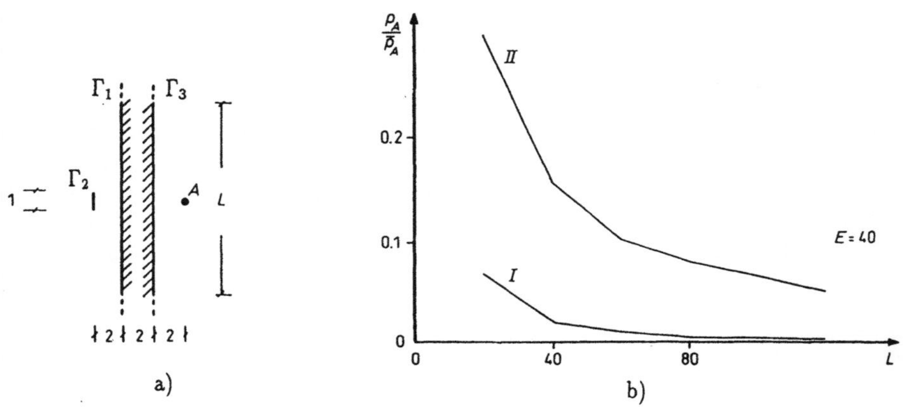

Fig. 5.2: a) Geometry of the system "loudspeaker - 2 walls", b) Relative pressure at point A behind one (I) and behind two walls (II), respectively. Dependence on the discretized wall length L

can lead to large equation systems and to increased computer-time, so that the question arises, whether it is possible to find another more efficient numerical method. For finding a simple, but correct numerical handling of such problems, a heuristic method proposed by the first author is applied. This method is inspired by the wave propagation theory [Wes][Höf]. According to Huygen's principle, each point reached by a wave produces again a new wave, a so-called "elementary wave". The front of all these propagating waves is formed by the envelope of the elementary waves (fig. 5.3a). This principle explains several phenomena, as e.g. the reflection or the interference of several wave fronts, but not all the wave phenomena, e.g. the fact that high-frequency waves, for instance the waves of light, propagate in a straight line. This appearing contradiction has been cleared up by the research of Fresnel (1788-1827) who introduced the Interference Principle. The experiments of Fresnel showed that the influence of waves propagating from a point B to a point A are almost exclusively produced by a small zone in the neighbourhood of the connection line AB. This "active" zone has only a radius of approximately the wave-length, which explains in some sense the ray character of the light.

The above physical conclusions permit the derivation of some heuristic rules which ameliorate the numerical performance of the B.E.M.

1st rule: If a connection line from one point on the boundary to another has a common point with the boundary, no cut-off of the boundary should exist in the "neighbourhood" (in these rules "neighborhood" means: "at a distance shorter than one wave-length") of this crossing.

2nd rule: In the "neighbourhood" of one cut-off no other cut-off should be placed (fig. 5.3c)

The first rule assures that the numerical influence of a source at a point B on the reactions at a point A will not be considerably affected by the cut-off of the boundary. Note, that in order to minimize the numerical error, the whole active zone in the neighbourhood of the intersection point S is needed. Moreover, a boundary cut-off produces the major error in its own near neighbourhood - an assumption which has been proved in a recent study [An87b, An87c].

By the second rule two cut-offs in a too short distance should be avoided in order to minimize the possibility that such errors create a together a much larger error.

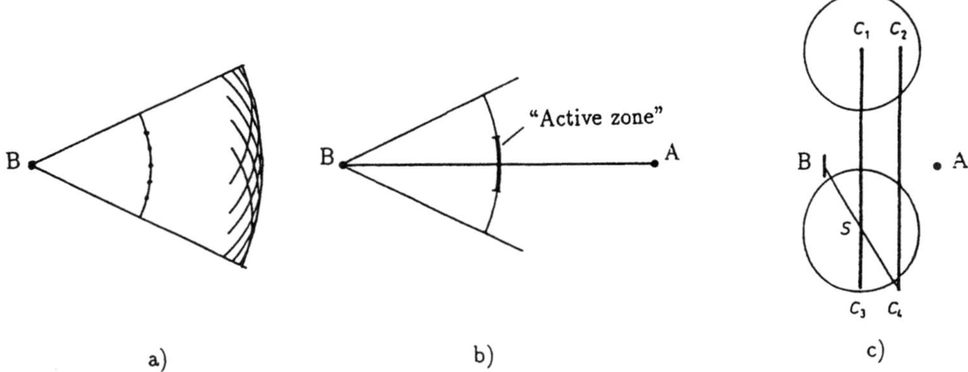

Fig. 5.3: a) Elementary waves and the wave front b) "Active zone" of sound wave propagation c) Unacceptable geometry relation between an intersection point S and the cut-offs C_1 and C_2 as well as C_3 and C_4

Let us fulfill now these two rules in the above described second application. Thus we have to approximate the second wall Γ_3 on a length shorter than the first wall Γ_2 about twice the wave length λ (fig. 5.4a). In fig. 5.4b is depicted the dependence of the sound wave pressure p_A at the point A on the discretization, i.e. on the number E of elements per wave-length λ. All three wall-systems are considered. It can be easily seen that, starting with $E = 40$ a certain convergence appears in all these three models. But, while model I shows excellent convergence to the exact value zero, in model II an unacceptable huge error appears which, moreover, seems to be independent of the discretization. This error is obviously generated by the two cut-offs which are too close to one another. This is confirmed by the results of model III, where the pressure ratio tends to $0,75\%$. This value is very good numerical approximation and verifies also the given heuristic rules.

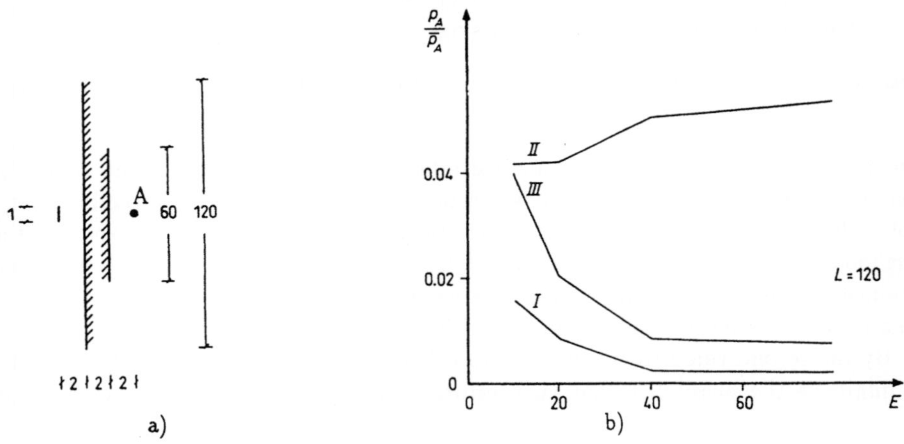

Fig. 5.4: a) Numerically correct model of a two-wall system b) Relative pressure at point A for models I, II, and III, dependence on E

The refraction of waves due to obstacles is a very interesting problem, which can be solved easily by the B.E.M. As an application we shall study how an excitation ahead of a wall of height $H = 5$ m will be registered at a point A behind the wall. Especially we deal with the question whether or not the excitation frequency f plays an important role in the refraction. The frequencies, which are considered, are in the rangeof $2.5\ Hz < f < 160\ Hz$, corresponding to wave-lengths between 136 m and 2.125 m. In order to obtain exact results it is necessary to consider thoroughly the modelling and the discretization of the boundary geometry. As we know from the previous examples around 40 elements per wave-length are sufficient to obtain numerically "exact" values. The wall is modelled with cut-offs in different height (fig. 5.5a) according to the two previous rules. In fig. 5.5b, the relative acoustic pressure $p_A(f)/\bar{p}_A(f)$ versus the excitation frequency f is plotted. It is apparent that the long waves, e.g. waves of length $\lambda = 136$ m can transport by refraction on the wall around 50 percentage of their pressure intensity to the point A behind the wall. Short waves of around $\lambda = 2$ m wave-length, however, are only able to transfer 2.5 percentage. Thus, the shorter the wave-length, the more "black" is the "shadow zone" which one may determine behind the wall. This fact is compatible with our experience.

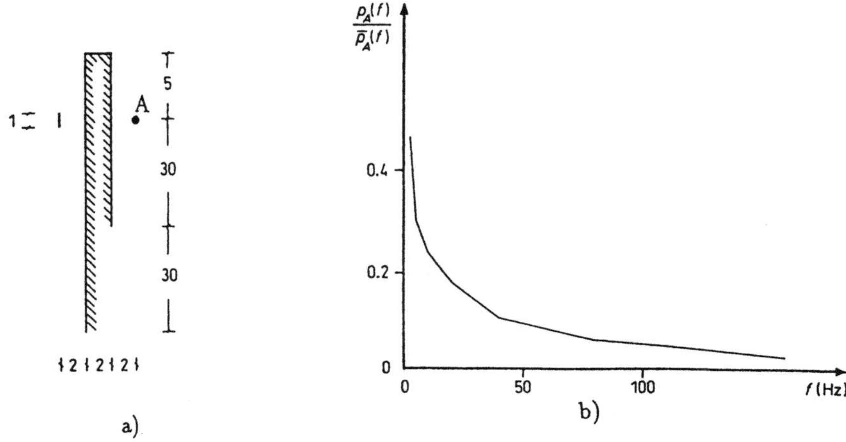

Fig. 5.5: a) Geometry and cut-offs of the wall model in order to get correct numerical results.   b) Relative pressure wave intensity at point A behind the wall versus frequency

5.3.2 Noise Distribution Around 2-D Barrier Models

Highway noise in urban areas is one source of pollution which can be rectified, e.g. by means of barriers. Generally, one is interested in reducing the traffic noise as effectively as possible; in many cases this reduction is more important in some special locations. From the point of view of environmental noise reduction, it is meaningless to describe the barrier's effect with precision. The only interesting result is to know whether the envelope of the maximum of sound field is "lowered" when a barrier is erected between the noise sources and the region to be protected. Furthermore, in many practical cases, there is a posibility to raise a long barrier which can be considered as almost straight. Thus, as shown in [Fil] a two-dimensional modelization can often be very efficient to predict the excess attenuation caused by a long screen in the presence of a line of point sources. In the following 2-D study (for more details, see [An91a]), two different types of barriers are examined: first, a thin hard reflecting edge wall (fig. 5.6), and second, a massive wedge-shaped barrier (fig. 5.7).

Let us consider a sound generated with a uniform intensity $\bar{q} = 1$ Pa/m along a vertical line of 1.5 m. It represents a vehicle noise including both the noises of the engine and of the tires. For protection against this noise, barriers of 5 m height have been erected such that their crest (of width 1 m) is at a distance of 10 m from the noise "producing line". As a first type, a thin edge wall of uniform thickness 1 m is considered and as a second type, a symmetric wedge-shaped wall with a basement width of 8.5 m. The soil surface behind the barriers is taken into account horizontally up to a distance of 50 m from the crests.

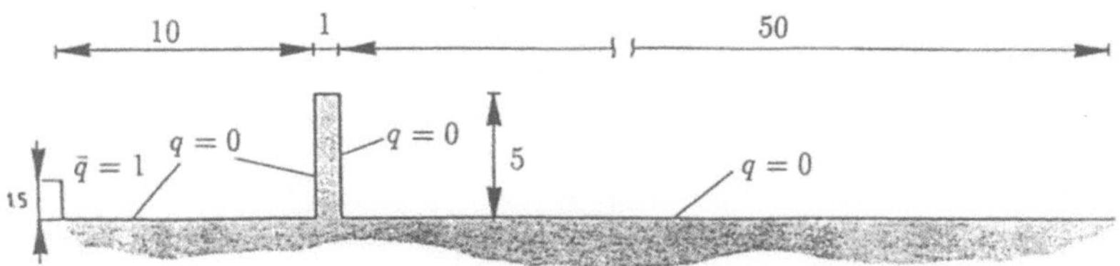

Fig. 5.6: Geometry of a thin edge barrier configuration

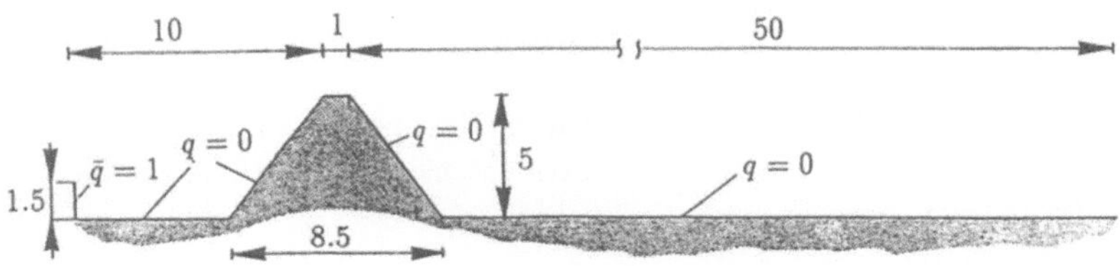

Fig. 5.7: Geometry of a wedge-shaped barrier configuration

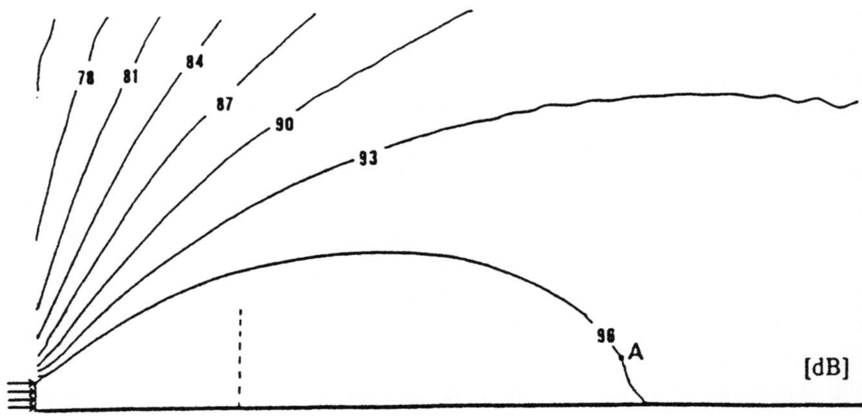

Fig. 5.8: Noise level distribution of undisturbed propagating sound: all surfaces are
hard reflecting (excitation frequency 100 Hz)

The numerical modelling uses linear approximation functions and elements of equal
length, but the discretization, i.e. the number and length of elements depends on the

wave length of the sound, i.e. on its frequency. Generally, a minimum of 3 elements per wavelength should be applied; however it is better to double this number. First, in the case of the thin edge wall, the dependence of the noise reduction effectiveness on the frequency of the noise is demonstrated. For a noise with frequency 100 Hz and assuming perfectly reflecting surfaces of both the ground and the barrier, the "undisturbed" level distribution (fig. 5.8) is compared to that when the barrier is erected (fig. 5.9).

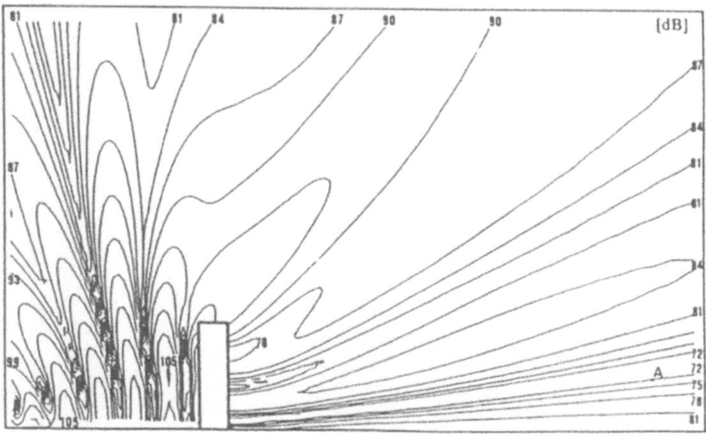

Fig. 5.9: Noise level distribution in front of and behind a edge barrier: all surfaces are hard reflecting (excitation frequency 100 Hz)

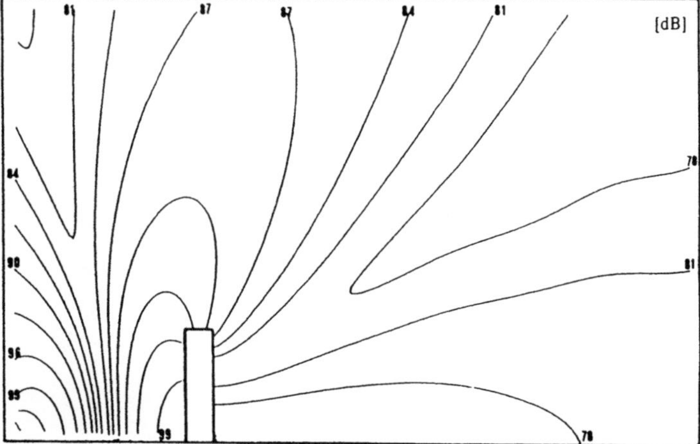

Fig. 5.10: Noise level distribution in front of and behind a edge barrier: all surfaces are hard reflecting (excitation frequency 20 Hz)

It is obvious that the "loss of transmission" caused by the barrier is rather high (about 20 dB at the reference point A), but this noise reduction is less in the case of lower frequencies - as shown for 20 Hz in fig. 5.1, and much greater for higher frequencies (see noise of 1000 Hz in fig. 5.13). Also, as known from experiments [Maek] and from observations on the diffraction theory [But], this loss of transmission is dependent on the absorptive property of both the barrier surface and the ground; more results about this may be found in [An91a].

Next, a similar study is performed for the symmetric wedge-shaped barrier (fig. 5.7). In fig. 5.11, we obtain on the assumption of perfect reflection at all surfaces, a transmission loss for a noise of 100 Hz. A control of this result gives less noise reduction behind this barrier than in the case of the thin edge barrier.

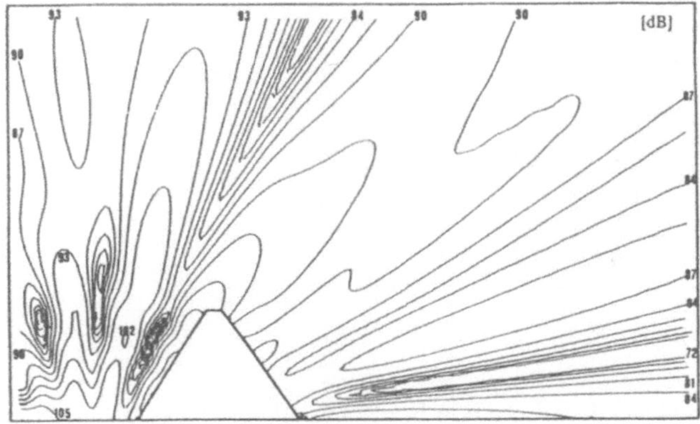

Fig. 5.11: Noise level distribution in front of and behind a wedge shaped barrier: all surfaces hard reflecting (excitation frequency 100 Hz)

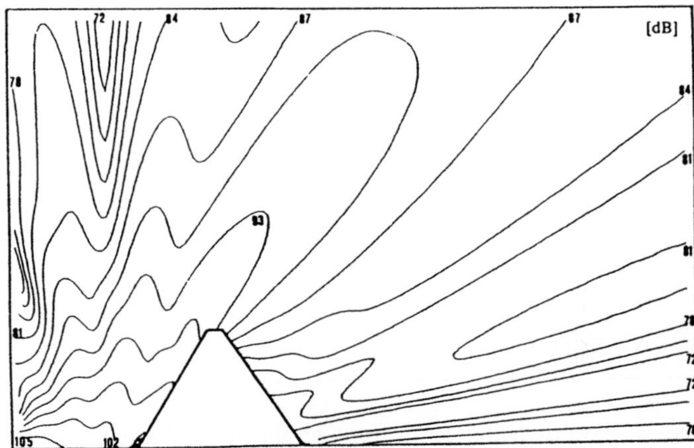

Fig. 5.12: Noise level distribution in front of and behind a wedge shaped barrier: barrier absorptive ($\alpha_r = 0.3$) (excitation frequency 100 Hz)

A more realistic barrier which model an earth wall, i.e. with a reflection factor of $\alpha_r = 0.3$, improves rather well the loss of transmission (fig. 5.12). Finally, the empirical result that "the higher the noise frequency, the greater the transmission loss of any barrier" is confirmed by studying the noise level distribution of an excitation of 1000 Hz (fig. 5.13).

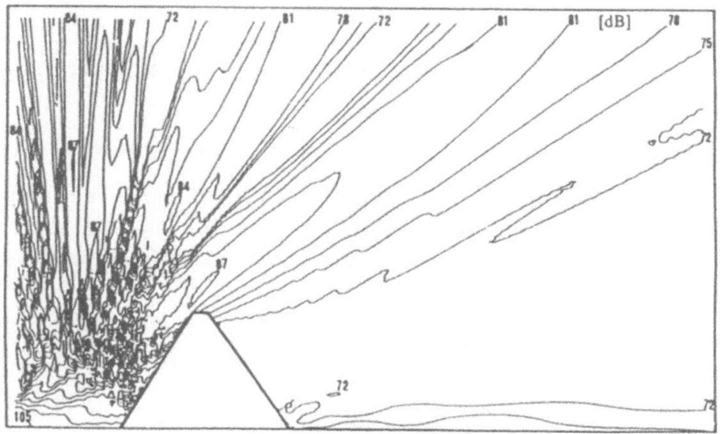

Fig. 5.13: Noise level distribution in front of and behind a wedge shaped barrier: barrier absorptive ($\alpha_r = 0.3$) (excitation frequency 1000 Hz)

5.4 Time Domain Formulation for Transient Problems.

5.4.1 Wave Propagation in 2-D Elastic Media.

Time domain formulations have the advantage over frequency domain solution procedures that they lead directly to the time history of transient response in an easier and more natural way, and that they create a basis for extension to nonlinear problems. Moreover, time-stepping solution procedures allow to incorporate directly any time-dependent variation of material data and boundary conditions, or to take into account even time-dependent changes of the boundary itself. In this Section the time-stepping boundary element procedures for determining the propagation of waves in 2-D elastic media as well as in 3-D compressible fluid domains are presented. We begin with the 2-D elastic media.

A time dependent integral representation of the dynamic initial- B.V.P. describing wave propagation in elastic media can be applying by employing Graffi's reciprocal theorem [Graf], which is a generalization of Betti's theorem for elastodynamics. As in the case of Betti's theorem, one has to consider two states which satisfy the basic differential equation system, i.e. the equations of motion (5.1). The first state includes the actual body force distribution $b_j(x,t)$ and the other a unit impulse in the direction x_i at time τ and point ξ. Finally, one obtains (the overdot "." indicates time differentiation, and $t' = t - \tau$)

$$d_{ij}(\xi)u_j(\xi,t) = \int_0^t \{ \int_\Omega \overset{\star(i)}{u_j}(x,\xi;t')b_j(x,\tau)d\Omega_x \tag{5.26}$$

$$+ \int_\Gamma [\overset{\star(i)}{u_j}(x,\xi;t')T_j(x,\tau) - \overset{\star(i)}{T_j}(x,\xi;t')u_j(x,\tau)]d\Gamma_x \}d\tau$$

$$+ \rho \int_\Omega [v_{j0}(x)\overset{\star(i)}{u_j}(x,\xi;t) - u_{j0}(x)\frac{\partial}{\partial t}\overset{\star(i)}{u_j}(x,\xi;t)]d\Omega_x.$$

On a smooth boundary ($\xi \in \Gamma$) we have $d_{ij} = 0.5\delta_{ij}$, while for interior points ($\xi \in \Omega$) $d_{ij} = \delta_{ij}$. The function $\overset{\star(i)}{u_j}(x,\xi;t')$ is known as the "time dependent fundamental solution of the full space", and gives the displacement response at a point x and time t due to a unit impuls located at point ξ and acting at the time τ in the direction x_i [Eri, Ma83a, Ma83b]. It reads

$$\overset{\star(i)}{u_j}(x,\xi;t') = \frac{1}{2\pi\rho}\{\frac{H(t'-\frac{r}{c})}{c_1 r^2}[(2R_1 + \frac{r^2}{R_1})r_{,i}r_{,j} - R_1\delta_{ij}] \tag{5.27}$$

$$- \frac{H(t'-\frac{r}{c})}{c_2 r^2}[(2R_2 + \frac{r^2}{R_2})r_{,i}r_{,j} - (R_2 + \frac{r^2}{R_2})\delta_{ij}]$$

where, $r = |x - \xi|$, $t' = t - \tau$ and

$$R_\alpha = (c_\alpha^2 t'^2 - r^2)^{1/2}; \quad \alpha = 1,2. \tag{5.28}$$

H is the Heaviside step-function, which guarantees the causality of the waves. The corresponding singular tractions $\overset{\star(i)}{T_j}$, determined by using the relation (5.5), contain strongly singular terms, and, in order to eliminate them, integration by parts with respect to time is carried out. The result of this regularization procedure is the following integro-differential equation [An85b, An87d]:

$$d_{ij}(\xi)u_j(\xi,t) = \int_0^t \{ \int_\Omega \overset{\star(i)}{u_j}(x,\xi;t')b_j(x,\tau)d\Omega_x + \int_\Gamma [\overset{\star(i)}{u_j}(x,\xi;t')T_j(x,\tau) \tag{5.29}$$

$$+ \overset{\star(i)}{P_j}(x,\xi;t')u_j(x,\tau) - \overset{\star(i)}{Q_j}(x,\xi;t')\dot{u}_j(x,\tau)]d\Gamma_x \}d\tau$$

$$+ \rho \int_\Omega [v_{j0}(x)\overset{\star(i)}{u_j}(x,\xi;t) - u_{j0}(x)\frac{\partial}{\partial t}\overset{\star(i)}{u_j}(x,\xi;t)]d\Omega_x,$$

where the regularized kernels $\overset{\star(i)}{P_j}$ and $\overset{\star(i)}{Q_j}$ have the forms

$$\overset{\star(i)}{P_j}(x,\xi;t') = \sum_{\alpha=1}^{2}(-1)^{\alpha}\frac{H(c_\alpha t'-r)}{2\pi c_\alpha R_\alpha}[2\,\overset{0}{a}_{ij}(r)\frac{2c_\alpha t'^2}{r^3} \tag{5.30a}$$

$$+\,\overset{\alpha}{a}_{ij}(r)\frac{r-c_\alpha t'}{R_\alpha^2} + \{\overset{1}{a}_{ij}(r)+\overset{2}{a}_{ij}(r)\}\frac{c_\alpha t}{r^2}]$$

$$\overset{\star(i)}{Q_j}(x,\xi;t') = \sum_{\alpha=1}^{2}(-1)^{\alpha}\frac{H(c_\alpha t'-r)}{2\pi c_\alpha^2}[\overset{\alpha}{a}_{ij}(r)\frac{1}{R_\alpha} + \{\overset{1}{a}_{ij}(r)+\overset{2}{a}_{ij}(r)\}\frac{R_\alpha}{r^2}] \tag{5.30b}$$

with

$$\overset{0}{a}_{ij}(r) = c_2^2(n_i r_{,j} + n_j r_{,i} + \delta_{ij}r_{,n} - 4r_{,i}r_{,j}r_{,n}) \tag{5.31a}$$

$$\overset{1}{a}_{ij}(r) = 2c_2^2 r_{,i}r_{,j}r_{,n} + (c_1^2 - 2c_2^2)n_j r_{,i} \tag{5.31b}$$

$$\overset{2}{a}_{ij}(r) = c_2^2(2r_{,i}r_{,j}r_{,n} - n_i r_{,j} - \delta_{ij}r_{,n}). \tag{5.31c}$$

The only singularities in these integro-differential equations are those that occur when r and t' tend to zero simultaneously. When only $r \to 0$ and t' does not tend to zero only pseudo-singularities occur which disappear, when contributions from similar terms concerning dilatational and shear waves are taken into account simultaneously in the calculations [Ma83a]. Thus, the above time-dependent equations can be integrated without difficulty, provided that appropriate shape-functions are used.

The domain integrals in eqs. (5.26) and (5.29), respectively, can be transformed to boundary integrals in the case of constant body force distributions, and if constant initial velocity $v_{j0}(x)$ and zero initial displacement $u_{j0}(y) = 0$ are prescribed [An90c]. One obtains for the initial velocity integral in (5.29) the expression

$$\rho v_{j0} \int_{\Omega} \overset{\star(i)}{u}_j(x,\xi;t,0)d\Omega_x = \tag{5.32}$$

$$= \rho v_{j0} \sum_{\alpha=1}^{2}\int_{\Gamma_{\alpha+}} [H(c_\alpha t-r)\,\overset{\alpha(i)}{R}_{jk}(x,\xi;t)]n_k(x)d\Gamma_x$$

$$+ \frac{v_{j0}t}{2\pi}[\Delta\varphi\delta_{ij} + \ln\frac{c_1}{c_2}\sin\Delta\varphi\,\Omega_{ij}(\varphi_1,\varphi_2)]$$

where

$$\overset{1(i)}{R}_{jk}(x,\xi;t) = \frac{1}{2\pi\rho}\frac{r_{,k}}{c_1 r}\{R_1(r_{,i}r_{,j} - \delta_{ij}) - c_1 t\ln(\frac{c_1 t + R_1}{r})[2r_{,i}r_{,j} - \delta_{ij}]\} \tag{5.33}$$

$$\overset{2(i)}{R}_{jk}(x,\xi;t) = \frac{-1}{2\pi\rho}\frac{r_{,k}}{c_2 r}\{R_2 r_{,i}r_{,j} - c_2 t\ln(\frac{c_2 t + R_2}{r})[2r_{,i}r_{,j} - \delta_{ij}]\}$$

and

$$\Omega_{ij}(\varphi_1,\varphi_2) = \begin{bmatrix} \cos(\varphi_1+\varphi_2) & \sin(\varphi_1+\varphi_2) \\ \sin(\varphi_1+\varphi_2) & -\cos(\varphi_1+\varphi_2) \end{bmatrix}. \tag{5.34}$$

On a smooth boundary, we have $\Delta\varphi = \pi$ which simply yields for the integral-free term $0.5v_{i0}t$. The boundary $\Gamma_{\alpha+} = \{x : r = |x - \xi| < c_\alpha t\}$ denotes the boundary of that part of the domain Ω which the p- and s-waves starting at point ξ have already traversed (see fig. 5.14). The total boundary of this domain is $\tilde{\Gamma} = \Gamma_{\alpha r} \cup \Gamma_{\alpha+} \cup \Gamma_\varepsilon$. When the wave fronts have reached all points x in the domain Ω obviously $\tilde{\Gamma}$ coincides with Γ. It should be mentioned that the evaluation of the integrals along $\Gamma_{\alpha r}$ gives no contribution; along Γ_ε the integrals are calculated analytically.

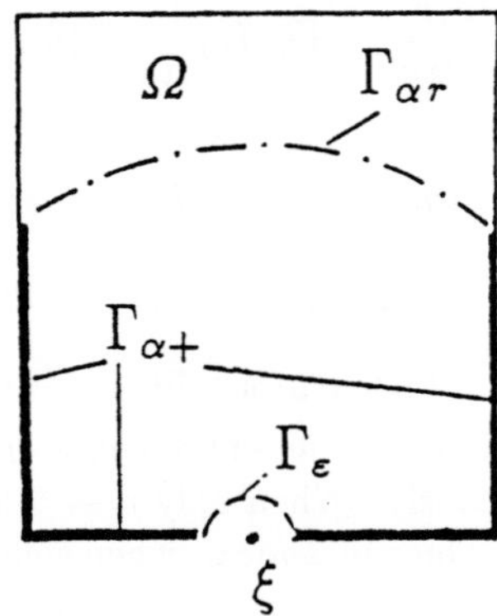

Fig. 5.14: The boundary parts $\Gamma_{\alpha+}, \Gamma_{\alpha r}$, and Γ_ε enclosing the domain covered by a wave starting at point ξ. $\Gamma_{\alpha r} := \{x | c_\alpha t = r\}$ $\Gamma_\varepsilon := \{x | r = \varepsilon\}$

After an additional time integration, the dead-weight domain integral gives rise to the following expression

$$\int_0^t b_{j0} \int_\Omega \overset{*(i)}{u_j} (x, \xi; t - \tau) d\Gamma_x d\tau \tag{5.35}$$

$$= b_{j0} \Big\{ \sum_{\alpha=1}^2 \int_{\Gamma_{\alpha+}} [H(c_\alpha t - r) \overset{\alpha(i)}{\tilde{r}_{jk}} (x, \xi; t)] n_k(x) d\Gamma_x$$

$$+ \frac{t^2}{4\pi\rho} \big[\Delta\varphi \delta_{ij} + \ln \frac{c_1}{c_2} \sin \Delta\varphi \Omega_{ij}(\varphi_1, \varphi_2) \big] \Big\}$$

where

$$\overset{1(i)}{\tilde{r}_{jk}} (x, \xi; t) = \frac{1}{2\pi\rho} \frac{r_{,k}}{r} \Big\{ t\sqrt{t^2 - \left(\frac{r}{c_1}\right)^2} (r_{,i} r_{,j} - \frac{3}{4} \delta_{ij}) \tag{5.36a}$$

$$+ \ln \frac{t + \sqrt{t^2 - (\frac{r}{c_1})^2}}{r/c_1} [\frac{1}{4}(\frac{r}{c_1})^2 \delta_{ij} - t^2 (r_{,i} r_{,j} - \frac{1}{2}\delta_{ij})] \Big\}$$

$$\overset{2(i)}{\tilde{r}_{jk}} (x, \xi; t) = \frac{-1}{2\pi\rho} \frac{r_{,k}}{r} \Big\{ t\sqrt{t^2 - \left(\frac{r}{c_2}\right)^2} (r_{,i} r_{,j} - \frac{1}{4} \delta_{ij}) \tag{5.36b}$$

$$- \ln \frac{t + \sqrt{t^2 - (\frac{r}{c_2})^2}}{r/c_2} [\frac{1}{4}(\frac{r}{c_2})^2 \delta_{ij} + t^2 (r_{,i} r_{,j} - \frac{1}{2}\delta_{ij})] \Big\}.$$

When r tends to zero, only pseudo-singularities occur in the boundary integral which vanish when the contributions from similar terms concerning the p-waves and the s-waves, i.e. to c_1 and c_2, are calculated simultaneously. Thus, this time-dependent boundary representation of the dead-weight integral can be integrated without difficulty. Then, the final, complete boundary integral formulation describing the wave propagation in a "heavy" structure Ω with a smooth boundary Γ, i.e. $\Delta\varphi = \pi$, and a constant initial velocity v_{j0} is given by

$$
\frac{1}{2}u_i(\xi, t) = \int_0^t \int_\Gamma \{ \overset{\star (i)}{u_j}(x, \xi; t - \tau)T_j(x, \tau) \tag{5.37}
$$

$$
+ (\overset{\star (i)}{P_j}(x, \xi; t - \tau)u_j(x, \tau) - \overset{\star (i)}{Q_j}(x, \xi; t - \tau)\dot{u}_j(x, \tau)) \} d\Gamma_x d\tau
$$

$$
+ \sum_{\alpha=1}^2 \int_{\Gamma_{\alpha+}} \{ b_{j0}r_{jk}^{(i)}(x, \xi; t) + \rho v_{j0}R_{jk}^{(i)}(x, \xi; t) \} n_k(x)d\Gamma_x
$$

$$
+ \frac{b_{i0}}{4\rho}t^2 + \frac{1}{2}v_{i0}t.
$$

This equation can be used to calculate the time dependent displacements in the interior of Ω as well as on the boundary Γ. However, one must first determine the values of the unknown boundary tractions T_j and displacements u_j. Since a general analytical solution is impossible, a numerical method is applied, which is based on a discretization of both, geometry and time:

(i)   the boundary is subdivided into a number k of line elements over which displacements and tractions are assumed to be constant

(ii)   the whole observation period is discretized by equal time increments Δt. Then linear (resp. constant) shape functions are used to approximate the transient behavior of the displacements (resp. tractions).

In order to arrive at systems of algebraic equations, the collocation method is applied at every node ξ_l and at all time steps $t_m = m\Delta t$. Then, integrations over each boundary element and over each time step have to be carried out, according to (i) and (ii), e.g. for $n \le m$

$$
U^{nm} = \{U_{jl\lambda}^{(i)nm}\} = \{ \int_{(n-1)\Delta t}^{n\Delta t} \int_{\Gamma_l} U_j^{(i)}(x, \xi_\lambda, m\Delta t - \tau)d\Gamma_x d\tau \}. \tag{5.38}
$$

It should be noted that, if all time steps Δt are the same, according to the causality condition[2] for the time dependent fundamental solution, the following equalities must hold for each $n \le m$

$$
U^{nm} = \overset{(n-m+1)}{U}, \quad T^{nm} = \overset{(n-m+1)}{T}. \tag{5.39}
$$

---

[2]It means that the disturbance is zero at the points, where the waves did not yet arrive.

Thus, for each time step t_m, it is necessary to determine only one extra block matrix $\overset{(\mu)}{U}$ and $\overset{(\mu)}{T}$, where the index $\mu = n - m + 1$ indicates the difference between the observation time $t_n = n\Delta t$ and the impulse time $t_m = m\Delta t$.

The integrations with respect to time have been performed analytically; the results may be found in [An85b]. The integrals over each boundary segment have been evaluated approximatively using Gaussian quadrature formulas. Finally, one obtains the discrete analogues of equations (5.37), which read

$$[0.5\boldsymbol{I} + \overset{(1)}{\boldsymbol{T}}] \cdot \boldsymbol{u}^{(n)} + \sum_{m=1}^{n-1} \overset{(n-m+1)}{\boldsymbol{T}} \cdot \boldsymbol{u}^{(m)} = \sum_{m=1}^{n} \overset{(n-m+1)}{\boldsymbol{U}} \cdot \boldsymbol{t}^{(m)} + \boldsymbol{r}^{(n)}, \qquad (5.40)$$

where $\boldsymbol{u}^{(m)}$ and $\boldsymbol{t}^{(m)}$ are vectors of all nodal displacements and tractions, respectively, at time step m, and the vector $\boldsymbol{r}^{(n)}$ contains terms due to non-zero body forces and/or non-zero initial conditions.

5.4.2 Sound Pressure Waves in 3-D Acoustics and Numerical Applications

Instead of describing the motion of a fluid by the partial differential equation (5.8) together with the boundary and initial conditions (5.9),(5.10), it is also possible to use an integral equation formulation. Its derivation is possible by means of the method of weighted residuals extended to time and space (cf. Sect. 2.2). After some integrations by parts the following reciprocal theorem results:

$$\int_0^t \left[\int_\Omega (p_{,ii} - \frac{1}{c^2}\ddot{p})p^\star d\Omega - \int_\Gamma qp^\star d\Gamma \right] d\tau + \frac{1}{c^2} \int_\Omega [\frac{\partial p}{\partial \tau}p^\star]|_{\tau=0}^t d\Omega \qquad (5.41)$$

$$= \int_0^t \left[\int_\Omega p(p_{,ii}^\star - \frac{1}{c^2}\frac{\partial^2 p^\star}{\partial \tau^2})d\Omega - \int_\Gamma pq^\star d\Gamma \right] d\tau + \frac{1}{c^2} \int_\Omega [p\frac{\partial p^\star}{\partial \tau}]|_{\tau=0}^t d\Omega,$$

which is the analogous of Graffi's theorem for elastodynamic problems [Graf], for the time-dependent scalar wave equation (5.8). Now, an appropriate weight function $p^\star$ must be specified, which is, in this formulation, the fundamental singular solution of eq. (5.8) in an infinite fluid medium due to a concentrated source given by the expression

$$\rho_f \dot{a}(x,t) = \delta(x - \xi)\delta(t - \tau). \qquad (5.42)$$

Here ξ and τ give the location of the point source and its starting time, respectively, and δ is the Dirac delta function. Then, the full space fundamental solution $p^\star(x, \xi; t')$ of eq. (5.8a) and its normal derivative $q^\star(x, \xi; t')$ are given by the formulae

$$p^\star = \frac{1}{4\pi}\frac{1}{r}\delta(t' - \frac{r}{c}) \qquad (5.43)$$

$$q^\star = \frac{1}{4\pi r}[-\frac{1}{r}\delta(t' - \frac{r}{c}) + \frac{\partial}{\partial r}\delta(t' - \frac{r}{c})]\frac{\partial r}{\partial n} \qquad (5.44)$$

where

$$r = |x - \xi| \quad \text{and} \quad R = \sqrt{c^2 t'^2 - r^2}, \tag{5.45}$$

and δ is again the Dirac function. Suppose now that we study the propagation of noise above horizontal hard reflecting surfaces. Then it is possible to use the half-space fundamental solution which is obtained by introducing the mirror source points (with respect to the reflecting surface) as additional singular points [Meis]. Using this singular solution as a special weight function in the reciprocal theorem (5.41), the following B.I.E results

$$d(\xi)p(\xi,t) + \int\limits_0^t \int\limits_\Gamma p(x,\tau)q^*(x,\xi;t,\tau)d\Gamma_x d\tau \tag{5.46}$$

$$= \int\limits_0^t \int\limits_\Gamma q(x,\tau)p^*(x,\xi;t,\tau)d\Gamma_x d\tau + \frac{1}{c^2}\int\limits_\Omega \Big[\bar{p}_0(x)\frac{\partial}{\partial t}p^*(x,\xi;t,0)$$

$$+\bar{b}_0(x)p^*(x,\xi,t,0)\Big]d\Omega_x + \rho\int\limits_0^t \int\limits_\Omega \frac{\partial}{\partial t}\gamma(x,\tau)p^*(x,\xi;t')d\Omega_x d\tau,$$

where the jump coefficient $d(\xi)$ is the same as in the case of the frequency domain equation (5.16). By means of the properties of the Dirac function and the relation

$$\frac{\partial}{\partial r}\delta(t' - \frac{r}{c}) = c\frac{\partial}{\partial \tau}\delta(t' - \frac{r}{c}) \tag{5.47}$$

the integral equation (5.46) can easily be transformed into an integro-differential equation. On the assumption of homogeneous initial conditions, this equation called Kirchhoff equation reads (the overdot denotes differentiation with respect to time)

$$d(\xi)p(\xi,t) - \int\limits_\Gamma \frac{\partial r}{\partial n}\frac{1}{4\pi r}\Big[\frac{1}{c}\dot{p}(x,t_r) + \frac{1}{r}p(x,t_r)\Big]d\Gamma_x \tag{5.48}$$

$$= \int\limits_\Gamma \frac{1}{4\pi}\Big[\frac{1}{r}q(x,t_r)\Big]d\Gamma_x + \rho\int\limits_\Omega \dot{\gamma}(x,t_r)\frac{1}{4\pi r}d\Omega_x,$$

where $t_r = t - \frac{r}{c}$ denotes the "retarded" time with the property $p(x,t_r) = q(x,t_r) = \dot{\gamma}(x,t_r) = 0$ for $t_r \leq 0$. As a special case, sources can be concentrated to any interior point x_l, i.e.

$$\gamma(x,t) = \sum_{l=1}^{L} g_l(t)\delta(x - x_l), \tag{5.49}$$

where $g_l(t)$ denotes the time dependent intensity of the l-th source. Then, the source term in eq. (5.48) becomes ($t_{l_r} = t - |x_l - \xi|/c$)

$$\rho\int\limits_\Omega \dot{\gamma}(x,t_r)\frac{1}{4\pi r}d\Omega_x = \sum_{l=1}^{L} \dot{g}_l(t_{l_r})\frac{\rho}{4\pi}\frac{1}{|x_l - \xi|}. \tag{5.50}$$

This expression is also valid when the position x_l of this point source is moving with a constant speed v_l, i.e. $x_l(t_r) = v_l t_r + x_l(0)$. The discrete form of eq. (5.48) is obtained by using point collocation and interpolation functions as simple as possible. Thus, for every point x on a plane triangular surface segment Γ^e

$$p(x,\tau) = \frac{1}{\Delta t}\big[(t_{m+1} - \tau)p^{em} + (\tau - t_m)p^{em+1}\big]$$

$$q(x,\tau) = q^{em+1}$$

$$(5.51)$$

are introduced in the time intervalls $[t_m, t_{m+1}]$ of length Δt. Finally, assuming that all time steps Δt are of the same duration yields the following system of algebraic equations as the discrete form of (5.48):

$$(\frac{1}{2}I + \overset{(1)}{Q}) \cdot p^{(n)} + \sum_{m=1}^{n-1} \overset{(n-m+1)}{Q} \cdot p^{(m)} = \sum_{m=1}^{n} \overset{(n-m+1)}{P} \cdot q^{(m)} + r^{(n)}. \qquad (5.52)$$

The block matrices $\overset{(m)}{Q}$ and $\overset{(m)}{P}, m = 1, 2, \ldots, n$, and, for point sources with constant intensity change $\dot{g}_l(t) = g_{l0}$, the source term vector $r^{(n)}$, are determined to be

$$\overset{(m)}{P} \doteq \left\{ \frac{1}{4\pi} \int_{\Gamma^e} \frac{1}{r_j} H(m - \frac{r_j}{c\Delta t})H(\frac{r_j}{c\Delta t} - m + 1)d\Gamma_x \right\} \qquad (5.53)$$

$$\overset{(m)}{Q} \doteq \left\{ \int_{\Gamma^e} \frac{n_i(x)}{4\pi r_j^2}\frac{\partial r_j}{\partial x_i} \left\{ \begin{array}{l} mH(m - \frac{r_j}{c\Delta t})H(\frac{r_j}{c\Delta t} - m + 1) - \\ -(m-2)H(m - 1 - \frac{r_j}{c\Delta t})H(\frac{r_j}{c\Delta t} - m + 2) \end{array} \right\} d\Gamma_x \right\} \qquad (5.54)$$

$$r^{(n)} \doteq \left\{ \frac{\rho}{4\pi} \sum_{l=1}^{L} g_{l0} \frac{1}{|x_l - \xi_j|} \right\}. \qquad (5.55)$$

In the numerical applications the integrals over the "non-singular" boundary elements have been calculated by using 21-point triangular Gaussian quadrature, while in the first time step the integrations for obtaining the matrix P have been carried out analytically [An87e, An88c, Meis] on plane elements containing the singular point ξ (the corresponding entries of Q are zero for plane elements). For two-dimensional problems, the respective integral formulations, fundamental solution, and boundary element matrix elements may be found in [An89b, An91b, Ma82].

We shall close this Section by means of a numerical application concerning the noise produced by moving sources. As we have studied in Sect. 5.2 the noise pollution can be rectified by erecting barriers, and the continuously produced traffic noise can be estimated by means of a simple, two-dimensional modelization. But often in residential areas, the traffic is not such big that long noise barriers make sense. Moreover, in these situations the traffic noise is usually produced by single driving cars or motorcycles. To study such noise problems a three-dimensional analysis in time domain is necessary because the noise source is moving in a real three-dimensional surrounding.

A sample of such moving noise source problems is examined in order to study the effects caused by the speed of the movement:

Starting two meters ahead of the front end of the barrier, a harmonic point source of intensity 1 mPa is passing a barrier of length 10 m and height 5 m on a course 1.5 m above the ground in a distance 5 m (see fig. 5.15) where

(i)   first, the harmonic noise point source has a frequency 170 Hz and is moving on straight line with a constant speed of 120 km/h.

(ii)   second, the point source is assumed to have the much lower frequency 20 Hz, but the much higher speed of 500 km/h.

Since the ground is assumed to be hard reflecting, the half-space fundamental solutions can be used. Therefore, only the boundary of the barrier must be discretized.

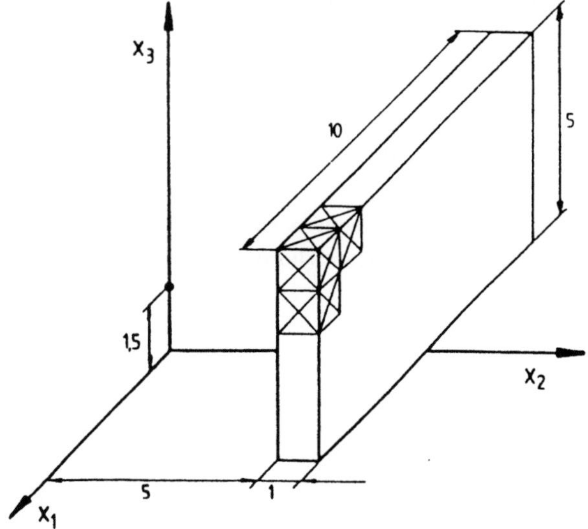

Fig. 5.15: Geometry and discretization of a 3-D edge barrier

First, the "slow" motion is studied: since the noise source starts two meters ahead of the front part of the barrier, portions of the first generated waves can pass undisturbed (see fig. 5.16). Several time steps later, when the noise source is passing directly in front of the barrier, due to the diffractions at the ends and over the top of the barrier, superposition of the waves can either diminish the noise (after 0.1239 secs $= 42\Delta t$, fig. 5.17) or amplify it (after 0.20945 secs $= 71\Delta t$, fig. 5.18).

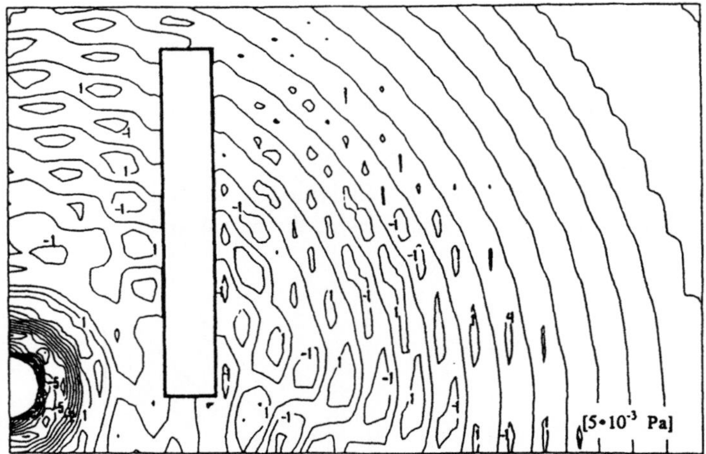

Fig. 5.16: Harmonic (170 Hz) point source with intensity 1 mPa moving with 120 km/h: Noise level distribution (horizontal section) after 0.0619 secs

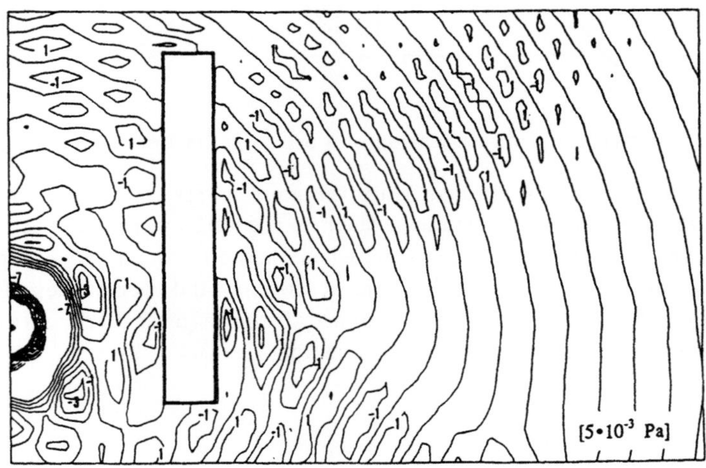

Fig. 5.17: Harmonic (170 Hz) point source with intensity 1 mPa moving with 120 km/h: Noise level distribution (horizontal section) after 0.1239 secs

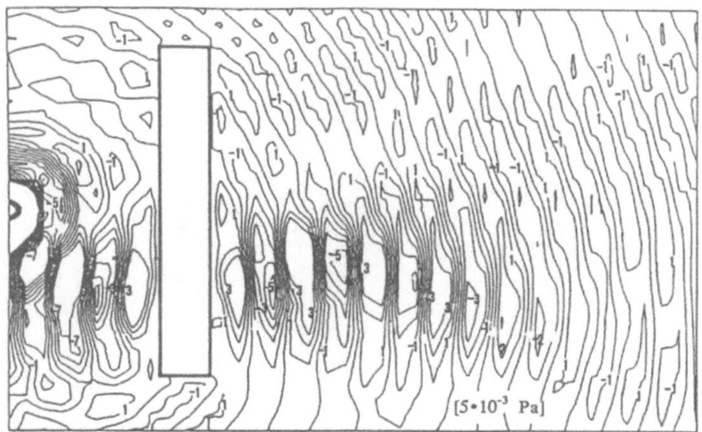

Fig. 5.18: Harmonic (170 Hz) point source with intensity 1 mPa moving with 120
km/h: Noise level distribution (horizontal section) after 0.20945 secs

Thus, as shown in fig. 5.17, the maximum noise measured at the observation points
during the entire considered time period T, that is

$$p_{\mathrm{max}}(x) = \max_{0 \leq t \leq T} |p(x, t)|$$

is not reduced behind the barrier.

When the point source is moving with the much higher speed of 500 km/h, this
fast movement creates the so-called Doppler effect. It can be observed (fig. 5.20) that
the lines of equal sound pressure are more dense ahead of the moving source than
behind, i.e. the wave length of the sound is shorter before the moving source than
after it. Moreover, due to the high speed of the driving noise source, almost none
of the generated waves can propagate undisturbed into the area behind the barrier.
Although the produced noise has the low frequency 20 Hz (which generally means
more diffractions [An91a]) the noise reduction due to the barrier is much greater than
for the slowly moving source (fig. 5.21).

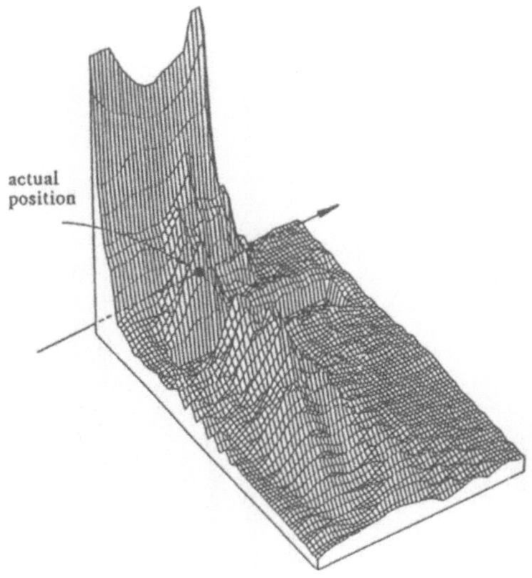

Fig. 5.19: Harmonic (170 Hz) point source with intensity 1 mPa moving with 120
km/h: distribution (horizontal section) of noise maxima during the period
$T = 0.2095$ secs

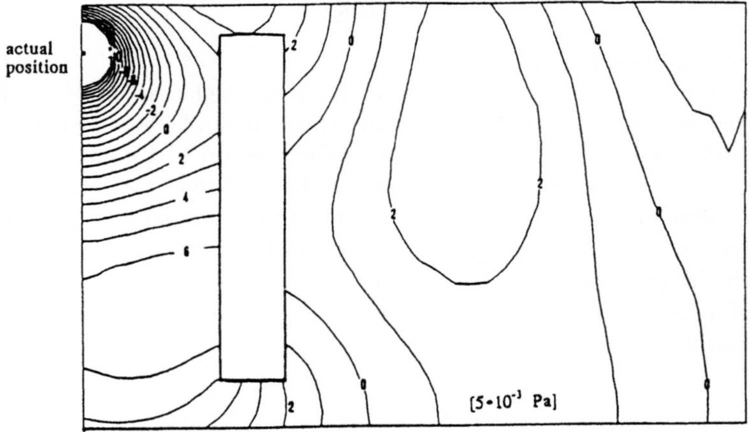

Fig. 5.20: Harmonic (20 Hz) point source with intensity 1 mPa moving with 500 km/h:
Noise level distribution (horizontal section) after 0.0885 secs

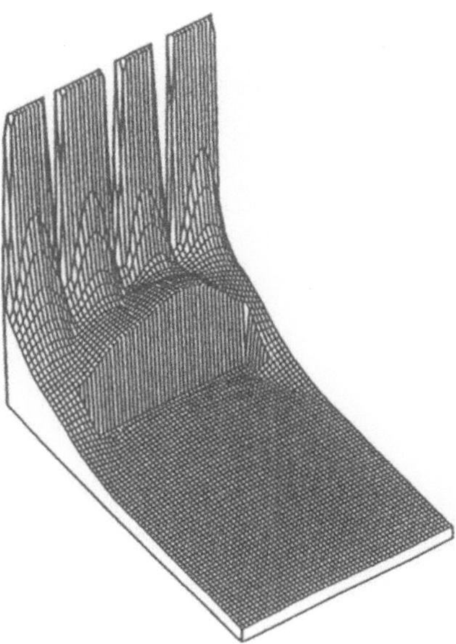

Fig. 5.21: Harmonic (20 Hz) point source with intensity 1 mPa moving with 500 km/h: distribution (horizontal section) of noise maxima during the period $T = 0.0885$ secs

Chapter 6
Dynamic Interaction Problems

6.1 Bilateral Coupling of Elastic Structures and Domains

The subject of dynamic interaction has been recognized as a very important aspect of the dynamic analysis and design of structures. When infinite or semi-infinite domains are involved, the most appropriate method to study relevant interaction effects is certainly the B.E.M. because it takes correctly into account the loss of energy due to radiation to infinity in a correct way. Moreover, the method is very convenient when the loading of the structure as well as the main unknowns are on the surface or on an interface, of the structure, since, in general, the B.E.M. works only on the discretized boundary.

In the analysis of the dynamic response of an elastic structure or a foundation on a layered soil or on a single layer - halfspace system, i.e. in the analyis of dynamic interaction problems, the equations of motion governing the coupled system must be solved simultaneously, and suitable conditions must be defined and satisfied along the interfaces. Assuming a contact zone without debonding, the following conditions have to be satisfied at each point of the interface Γ_i between two deforming subregions Ω_A and Ω_B

(i)   compatibility of the displacements, i.e. the components of the displacements on the discretized interface between the subregion Ω_A and the subregion Ω_B must coincide, i.e.

$$u_{A_i} = u_{B_i} = u_i \qquad \text{in cartesian coordinates} \qquad (6.1)$$

$$\left. \begin{aligned} u_{A_n} &= -u_{B_n} = u_n \\ u_{A_s} &= -u_{B_s} = u_s \end{aligned} \right\} \quad \text{in intrinsic coordinates } (n = \text{normal}, s = \text{tangential})$$

(ii) equilibrium of the tractions, i.e. the components of the tractions acting along the discretized interface Γ_i on the subregion Ω_A must be opposite to those on the subregion Ω_B, i.e.

$$t_{A_i} = -t_{B_i} = t_i \qquad \text{in cartesian co-ordinates} \qquad (6.2)$$

$$\left. \begin{aligned} t_{A_n} &= t_{B_n} = t_n \\ t_{A_s} &= t_{B_s} = t_s \end{aligned} \right\} \quad \text{in intrinsic co-ordinates .}$$

In order to incorporate these conditions in the boundary element formulation of the interaction problem, both the boundary element equation systems concerning the subregion Ω_A and the subregion Ω_B have to be adequately combined. Here, this coupling will be explained in detail for the time-stepping boundary element equations (5.40).

First, both systems of equations have to be rearranged in order to distinguish between the collocation and nodal points on the "outer" boundaries (Γ_0^A and Γ_0^B, subscript 0) and on the interface (Γ_i^A and Γ_i^B, subscript i). The equations for the region Ω_A read

$$
\begin{bmatrix} \overset{(1)}{\hat{\boldsymbol{T}}}_{A_{ii}} & \overset{(1)}{\boldsymbol{T}}_{A_{i0}} \\ \overset{(1)}{\boldsymbol{T}}_{A_{0i}} & \overset{(1)}{\hat{\boldsymbol{T}}}_{A_{00}} \end{bmatrix} \cdot \begin{bmatrix} \boldsymbol{u}_{A_i}^{(n)} \\ \boldsymbol{u}_{A_0}^{(n)} \end{bmatrix} = \begin{bmatrix} \overset{(1)}{\boldsymbol{U}}_{A_{ii}} & \overset{(1)}{\boldsymbol{U}}_{A_{i0}} \\ \overset{(1)}{\boldsymbol{U}}_{A_{0i}} & \overset{(1)}{\boldsymbol{U}}_{A_{00}} \end{bmatrix} \cdot \begin{bmatrix} \boldsymbol{t}_{A_i}^{(n)} \\ \boldsymbol{t}_{A_0}^{(n)} \end{bmatrix}
$$
$$
+ \sum_{m=1}^{n-1} \{ \overset{(n-m+1)}{\boldsymbol{U}}_A \cdot \boldsymbol{t}_A^{(m)} - \overset{(n-m+1)}{\boldsymbol{T}}_A \cdot \boldsymbol{u}_A^{(m)} \} - \boldsymbol{r}_A^{(n)}
$$
(6.3)

and for the region Ω_B

$$
\begin{bmatrix} \overset{(1)}{\hat{\boldsymbol{T}}}_{B_{ii}} & \overset{(1)}{\boldsymbol{T}}_{B_{i0}} \\ \overset{(1)}{\boldsymbol{T}}_{B_{0i}} & \overset{(1)}{\hat{\boldsymbol{T}}}_{B_{00}} \end{bmatrix} \cdot \begin{bmatrix} \boldsymbol{u}_{B_i}^{(n)} \\ \boldsymbol{u}_{B_0}^{(n)} \end{bmatrix} = \begin{bmatrix} \overset{(1)}{\boldsymbol{U}}_{B_{ii}} & \overset{(1)}{\boldsymbol{U}}_{B_{i0}} \\ \overset{(1)}{\boldsymbol{U}}_{B_{0i}} & \overset{(1)}{\boldsymbol{U}}_{B_{00}} \end{bmatrix} \cdot \begin{bmatrix} \boldsymbol{t}_{B_i}^{(n)} \\ \boldsymbol{t}_{B_0}^{(n)} \end{bmatrix}
$$
$$
+ \sum_{m=1}^{n-1} \{ \overset{(n-m+1)}{\boldsymbol{U}}_B \cdot \boldsymbol{t}_B^{(m)} - \overset{(n-m+1)}{\boldsymbol{T}}_B \cdot \boldsymbol{u}_B^{(m)} \} - \boldsymbol{r}_B^{(n)},
$$
(6.4)

the relations (6.3) and (6.4) together yield the coupled system

$$
\begin{bmatrix} \overset{(1)}{\hat{\boldsymbol{T}}}_{A_{00}} & \overset{(1)}{\boldsymbol{T}}_{A_{0i}} & \boldsymbol{0} \\ \overset{(1)}{\boldsymbol{T}}_{A_{i0}} & \overset{(1)}{\hat{\boldsymbol{T}}}_{A_{ii}} & \boldsymbol{0} \\ \boldsymbol{0} & -\overset{(1)}{\hat{\boldsymbol{T}}}_{B_{ii}} & \overset{(1)}{\boldsymbol{T}}_{B_{i0}} \\ \boldsymbol{0} & -\overset{(1)}{\boldsymbol{T}}_{B_{0i}} & \overset{(1)}{\hat{\boldsymbol{T}}}_{B_{00}} \end{bmatrix} \cdot \begin{bmatrix} \boldsymbol{u}_{A_0}^{(n)} \\ \boldsymbol{u}_i^{(n)} \\ \boldsymbol{u}_{B_0}^{(n)} \end{bmatrix} = \begin{bmatrix} \overset{(1)}{\boldsymbol{U}}_{A_{00}} & \overset{(1)}{\boldsymbol{U}}_{A_{0i}} & \boldsymbol{0} \\ \overset{(1)}{\boldsymbol{U}}_{A_{i0}} & \overset{(1)}{\boldsymbol{U}}_{A_{ii}} & \boldsymbol{0} \\ \boldsymbol{0} & \overset{(1)}{\boldsymbol{U}}_{B_{ii}} & \overset{(1)}{\boldsymbol{U}}_{B_{i0}} \\ \boldsymbol{0} & \overset{(1)}{\boldsymbol{U}}_{B_{0i}} & \overset{(1)}{\boldsymbol{U}}_{B_{00}} \end{bmatrix} \cdot \begin{bmatrix} \boldsymbol{t}_{A_0}^{(n)} \\ \boldsymbol{t}_i^{(n)} \\ \boldsymbol{t}_{B_0}^{(n)} \end{bmatrix}
$$
(6.5)
$$
+ \begin{bmatrix} \sum_{m=1}^{n-1} \{ \overset{(n-m+1)}{\boldsymbol{U}}_A \cdot \boldsymbol{t}_A^{(m)} - \overset{(n-m+1)}{\boldsymbol{T}}_A \cdot \boldsymbol{u}_A^{(m)} \} - \boldsymbol{r}_A^{(n)} \\ \sum_{m=1}^{n-1} \{ \overset{(n-m+1)}{\boldsymbol{U}}_B \cdot \boldsymbol{t}_B^{(m)} - \overset{(n-m+1)}{\boldsymbol{T}}_B \cdot \boldsymbol{u}_B^{(m)} \} - \boldsymbol{r}_B^{(n)} \end{bmatrix} .
$$

These relations must be reordered accordingly to the actual "outer" boundary conditions of the problem. For instance, if $\boldsymbol{t}_{A_0}$ and $\boldsymbol{t}_{B_0}$ are given, then we write (6.5) in the form

$$
\begin{bmatrix} \overset{(1)}{\hat{\boldsymbol{T}}}_{A_{00}} & \overset{(1)}{\boldsymbol{T}}_{A_{0i}} & -\overset{(1)}{\boldsymbol{U}}_{A_{0i}} & \boldsymbol{0} \\ \overset{(1)}{\boldsymbol{T}}_{A_{i0}} & \overset{(1)}{\hat{\boldsymbol{T}}}_{A_{ii}} & -\overset{(1)}{\boldsymbol{U}}_{A_{ii}} & \boldsymbol{0} \\ \boldsymbol{0} & -\overset{(1)}{\hat{\boldsymbol{T}}}_{B_{ii}} & -\overset{(1)}{\boldsymbol{U}}_{B_{ii}} & \overset{(1)}{\boldsymbol{T}}_{B_{i0}} \\ \boldsymbol{0} & -\overset{(1)}{\boldsymbol{T}}_{B_{0i}} & -\overset{(1)}{\boldsymbol{U}}_{B_{0i}} & \overset{(1)}{\hat{\boldsymbol{T}}}_{B_{00}} \end{bmatrix} \cdot \begin{bmatrix} \boldsymbol{u}_{A_0}^{(n)} \\ \boldsymbol{u}_i^{(n)} \\ \boldsymbol{t}_i^{(n)} \\ \boldsymbol{u}_{B_0}^{(n)} \end{bmatrix} = \begin{bmatrix} \overset{(1)}{\boldsymbol{U}}_{A_{00}} & \boldsymbol{0} \\ \overset{(1)}{\boldsymbol{U}}_{A_{i0}} & \boldsymbol{0} \\ \boldsymbol{0} & \overset{(1)}{\boldsymbol{U}}_{B_{i0}} \\ \boldsymbol{0} & \overset{(1)}{\boldsymbol{U}}_{B_{00}} \end{bmatrix} \cdot \begin{bmatrix} \boldsymbol{t}_{A_0}^{(n)} \\ \boldsymbol{t}_{B_0}^{(n)} \end{bmatrix}
$$
(6.6)

$$+ \left[\begin{array}{c} \sum_{m=1}^{n-1} \left\{ \overset{(n-m+1)}{\boldsymbol{U}_A} \cdot t_A^{(m)} - \overset{(n-m+1)}{\boldsymbol{T}_A} \cdot u_A^{(m)} \right\} - r_A^{(n)} \\[2ex] \sum_{m=1}^{n-1} \left\{ \overset{(n-m+1)}{\boldsymbol{U}_B} \cdot t_B^{(m)} - \overset{(n-m+1)}{\boldsymbol{T}_B} \cdot u_B^{(m)} \right\} - r_B^{(n)} \end{array} \right] .$$

For steady state or harmonic interaction problems, the treatment of the coupled conditions is analogous. There, all indices indicating the time steps can be neglected, and, moreover, all terms giving the influence of the past steps can be cancelled. Thus, we obtain the much simpler system

$$\left[\begin{array}{cccc} \hat{\boldsymbol{T}}_{A00} & \boldsymbol{T}_{A0i} & -\boldsymbol{U}_{A0i} & \mathbf{0} \\ \boldsymbol{T}_{Ai0} & \hat{\boldsymbol{T}}_{Aii} & -\boldsymbol{U}_{Aii} & \mathbf{0} \\ \mathbf{0} & -\hat{\boldsymbol{T}}_{Bii} & -\boldsymbol{U}_{Bii} & \boldsymbol{T}_{Bi0} \\ \mathbf{0} & -\boldsymbol{T}_{Boi} & -\boldsymbol{U}_{Boi} & \hat{\boldsymbol{T}}_{B00} \end{array} \right] \cdot \left[\begin{array}{c} u_{A0} \\ u_i \\ t_i \\ u_{B0} \end{array} \right] = \left[\begin{array}{cc} \boldsymbol{U}_{A00} & \mathbf{0} \\ \boldsymbol{U}_{Ai0} & \mathbf{0} \\ \mathbf{0} & \boldsymbol{U}_{Bi0} \\ \mathbf{0} & \boldsymbol{U}_{B00} \end{array} \right] \cdot \left[\begin{array}{c} t_{A0} \\ t_{B0} \end{array} \right] . \tag{6.7}$$

This technique of coupling subregions is necessary when the material data are different in parts of the problem domain. But, it can also be used in the case of homogeneous domains in order to have some advantages in data processing. For instance, in the "substructuring technique" homogeneous domains are artificially subdivided into subregions which then are coupled again. This has several advantages, e.g. the easier application of suitable background storage techniques. Moreover, one obtains a well-pronounced diagonal band structures which diminish, at least partially, the disadvantage of the fully occupied matrices.

The methods presented in Sect. 5.3.1 and the previous relations are now used to solve some problems concerning the response of foundations. Especially in layers of different soil material the occuring wave reflections and refractions influence the dynamic response of a rigid surface foundation [An87d, An89c].

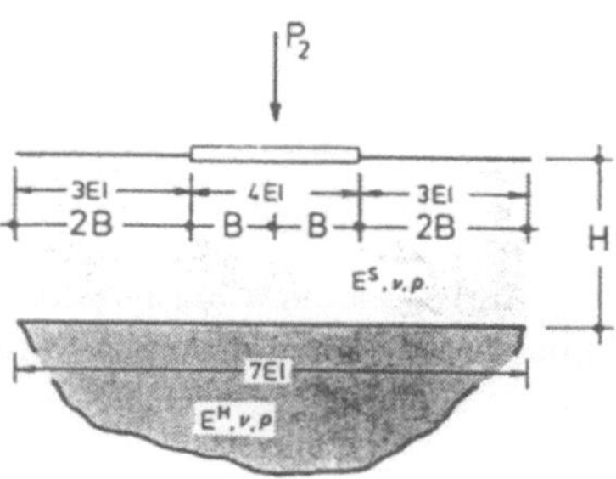

Fig. 6.1: Geometry and discretization of a surface foundation on an elastic layer (E^S) over the elastic halfspace (E^H)

Let us consider a rigid, massless surface strip foundation (the foundation is represented by elements which are combined such that they are only able to move like a rigid body; for more details see [An88c]) of width $2B = 2$ m on a homogeneous elastic soil-layer of thickness H perfectly bonded to an elastic halfspace. The material of the layer has the Young's modulus $E^S = 2.66 \cdot 10^5 \, KN/m^2$, while that of the underlying halfspace E^H varies in a certain range. The rest of the material data, i.e. the mass density $\rho = 2000 \, kg/m^3$, and the Poisson's ratio $\nu = 0.33$ is taken to be the same for both the layer and the halfspace. The discretization consists of four elements for the foundation, and three elements of equal length at both sides of the foundation, while the interface between the elastic layer and the halfspace is modelled by seven equal elements (see fig. 6.1). The duration of one time step is taken to be $\Delta t = 1.3516 \cdot 10^{-3}$ secs. In order to study the dependence of the reflections and refractions generated by the layer - halfspace interface on the material data, i.e here, on the Young's modulus ratio E^H/E^S, waves are created by a vertical impulsive force of intensity $740 \, KN/m$ acting on the foundation during the time interval 0.0013516 secs $(= \Delta t)$.

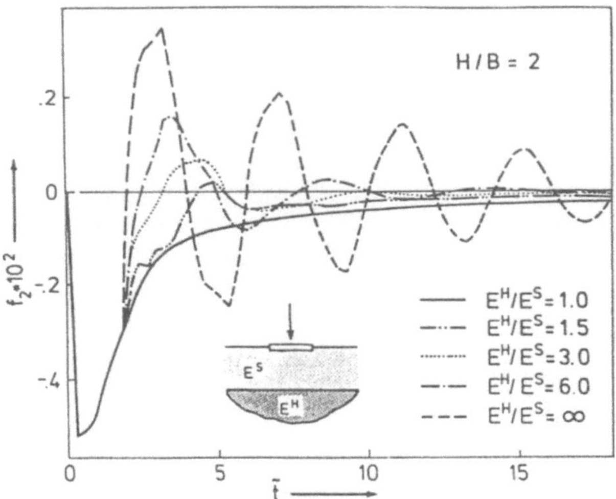

Fig. 6.2: Time history of the vertical foundation response subjected to a vertical impulsive force: Influence of Young's moduli ratio E^H/E^S and of the layer thickness $H = 2$ m. $(\tilde{t} = c_2^s t/B;\ f_\alpha = G^s u_\alpha)$

By comparing the time dependent vertical response of the foundation, when the later simply rests on the homogeneous halfspace, i.e. if $E^H/E^S = 1.0$, with the response observed when the material data of the layer and the underlying halfspace are different, certain important effects become apparent. Depending on the thickness of the layer – $H = 2$ m in fig. 6.1 and $H = 5$ m in fig. 6.2 – after 0.001 secs and 00.022 secs, respectively, vibrating motions of the rigid foundation are induced in the cases of different Young's moduli in the layer and in the underlying supporting halfspace.

This time is exactly that required by the p-waves to traverse the distance from the foundation to the interface and back. Morever, as expected, the amplitudes of these vibrations are related to the stiffness of the supporting halfspace. Indeed the softer the halfspace, the less is the reflected part of the waves and thus, the smaller are the amplitudes. Besides, in any case, even for a rigid bedrock ($E^H/E^S = \infty$) the response to the transient excitation is damped by the radiation of the waves in both the horizontal directions.

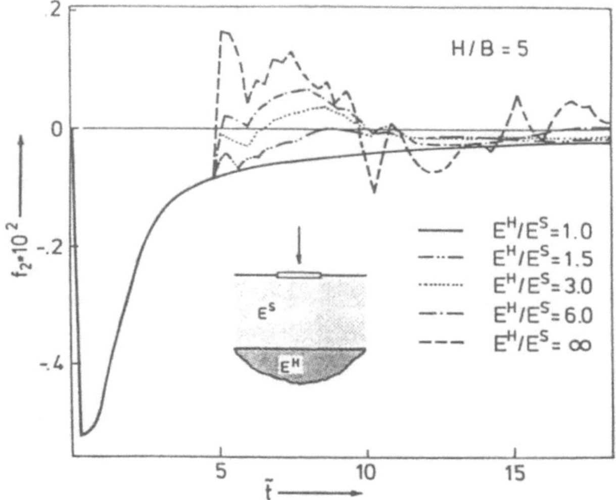

Fig. 6.3: Time history of the vertical foundation response to a vertical impulsive force: Influence of Young's moduli ratio E^H/E^S and of the layer thickness $H = 5$ m ($\tilde{t} = c_2^s t/B$; $f_\alpha = G^s u_\alpha$)

Another application concerns stress waves in composite strips. One special example will be given now which demonstrates quite impressively the applicabilty and accuracy of the time-stepping indirect BE procedure, because its results can be compared to those of a calculation using the traditional "exact" method of "Bicharacteristics" [Ball]. A semi-infinite strip of breadth two times 700 mm (see fig. 6.4a) is composed of two parts, of aluminium and of steel. The respective material data are for the aluminium a Young's modulus $E^A = 0.72 \cdot 10^8 KN/m^2$, a shear modulus $G^A = 0.27 \cdot 10^8 KN/m^2$, and a density $\rho^A = 2700 kg/m^3$, and for the steel a Young's modulus $E^S = 2.1 \cdot 10^8 KN/m^2$, a shear modulus $G^S = 0.8 \cdot 10^8 KN/m^2$, and a density $\rho^S = 7800 kg/m^3$. Under conditions of plane stress, the pressure wave speeds are $c_1^A = 5477 m/s$ and $c_1^S = 5462 m/s$, while those of the shear waves are $c_2^A = 3162 m/s$ and $c_2^S = 3202$ m/s. The response of the structure, when an explosion takes plate at the point A, is examined.

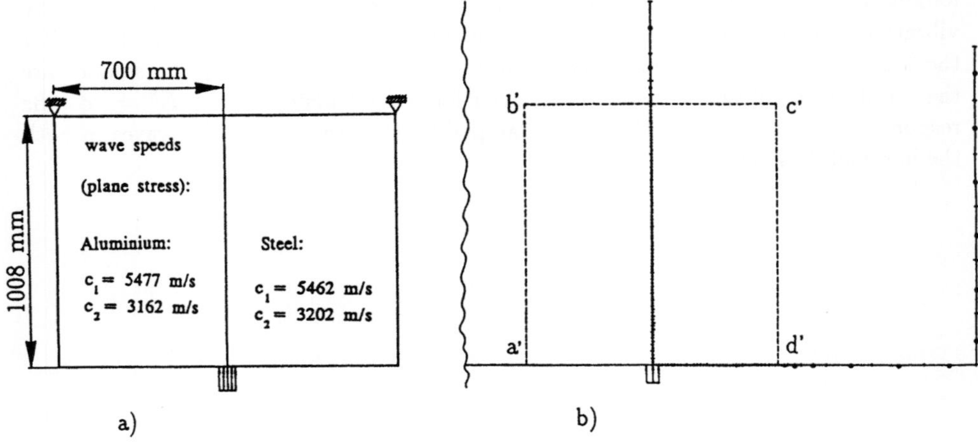

Fig. 6.4: a) Geometry and b) discretization of a semi-infinite composite strip. Here $a'b'c'd'$ denotes the domain appearing in fig. 6.6

The numerical treatment for the indirect B.E.M formulation consists of the boundary discretization by a total number of 152 elements where the infinite strip is numerically modelled only on a length of 1008 mm (see fig. 6.4a). The boundary elements are chosen to be of two different lengths: the parts of the boundary in the neighborhood of the explosion are modeled by much smaller elements than the parts in a larger distance. Thus, along both parts of the interface 40 small and 4 large elements, and along the "outer" boundaries again 40 small and 6+6 large elements on each side are used (see fig. 6.4b)

The two central elements near A are subjected to a transient "explosive" loading whose time dependent strength distribution during the whole action period of $135\,\mu$secs is given in fig. 6.5. The time step duration is $\Delta t = 2.72 \cdot 10^{-6}$ secs.

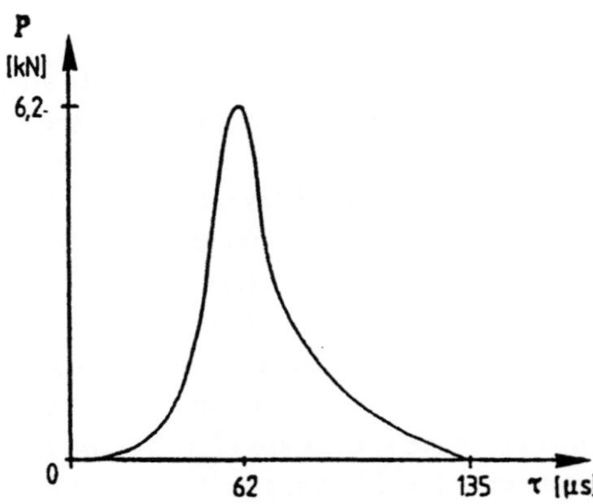

Fig. 6.5: Time dependence of the explosive load intensity

In fig. 6.6, the distribution of the principal shear stress $\tau_{\mathbf{max}}$ at the time 125μsecs, i.e. $46\Delta t$ after the start of "explosion", is plotted for both methods, the method of Bi-Characteristics [Ball] and the Indirect BEM in the time domain.

It is obvious, that due to the dynamic interactions across the interface the so-called Stonely waves are generated along the interface. Moreover, besides the p-waves and the s-waves in both parts of the composite strip, Rayleigh-waves have been developed along the free boundaries. The agreement of both the results determined via completely different methods is impressive, not only in the axonometric plottings (fig. 6.6) but also when comparing the corresponding contour line plottings of this principle shear stress distribution (fig. 6.7).

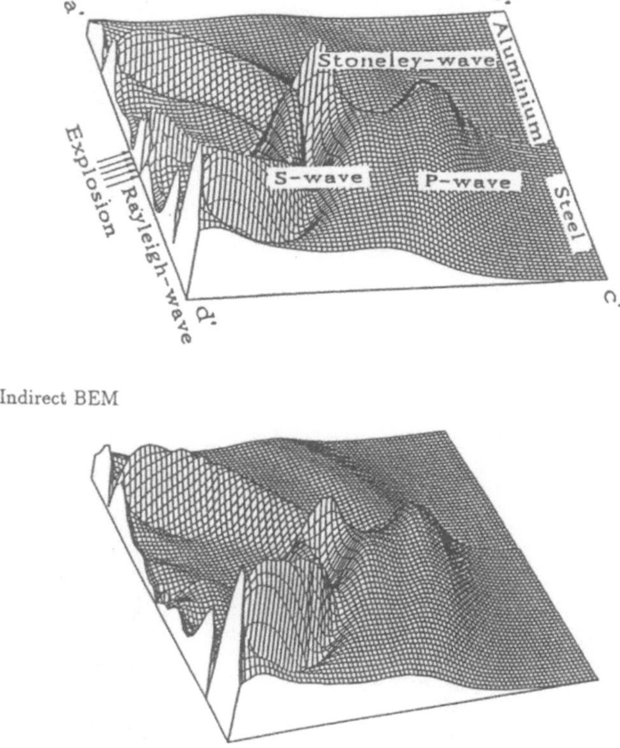

Fig. 6.6: Plots of the principle shear stress distribution after 125 μsecs: comparison of the results obtained by the method of Bi-Characteristics and by the Indirect B.E.M.

Method of BI-CHARACTERISTICS

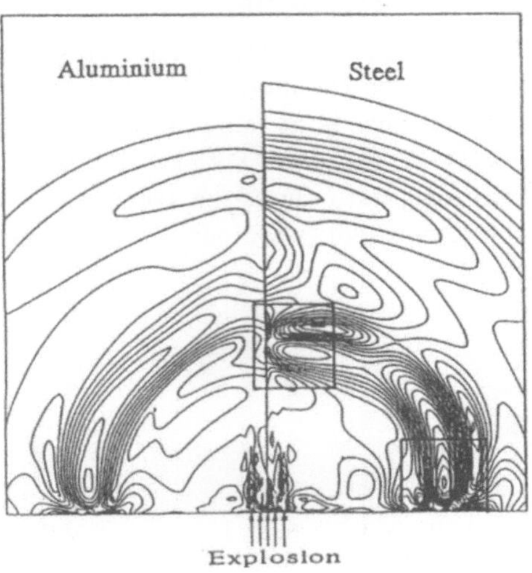

INDIRECT BEM

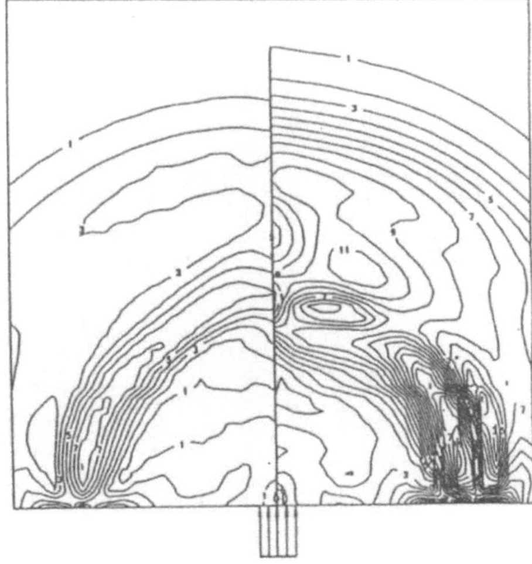

Fig. 6.7: Contour lines of the principle shear stress distribution after 125μsecs: Comparison of results found by the method of Bi-Characteristics and by the Indirect B.E.M.

6.2 Fluid-Structure Interaction

A very important question in fluid mechanics concerns the effect of hydrodynamic pressure on dams during earthquakes. Since 1933 when Westergaard [West] first published his work on rigid dams, a large number of analytical and some numerical studies have been carried out to analyze the dynamic interaction problem of dam reservoir systems (cf. e.g. [Cho68, Cho73, Cho83]. The important conclusions obtained from these studies are that the water compressibility cannot be neglected in the analysis, and that the dam deformation may significantly alter the pressure.

In order to study hydrodynamic pressure waves in semi-infinite domains, e.g. in large water reservoirs, the most appropriate method is certainly the B.E.M. Although in most cases the soil properties will not be uniform, the assumption of a homogeneous isotropic linearly elastic halfspace is a natural and appropriate idealization for the bottom of the reservoir. The dam itself, e.g. an earth-dam or a concrete gravity dam, is actually an inhomogeneous structure which should be better numerically modelled by means of the Finite Element Method [An91b]. However certain effects on the dam caused by the contact with the water, like the damping, can also be studied by considering a homogeneous dam medium with elastic behavior. Here a fluid-structure interaction problem is studied by modelling all subdomains of the problem, i.e. the water domain, the soil, and the dam, by boundary elements. In the time domain analysis (resp. the frequency domain analysis) of the dynamic behaviour of coupled fluid- structure systems the eqs (5.29) and (5.48), (resp. (5.16)) and the corresponding ones for elastic regions (see [An86c]), must be solved simultaneously. Then suitable coupling conditions have to be prescribed along the fluid-structure interfaces. If one assumes perfect sliding of the fluid particles along the elastic structure, the following conditions have to be satisfied at each point of the interface Γ_i which is considered to be always submerged.

(i)   the tangential component of the traction is zero.

(ii)   the hydrodynamic pressure on the structure is equal to the normal component of the external traction.

(iii)   the acceleration components of the fluid particles in the normal direction must coincide with those of the structure.

Thus, on the parts of the structure which are in contact with the fluid, these conditions imply using for the fluid either the (p, q) or the (Φ, Ψ) formulation, (cf. Sect. 5.1) that in the time domain for $t > 0$ on Γ_i

$$\text{(i)} \quad T_j(x,t)s_j(x) = 0 \tag{6.8a}$$

$$\left. \begin{array}{l} \text{(ii)} \quad p(x,t) = -T_j(x,t)n_j(x) \\ \text{or (ii)} \quad \rho_f \dot{\Phi}(x,t) = -T_j(x,t)n_j(x) \end{array} \right\} \tag{6.9a}$$

$$\left. \begin{array}{l} \text{(iii)} \quad q(x,t) = \rho_f \ddot{u}_j(x,t)n_j(x) \\ \text{or (iii)} \quad \Psi = \dot{u}_j(x,t)n_j(x) \end{array} \right\} \tag{6.10a}$$

whereas in the frequency domain on Γ_i

$$\text{(i)} \quad \hat{T}_j(x) s_j(x) = 0 \tag{6.8b}$$

$$\left.\begin{array}{ll}\text{(ii)} & \hat{p}(x) = -\hat{T}_j(x) n_j(x) \\ \text{or (ii)} & \rho_f s \hat{\Phi}(x) = -\hat{T}_j(x)\end{array}\right\} \tag{6.9b}$$

$$\left.\begin{array}{ll}\text{(iii)} & \hat{q}(x) = \rho_f s^2 \hat{u}_j(x) n_j(x) \\ [or(iii)] & \hat{\Psi} = s\hat{u}_j(x) n_j(x)\end{array}\right\} \tag{6.10b}$$

Here, $s_j(x)$ and $n_j(x)$ indicate the tangential and the outward normal to the boundary of the elastic structure, respectively. The formulation of the third condition (6.10a) results from equation (5.6) and the definition (5.11b) of the flux q:

$$q = p_{,j} n_j^f = -(\rho_f \dot{v}_j^f) n_j^f = -(\rho_f \ddot{u}_j^f) n_j^f = -\rho_f \ddot{u}_j(-n_j). \tag{6.11}$$

In the time domain, the choice between these two alternative descriptions of the wave propagation in fluids will be strongly influenced by these coupling conditions. If the (p,q) - formulation is used, it is necessary to use shape functions for $u_j(\mathbf{x}, t)$ which are twice differentiable with respect to time, whereas in the other case it is possible to take interpolation function which are only piecewise linear in time.

The above coupling conditions have to be encorporated in both the systems of boundary element equations, for the fluid and the elastic domain, i.e. in

$$\overset{(n-m+1)}{\boldsymbol{T}} \cdot (\boldsymbol{n} \cdot \boldsymbol{u}_n^{(m)} + \boldsymbol{s} \cdot \boldsymbol{m}_s^{(m)}) = \overset{(n-m+1)}{\boldsymbol{U}} \cdot (\boldsymbol{n} \cdot \boldsymbol{t}_n^{(m)} + \boldsymbol{s} \cdot \boldsymbol{t}_s^{(m)}), \tag{6.12}$$

$$\overset{(n-m+1)}{\boldsymbol{G}} \cdot \boldsymbol{\Phi}^{(m)} = \overset{(n-m+1)}{\boldsymbol{H}} \cdot \boldsymbol{\Psi}^{(m)}. \tag{6.13}$$

Here, in the time-dependent case, summation over the time steps m is taking place, while the multiplication with $\boldsymbol{n}$ and $\boldsymbol{s}$ is performed, because intrinsic reference coordinates are used in the coupling conditions. Imposing the interface conditions (6.9) and (6.10) in the time- and frequency domain, respectively, one obtains

$$\left.\begin{array}{lll}t_{ni}^{(m)} & = & -\rho_f \dot{\Phi}_i^{(m)} \cong -\rho_f \frac{1}{\Delta t}(\Phi_i^{(m)} - \Phi_i^{(m-1)}) \\ t_{ni} & = & -\rho_f s \Phi_i\end{array}\right\} t_{ni} = -\rho_f \boldsymbol{D} \cdot \Phi_i \tag{6.14}$$

$$\left.\begin{array}{lll}\boldsymbol{\Psi}_i^{(m)} & = & \dot{u}_{ni}^{(m)} \cong \frac{1}{\Delta t}(u_{ni}^{(m)} - u_{ni}^{(m-1)}) \\ \boldsymbol{\Psi}_i & = & s u_{ni}\end{array}\right\} \boldsymbol{\Psi}_i = \boldsymbol{D} \cdot u_{ni} \tag{6.15}$$

where the index i indicates variables along the "inner" boundary, i.e. along the interfaces.

Taking the potential of the fluid particles and the displacement components of the elastic medium as the unknown values at the interface nodes, the entire set of equations can be reordered and expressed (at each time step) in the coupled matrix form (6.16). In this equation, $\Gamma^{ei} = \Gamma^{fi}$ denotes the fluid-elastic body interface and Γ^{e0} and Γ^{f0} stand for the non-common "outer" boundaries of the elastic and the fluid domains. Accordingly, the variables $\boldsymbol{u}_{n0}, \boldsymbol{u}_{s0}$, and $\boldsymbol{\Psi}_0$ denote the unknown displacements and the unknown flux along these outer boundaries. The traction components $\bar{t}_{no}$ and $\bar{t}_{s0}$, and the potential Φ_0 are assumed to be prescribed along the outer boundary of the elastic domain and the fluid, respectively:

$$
\begin{bmatrix}
T_{00} \cdot n^0 & T_{00} \cdot s^0 & T_{ni} \cdot n^i & T_{0i} \cdot s^i & \rho_f U_{0i} \cdot n^i \cdot D & 0 \\
T_{i0} \cdot n^0 & T_{i0} \cdot s^0 & T_{ii} \cdot n^i & T_{ii} \cdot s^i & \rho_f U_{ii} \cdot n^i \cdot D & 0 \\
0 & 0 & -H_{ii} \cdot D & 0 & G_{ii} & -H_{i0} \\
0 & 0 & -H_{0i} \cdot D & 0 & G_{0i} & -H_{i0}
\end{bmatrix}
\cdot
\begin{bmatrix}
u_{n0} \\ u_{s0} \\ u_{ni} \\ u_{si} \\ \Phi_i \\ \Psi_0
\end{bmatrix}
$$

$$
=
\begin{bmatrix}
U_{00} \cdot n^0 & U_{n0} \cdot s^0 & 0 \\
U_{i0} \cdot n^0 & U_{i0} \cdot s^0 & 0 \\
0 & 0 & -G_{i0} \\
0 & 0 & -G_{00}
\end{bmatrix}
\cdot
\begin{bmatrix}
\bar{t}_{n0} \\ \bar{t}_{s0} \\ \bar{\Phi}_0
\end{bmatrix}
\quad \text{for}
\begin{array}{ll}
\xi & \text{on } \Gamma^{e0} \\
\xi & \text{on } \Gamma^{ei} \\
\xi & \text{on } \Gamma^{fi} \\
\xi & \text{on } \Gamma^{f0} .
\end{array}
\tag{6.16}
$$

The importance of the consideration of the dynamic fluid-structure interaction effects becomes obvious when one studies, for example, the dependence of the total hydrodynamic force (acting on the upstream face of a dam, see fig. 6.8) on the material of the reservoir bottom.

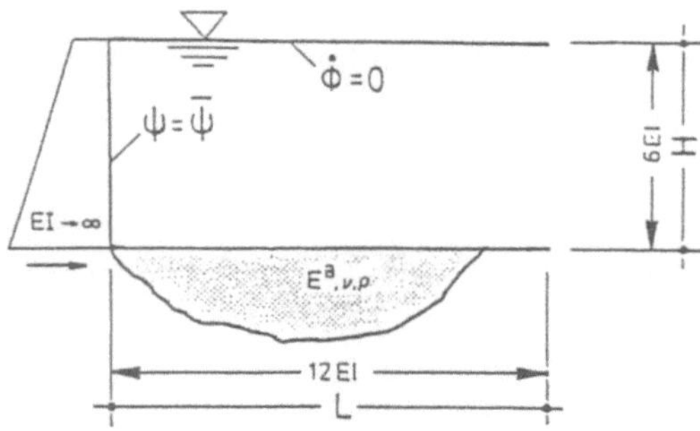

Fig. 6.8: Geometry and discretization of an infinite reservoir

Let us give an application. Consider a 2-D model of a dam of height $H = 100$ m storing a reservoir of the same depth. The velocity of sound in water is taken as 1438 m/s. Let this dam now be subjected to a horizontal ground motion. In the time domain this motion is taken as a unit impulse (here approximated by an impulse $1/\Delta t$ of duration Δt). The soil of the reservoir bottom is taken to be elastic, a plane strain assumption is adopted, and the mass density ρ and Poisson's ratio ν have the following values: $\rho = 2000 \ kg/m^3$ and $\nu = 0.4$. In the following parametric study the value of Young's modulus E varies. First, the case of a rigid dam storing an 'infinite' reservoir is examined (fig. 6.8). In order to show the influence of the coupling, the bottom of the reservoir is taken to be both rigid and elastic, and in the second case two different Young's moduli are considered. The horizontal impulse $1/\Delta t$ acts during the first time step of $\Delta t = 0.0348$ secs. Fig. 6.9 presents the difference in the time history of the total hydodynamic force (relative to the total static load F_{stat}) resulting from the interaction of the liquid with different soil media.

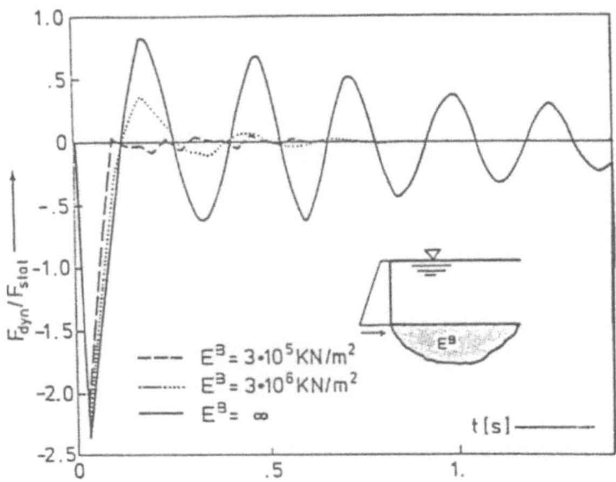

Fig. 6.9: Time history of relative total hydrodynamic force on the dam: infinite reservoir with horizontal elastic and rigid beds

Second, a rigid dam is considered under a horizontal impulse $1/\Delta t$ of duration $\Delta t = 0.0232$ secs. The dam stores a finite reservoir of length $L = 200$ m. Both the effect of an interaction with an elastic bottom (fig. 6.10) and with an elastic far end (fig. 6.11) are examined.

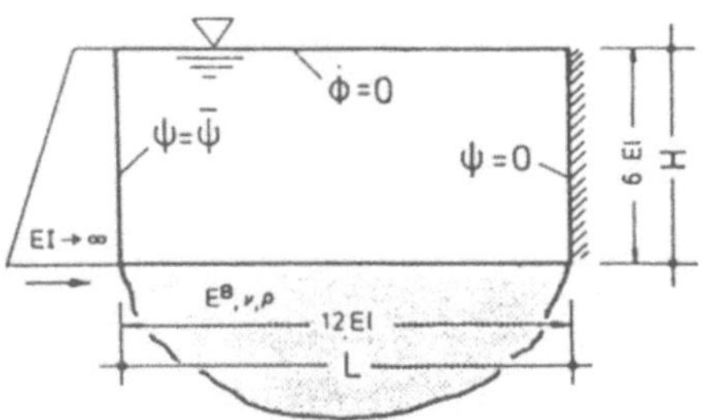

Fig. 6.10: Geometry and discretization of a finite reservoir with elastic bed

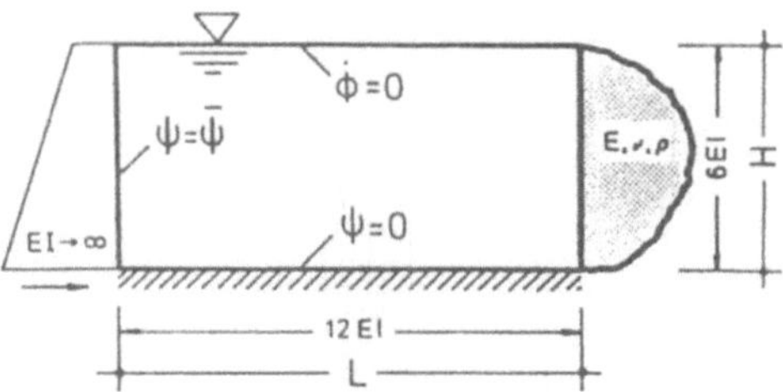

Fig. 6.11: Geometry and discretization of a finite reservoir with elastic far end

As in the previous case of the infinite reservoir, the damping effect due to the dynamic interaction increases with a decreasing Young's modulus (fig. 6.12). Similar behaviour can be observed in fig. 6.13 which gives the comparison of the time history of the relative total hydrodynamic force for a finite reservoir with rigid walls to that for a reservoir with an elastic ($E^B = 3 \times 10^5 KN/m^2$) far end to the right. The size Δt of the time steps has been choosen such that the p-waves travel a distance equal to β-times the length of a boundary element. The factor β can be any real number ranging from 0.5 to 2.0 for the waves in the fluid, whereas the computations of wave propagation in the soil are numerically more sensitive so that there the upper limit for β is approximately 1.2. In the case of the finite reservoir with elastic far end (fig. 6.11) the factors $\beta_f = 2.0$ and $\beta_s = 0.788$ have been chosen for the fluid and the soil domain, respectively.

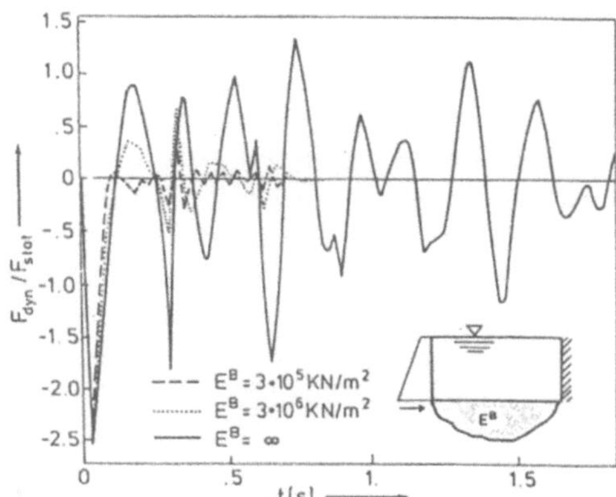

Fig. 6.12: Time history of the relative total hydrodynamic force on the dam: finite reservoir with rigid and elastic beds

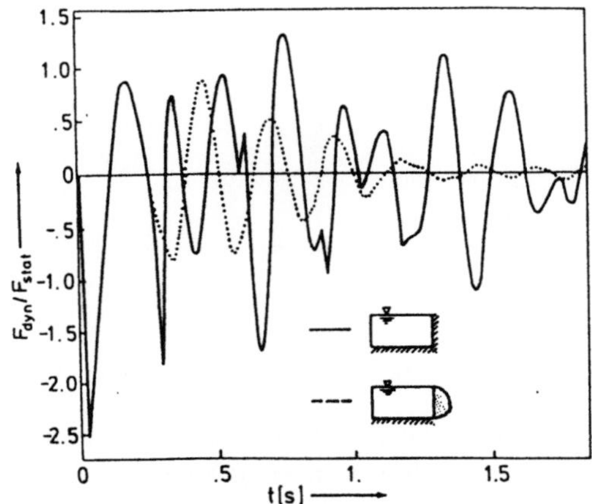

Fig. 6.13: Time history of the relative total hydrodynamic force on the dam: finite reservoir with rigid and elastic far end to the right

6.3 Unilateral Contact Problems

6.3.1 Dynamical Problems. The Trial and Error Method

The analysis of elastic bodies in unilateral contact is a common problem for engineers. Structural elements which are in contact or may come in contact with other parts of the structure or other structures, are generally subjected to quasi-static loadings, i.e. to progressively applied loads, or to transient dynamic excitations, or to incident waves. In any case, the study of the load transfer is important: either because high peak stresses and strains can arise at the contact locations and the associated large fatigue can remarkably shorten the structural lifetime, or because partial uplift and sliding can occur for small or large time intervals and this behavior can be responsible for the good (resp. bad) performance of a seemingly unstable (resp. stable) structure. Obviously, the B.E.M is particularly well suited for solving contact problems since the variables of interest of both the problem and the solution procedure are on the boundaries. Moreover, the boundary tractions are calculated with the same accuracy as the boundary displacements, and the normal and tangential tractions can appear in a coupled form directly in the equation system. Anderson et al. [And80] were among the first who obtained numerical results by means of the direct B.E.M for two-dimensional frictionless static problems with debonding by using constant elements. Later they extended their formulation and they applied it to the frictional contact [And81] problem with linear and parabolic elements [And83]. Also, discontinuous linear elements have been used to model typical friction contact problems [Par], and, moreover, a non-classical friction law has already been employed [Jin]. The common characteristic of all these efforts is the "trial and error" procedure. Indeed since the contact and debonding regions are not a priori known, as well as the adhesive friction and the sliding friction regions, a trial and error procedure was applied in order to determine the arising free boundaries between the aforementioned regions. Note that the same procedure was applied to solve several mechanical problems in elastostatics (e.g. the effects of inclusions embedded in an elastic medium [Sel] or the contact between plates [Bez84]), in thermoelastic analysis and in several other engineering problems. Note that these trial and error methods can lead to erroneous results (cf. e.g. [Pan80,Pan85,Tal,Bis86] and that certain theoretical approaches to the correct formulation of the inequality problem as a multivalued B.I.E. can be found in [Laz] and [Pan87] concerning the numerical approach. The second author developed for both the Direct and the Indirect B.E.M. [Pa85, Pan85] for unilateral contact and friction problems [Pa87b] suitable multivalued B.I.Es or equivalently boundary variational "principles" that allow to use very effective quadratic programming algorithms for the numerical treatment. Analogous boundary variational formulations have been derived independently by Tralli et al. [Tra89, Tra90] by means of standard direct B.E. procedures. In the next Sections we shall study these new formulations. In this Section and in the next one, we give some numerical results concerning the numerical study of dynamic contact problems by the B.E.M. combined with the aforementioned trial end error methods for the determination of the free boundary between contact and noncontact regions. Although most of the real contact problems are time history dependent, adequate, even trial and error, B.E. formulations acting directly in the time domain are rare. Besides some investigations pertinent to foundations (for a survey see [Spy]) and other frictionless uplifting problems ([An88d, An89c]), only recently, almost at the same time, but independently Dominguez et al. [Dom] and Antes et al. [An90d, An90e] developed iterative time-stepping direct B.E. techniques to solve the dynamic frictional contact problem, i.e. to determine which part of an interface remains sticking, or slides, looses the contact by uplifting, or returns to sliding etc.

Two homogeneous and linearly elastic structures occupying the domains A and B in their undeformed state may have contact along the interface Γ_I. A convenient boundary element mesh is defined on the boundary Γ^A of A and Γ^B of B, where each pair of boundary nodes of the interface zone is initially at the same geometric location. Besides, the following assumptions are made:

1. The gravity effect, i.e. the dead-weight of a structure is taken into account, while that of the supporting soil is ignored.

2. The frictional forces acting at the contact interface follow the Coulomb type criterion for dry friction, i.e. during sliding the Coulomb coefficient μ determines the ratio between the tangential shear and normal compressive stress along the interface.

At the contact region between the two bodies forces are transmitted from one body to the other by means of normal compressive stresses, and shear stress if friction occurs. Frictional contact may be of either the sticking (adhesive) or sliding type, depending on whether or not relative tangential displacements occur. Indeed, a particular contact region may contain zones of each type, the extents of which are not known in advance and, moreover, which are time-dependent in the case of dynamic loadings. Contact boundary conditions can be expressed in terms of the displacements u and tractions T, referred to a local coordinate system consisting of the outward normal vector and the unit vectors corresponding to the tangential directions relative to the contact surface. Let us indicate the normal direction by the subscript n and the two tangential directions by subscript s, respectively. Also, indices A and B are used to distinguish between corresponding values of nodal points (a nodepair) on the boundaries of the two contacting bodies. For a prescribed static loading on the system, a node pair may have one of the following four modes of behaviour. In a dynamic loading as it is the case in our problem a node pair may enter from one behaviour mode to the other.

(i) Stick Mode:

   As long as no tensile stresses occur and the tangential stresses do not reach the Coulomb limit, traction equilibrium and displacement compatibility require, as in the bilateral contact, i.e. the contact without debonding $(n^I = n^A = -n^B, s^I = s^A = -s^B)$, that

$$T_s^A = T_s^B = T_s^I \quad \text{or} \quad T_s^A - T_s^B = 0 \tag{6.17}$$

$$T_n^A = T_n^B = T_n^I \quad \text{or} \quad T_n^A - T_n^B = 0 \tag{6.18}$$

$$u_s^A = -u_s^B = u_s^I \quad \text{or} \quad u_s^A + u_s^B = 0 \tag{6.19}$$

$$u_n^A = -u_n^B = u_n^I \quad \text{or} \quad u_n^A + u_n^B = 0 \tag{6.20}$$

(ii) Coulomb Slip Mode:

   If no tensile normal stresses occur, but the level of tangential stress reaches the limit prescribed by Coulomb's law (the dependency of the friction coefficient μ from the total effective slip [And81] will be neglected), slipping will occur in the tangential direction, i.e. relative tangential displacements $(u_s^A - u_s^B) \neq 0$ are permitted; by the condition of energy dissipation, the direction of the "friction force" T_s^I has to be opposite to the direction of the relative movements (". " denotes the time derivative):

$$|T_s^I| = \mu|T_n^I| \quad \text{with} \quad T_s^I(\dot{u}_s^A + \dot{u}_s^B) < 0 \tag{6.21}$$

In the normal direction, equilibrium and compatibility still require:

$$T_n^A = T_n^B = T_n^I \quad (<0) \tag{6.22}$$

$$u_n^A = -u_n^B = u_n^I. \tag{6.23}$$

Note that instead of the (6.21) we may write

$$\text{If} \quad |T_s^I| = \mu|T_n^I| \quad \text{there exist} \quad \lambda \geq 0 \quad \text{such that} \quad T_s^I = -\lambda(\dot{u}_s^A + \dot{u}_s^B). \tag{6.21a}$$

(iii) Separation Mode:

   If tensile normal stresses occur in some region of the interface, such parts debond and a gap opens. There, both boundaries are free of stresses, i.e.

$$T_n^A = T_s^A = 0 \tag{6.24}$$

$$T_n^B = T_s^B = 0, \tag{6.25}$$

and they can move independently, but under certain constraints, i.e. their relative displacements have to be compatible. Neglecting relative rotations, this gap is given by

$$\Delta u_n^I := -u_n^A - u_n^B > 0. \tag{6.26}$$

If an opened gap Δu_n^I tends to zero again, either sliding or sticking of the two contacting structures can take place, i.e either the Coulomb slip mode conditions (ii) may occur again, or the conditions of the socalled rebonding.

(iiii) Rebonding mode:

If separation or sliding contact returns to stick contact, for instance at the n-th time step, the condition of tangential compatibility (6.19) must be changed to the condition

$${}^{(n)}u_s^A + {}^{(n)}u_s^B \,\hat{=}\, {}^{(n)}g_s = {}^{(n-1)}g_s = {}^{(n-1)}u_s^A + {}^{(n-1)}u_s^B, \tag{6.27}$$

where ${}^{(n-1)}g_s$ is the total tangential displacement at time $(n-1)\Delta t$, i.e. the total effective slip which has occured during the separation and the sliding process untill the node returns to a nonseparating and nonsliding contact. By means of the "new" tangential displacements

$${}^{(n)}u_{sr}^A \,\hat{=}\, {}^{(n)}u_s^A - \alpha\, {}^{(n)}g_s \quad \text{and} \quad {}^{(n)}u_{sr}^B \,\hat{=}\, {}^{(n)}u_s^B - (1-\alpha)\, {}^{(n)}g_s \tag{6.28}$$

we may write the generalized tangential compatibility condition in the form

$${}^{(n)}u_{sr}^A + {}^{(n)}u_{sr}^B = 0 = {}^{(n)}u_s^A + {}^{(n)}u_s^B - {}^{(n)}g_s . \tag{6.29}$$

This condition replaces the condition (6.19). All other conditions of the stick mode, i.e. (6.17), (6.18), and (6.20), hold in the same form as before.

The dynamic interaction analysis of elastic structures, e.g. Ω^A and Ω^B in unilateral contact with friction, requires the generation (as described in Sect. 5.3.1) of a system of boundary element equations of the form

$$\left[0.5\boldsymbol{I} + {}^{(1)}\boldsymbol{T}\right] \cdot {}^{(n)}\boldsymbol{u} + \sum_{m=1}^{n-1} {}^{(n-m-1)}\boldsymbol{T} \cdot {}^{(m)}\boldsymbol{u} = \sum_{m=1}^{n} {}^{(n-m-1)}\boldsymbol{U} \cdot {}^{(m)}\boldsymbol{t} + {}^{(n)}\boldsymbol{r} \tag{6.30}$$

for each structure. Then we have to combine these systems of algebraic equations by imposing the adequate interface conditions. In (6.30), the vectors ${}^{(m)}\boldsymbol{u}$ and ${}^{(m)}\boldsymbol{t}$ contain all nodal displacements and tractions, respectively, of time step m, while the vector ${}^{(n)}\boldsymbol{r}$ gives the influence of non-zero body forces. Thus, for any coupling, the algebraic equations contain within each time step n boundary values of the heavy, elastic structure Ω^A, e.g. an elastic massive strip foundation, and of the domain Ω^B, e.g. the supporting soil. The eqs. (6.30) have to be subdivided and rearranged according to the position of the collocation and observation points at the contact and "outer" boundary, respectively. For instance, coupling with the stick contact conditions, and then rearranging according to the unknown values, the prescribed ones, and the values already known or found in the previous time steps, yield the following system of equations [An88d, An90d]

$$
\overset{(1)}{[Q]} \cdot
\begin{Bmatrix}
\overset{(n)}{u^A_o} \\
\overset{(n)}{u^A_c} \\
\overset{(n)}{t^A_c} \\
\overset{(n)}{u^B_o}
\end{Bmatrix}
= \overset{(1)}{[P]} \cdot \{ \overset{(n)}{\bar{t}^A_o} \} +
\begin{Bmatrix}
\overset{(n)}{R^A_o} \\
\overset{(n)}{R^A_c} \\
\overset{(n)}{R^B_c} \\
\overset{(n)}{R^B_o}
\end{Bmatrix} ,
\tag{6.31}
$$

where the matrices $\overset{(1)}{Q}$ and $\overset{(1)}{P}$ are explicitly given by the formula (we use the notation $\overset{\star}{T} = \tfrac{1}{2} I + T$)

$$
\overset{(1)}{Q} \;\hat{=}\;
\begin{bmatrix}
\overset{(1)}{\overset{\star}{T}}{}^A_{oo} & \overset{(1)}{T}{}^A_{oc} & -\overset{(1)}{U}{}^A_{oc} & 0 \\
\overset{(1)}{T}{}^A_{co} & \overset{(1)}{\overset{\star}{T}}{}^A_{cc} & -\overset{(1)}{U}{}^A_{cc} & 0 \\
0 & -\overset{(1)}{\overset{\star}{T}}{}^B_{cc} & -\overset{(1)}{U}{}^B_{cc} & \overset{(1)}{T}{}^B_{co} \\
0 & -\overset{(1)}{T}{}^B_{oc} & -\overset{(1)}{U}{}^B_{oc} & \overset{(1)}{\overset{\star}{T}}{}^B_{oo}
\end{bmatrix} ,
\qquad
\overset{(1)}{P} \;\hat{=}\;
\begin{bmatrix}
\overset{(1)}{U}{}^A_{oo} \\
\overset{(1)}{U}{}^A_{co} \\
0 \\
0
\end{bmatrix} .
\tag{6.32}
$$

In these matrices the first subscript (resp. second) shows, where the collocation (resp. the nodal) point is located and "o" denotes the "outer" boundary whereas "c" denotes the contact boundary. The two central block-rows of the matrix $\overset{(1)}{Q}$ in (6.32) are denoted by $\overset{(1)}{C}$ in the sequel. The terms $\overset{(n)}{R^A}$ and $\overset{(n)}{R^B}$ represent the influence of all the previous time steps ($m < n$) and, of the dead weight of a "heavy" structure if this weight is taken into account. For instance

$$
\{ \overset{(n)}{R^A_c} \} = \{ \overset{(n)}{R^A_c}(\overset{(m)}{t}{}^A, \overset{(m)}{u}{}^A)\} = \sum_{m=1}^{n-1} [\; \overset{(n-m+1)}{U^A_c} \cdot \overset{(m)}{t}{}^A - \overset{(n-m+1)}{T^A_c} \cdot \overset{(m)}{u}{}^A] + \overset{(n)}{r}{}^A
\tag{6.33}
$$

In the formulation of the eq. (6.31), it has been assumed that the non-contacting surface of the soil Ω^B is free of stresses ($t^B_o = 0$), and that the dynamic loadings t^A_o are prescribed along the outer boundary of the structure Ω^A.

Stepping from time $(n-1)\Delta t$ to time $n\Delta t$ is achieved only by changing the current prescribed boundary values, and by updating the terms $\{ \overset{(n)}{R^A} \}$ and $\{ \overset{(n)}{R^B} \}$. Moreover, in order to verify the sliding and the debonding on certain interface elements, nothing but the part $\overset{(1)}{C}$ of the matrix $\overset{(1)}{Q}$ has to be changed, while for the rebonding mode the relative slips have to be taken into account additionally in the right hand side of eq. (6.31). When within a certain time step n, sliding takes place on some interface elements, the tangential and the normal components of those nodal values behave in a different way: while in the normal direction equilibrium and compatibility are still preserved, tangentially relative displacements are permitted and the "friction force" is determined by the normal compressive stress via the relation (6.21). To find the right sign of the friction force for each sliding node, it is first assumed, that its direction is the same as under sticking condition just before:

$$\overset{(n)I}{T_s} = \operatorname{sign}\left(\overset{(n-1)I\text{stick}}{T}_s\right)\mu\, \overset{(n)I}{T_n} .\tag{6.34}$$

But in some cases the direction may change. Therefore one must check whether the relation (6.21a) is satisfied, i.e.

$$\operatorname{sign}(\overset{(n)I}{T_s}) = -\operatorname{sign}(\overset{(n)A}{u}_n + \overset{(n)B}{u}_n) \cong \operatorname{sign}(\overset{(n-1)A}{u_s} + \overset{(n-1)B}{u_s}) - \operatorname{sign}(\overset{(n)A}{u}_s + \overset{(n)B}{u}_s).\tag{6.35}$$

If sliding does not occur, sticking is again assumed in this time step for the element under consideration, and the iteration cycle starts again. When separation takes place – the debonding criterion is that "the normal traction in certain interface elements becomes tensile" – both traction components must disappear in the debonded elements and both faces move separately. Such debonded elements are subjected to the debonding constraint $\Delta u_n \geq 0$, which is numerically approximated by $|\Delta u_n| \leq |\epsilon_n|$, where ϵ_n represents a small "admissible" incompatibility, i.e. either a tiny interpenetration or a little gap, at the interface may occur as a result of the trial and error procedure and the approximation error. Moreover the criterion of the debonding which we have put in quotation marks is responsible for an erroneous equilibrium position, if the time step is not small enough. (cf. e.g. [Pan80, Pan85, Tal, Bi89]). All the aforementioned disadvantages of the trial end error method as well as the law level of automatization of the calculations are avoided in the quadratic programming approach to the unilateral contact problem with friction which will be developed in the next Chapters. Thus, the last types of contact modes cause typical changes in the contact related part $\overset{(1)}{C}$ of the matrix $\overset{(1)}{Q}$ (see definition (6.32)). We have for sliding elements

$$\overset{(1)}{C} =
\begin{bmatrix}
\overset{(1)}{T}{}^A_{noc} & \overset{(1)}{T}{}^A_{soc} & 0 & -\overset{(1)}{U}{}^A_{noc} + \mu\operatorname{sign}(\overset{(n)}{T}{}^I_s)\,\overset{(1)}{U}{}^A_{soc} \\[4pt]
\overset{(1)}{\overset{\star}{T}}{}^A_{ncc} & \overset{(1)}{\overset{\star}{T}}{}^A_{scc} & 0 & -\overset{(1)}{U}{}^A_{ncc} + \mu\operatorname{sign}(\overset{(n)}{T}{}^I_s)\,\overset{(1)}{U}{}^A_{scc} \\[4pt]
-\overset{(1)}{\overset{\star}{T}}{}^B_{ncc} & 0 & \overset{(1)}{\overset{\star}{T}}{}^B_{scc} & -\overset{(1)}{U}{}^B_{ncc} + \mu\operatorname{sign}(\overset{(n)}{T}{}^I_s)\,\overset{(1)}{U}{}^B_{scc} \\[4pt]
-\overset{(1)}{T}{}^B_{noc} & 0 & \overset{(1)}{T}{}^B_{soc} & -\overset{(1)}{U}{}^B_{noc} + \mu\operatorname{sign}(\overset{(n)}{T}{}^I_s)\,\overset{(1)}{U}{}^B_{soc}
\end{bmatrix},\tag{6.36}$$

for debonded elements

$$\overset{(1)}{C} =
\begin{bmatrix}
\overset{(1)}{T}{}^A_{noc} & 0 & \overset{(1)}{T}{}^A_{soc} & 0 \\[4pt]
\overset{(1)}{\overset{\star}{T}}{}^A_{ncc} & 0 & \overset{(1)}{\overset{\star}{T}}{}^A_{scc} & 0 \\[4pt]
0 & \overset{(1)}{\overset{\star}{T}}{}^B_{ncc} & 0 & \overset{(1)}{\overset{\star}{T}}{}^B_{scc} \\[4pt]
0 & \overset{(1)}{T}{}^B_{noc} & 0 & \overset{(1)}{T}{}^B_{soc}
\end{bmatrix},\tag{6.37}$$

for bonded or rebonded elements (see also definition (6.32))

$$
\overset{(1)}{C} =
\begin{bmatrix}
\overset{(1)}{T}{}^A_{noc} & \overset{(1)}{T}{}^A_{soc} & -\overset{(1)}{U}{}^A_{noc} & -\overset{(1)}{U}{}^A_{soc} \\[2mm]
\overset{(1)}{\overset{\star}{T}}{}^A_{ncc} & \overset{(1)}{\overset{\star}{T}}{}^A_{scc} & -\overset{(1)}{U}{}^A_{ncc} & -\overset{(1)}{U}{}^A_{scc} \\[2mm]
-\overset{(1)}{\overset{\star}{T}}{}^B_{ncc} & -\overset{(1)}{\overset{\star}{T}}{}^B_{scc} & -\overset{(1)}{U}{}^B_{ncc} & -\overset{(1)}{U}{}^B_{scc} \\[2mm]
-\overset{(1)}{T}{}^B_{noc} & -\overset{(1)}{T}{}^B_{soc} & -\overset{(1)}{U}{}^B_{noc} & -\overset{(1)}{U}{}^B_{soc}
\end{bmatrix},
\tag{6.38}
$$

where the respective unknown node values are

$$
\left[\ \overset{(n)}{u}{}^A_n,\ \overset{(n)}{u}{}^A_s,\ \overset{(n)}{u}{}^B_s,\ \overset{(n)}{t}{}^A_n\ \right] \quad \text{in the sliding elements,}
\tag{6.39}
$$

$$
\left[\ \overset{(n)}{u}{}^A_n,\ \overset{(n)}{u}{}^B_n,\ \overset{(n)}{u}{}^A_s,\ \overset{(n)}{u}{}^B_s\ \right] \quad \text{in the debonded elements,}
\tag{6.40}
$$

$$
\left[\ \overset{(n)}{u}{}^A_n,\ \overset{(n)}{u}{}^A_{sr},\ \overset{(n)}{t}{}^A_n,\ \overset{(n)}{t}{}^A_s\ \right] \quad \text{in the debonded elements.}
\tag{6.41}
$$

If an interface node returns from sliding or separation to the stick mode, the relative slip $\{\overset{(n)}{g}\}$ of the interface elements has to be taken into account, i.e. it has to be added at the right hand side of eq. (6.31) :

$$
\overset{(1)}{[Q]} \cdot
\left\{
\begin{array}{c}
\overset{(n)}{u}{}^A_o \\[1mm]
\overset{(n)}{u}{}^A_c \\[1mm]
\overset{(n)}{t}{}^A_c \\[1mm]
\overset{(n)}{u}{}^B_o
\end{array}
\right\}
= \overset{(1)}{[P]} \cdot \{\overset{(n)}{\bar{t}}{}^A_o\} -
\begin{bmatrix}
\alpha\,\overset{(1)}{T}{}^A_{soc} \\[2mm]
\alpha\,\overset{(1)}{\overset{\star}{T}}{}^A_{scc} \\[2mm]
(1-\alpha)\,\overset{(1)}{\overset{\star}{T}}_{scc} \\[2mm]
(1-\alpha)\,\overset{(1)}{T}{}^B_{scc}
\end{bmatrix}
\cdot \{\overset{(n)}{g}{}_s\} +
\left\{
\begin{array}{c}
\overset{(n)}{R}{}^A_o \\[1mm]
\overset{(n)}{R}{}^A_c \\[1mm]
\overset{(n)}{R}{}^B_c \\[1mm]
\overset{(n)}{R}{}^B_o
\end{array}
\right\} .
\tag{6.42}
$$

The value of the factor α can be chosen arbitrarily in the range 0 to 1. This statement (see also [And81]) has been checked in running some parametric studies, e.g. a heavy surface foundation. Trial values $\alpha = 0.0, 0.5$, and 1.0 gave almost identical interface displacements and tractions. The examples in the next Section are calculated with $\alpha = 0.0$, i.e. $u^A_{sr} = u^A_s$ and $u^B_{sr} = u^B_s - g_s$. Finally, after the matrix $\overset{(1)}{Q}$ is appropriately modified according to the current situation along the interface, the eqs. (6.31) or (6.42) must be solved again in order to provide the unknowns of the new, i.e. partially slidden, debonded or rebonded state of the interface.

6.3.2 Examples: Elastic Massive Foundations on Elastic Soil

Here the results of two studies are presented in order to demonstrate the applicability of the time-stepping trial and error B.E.M procedure, and to show the importance of considering unilateral frictional contact along foundation-soil interfaces.

First, let us consider a rectangular shaped, massive, and elastic strip foundation (see fig. 6.14) of height $H = 1.0$ m and width $2B = 4.0$ m resting on the free surface of a supporting homogeneous soil medium. Adopting plane strain assumptions, the soil material is assumed to have a mass density $\rho_s = 2000 \ kg/m^3$, Poisson's ratio $\nu_s = 0.33$, and Young's modulus $E_s = 4.788 \cdot 10^6 \ KN/m^2$, while the foundation material is considered to have larger stiffness with a mass density $\rho_f = 2500 \ kg/m^3$, Poisson's ratio $\nu_f = 0.20$, and Young's modulus $E_f = 1.968 \cdot 10^7 \ KN/m^2$. Hence, the velocities of the pressure wave and the shear wave are $c_{1s} = 1833.4 \ m/s$ and $c_{2s} = 948.7 \ m/s$ in the soil medium, and $c_{1f} = 2957.5 \ m/s$ and $c_{2f} = 1811.1 \ m/s$ in the foundation, respectively.

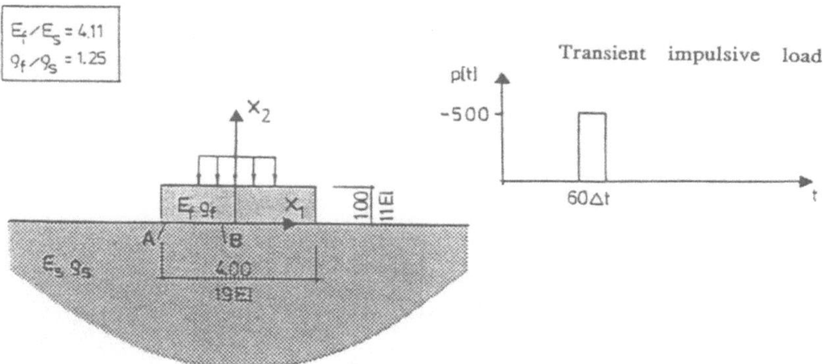

Fig. 6.14: Geometry, discretization, and excitations of a flexible massive strip foundation resting on the elastic halfspace

The foundation boundary is discretized by using 11 elements on both free side faces, 13 elements on the top surface, and 19 elements along the interface, i.e. by 54 elements in the whole. The length of these elements varies; it is chosen to decrease towards the expected stress concentration points, e.g. along the interface from 0.45 m for both the "central" elements to 0.01 m for the "corner" elements. In addition to the 19 interface elements, the free soil surface discretization with 11 unequal elements on each side (the first, next the foundation, is 0.055 m, the last is 3.0 m long) extends to a distance of 10.0 m. In order to evaluate the foundation dead weight integral numerically using Gaussian quadrature formula, the foundation domain has been subdivided into interior cells. The time increment used is $\Delta t = 2.258 \cdot 10^{-4}$ secs; the whole observation period is $177\Delta t = 0.04$ secs.

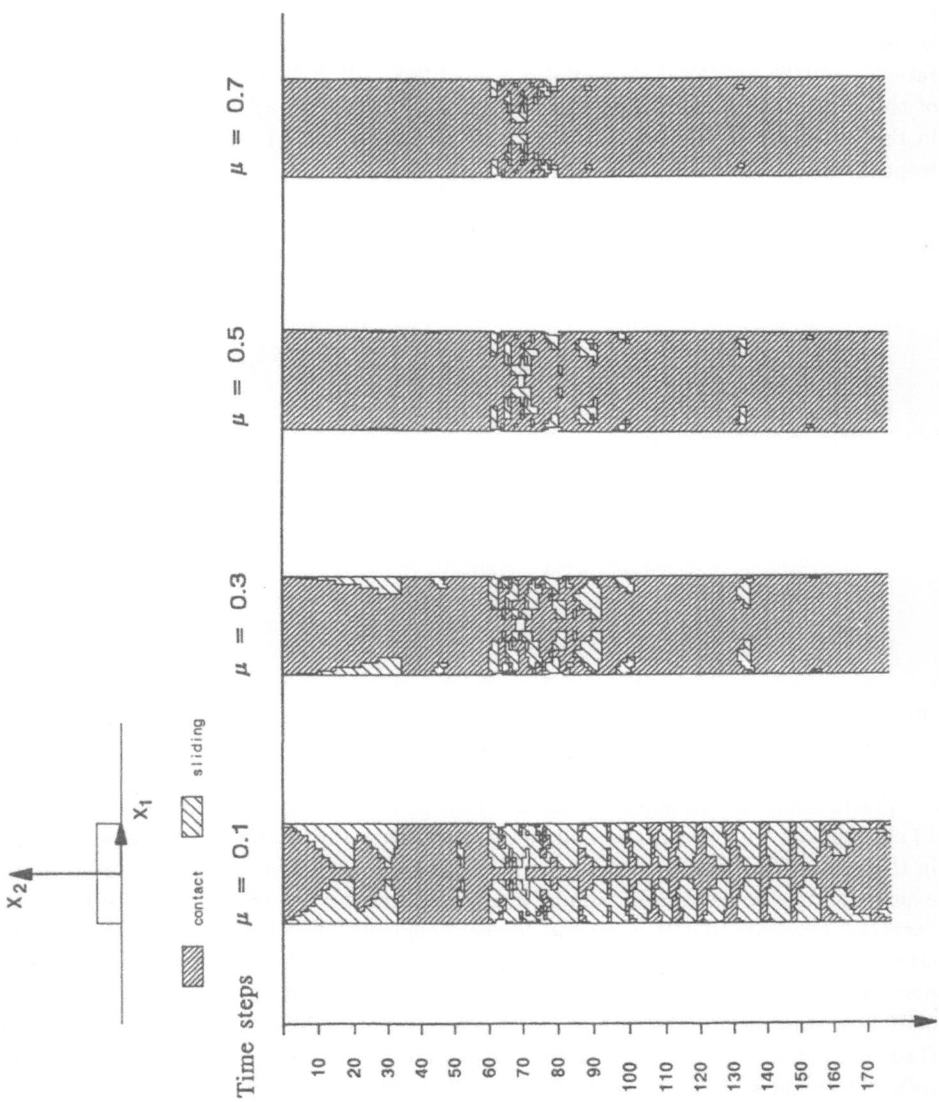

Fig. 6.15: Time-dependent distribution of the bonded, the sliding, and the uplifted interface areas: influence of the Coulomb friction coefficient

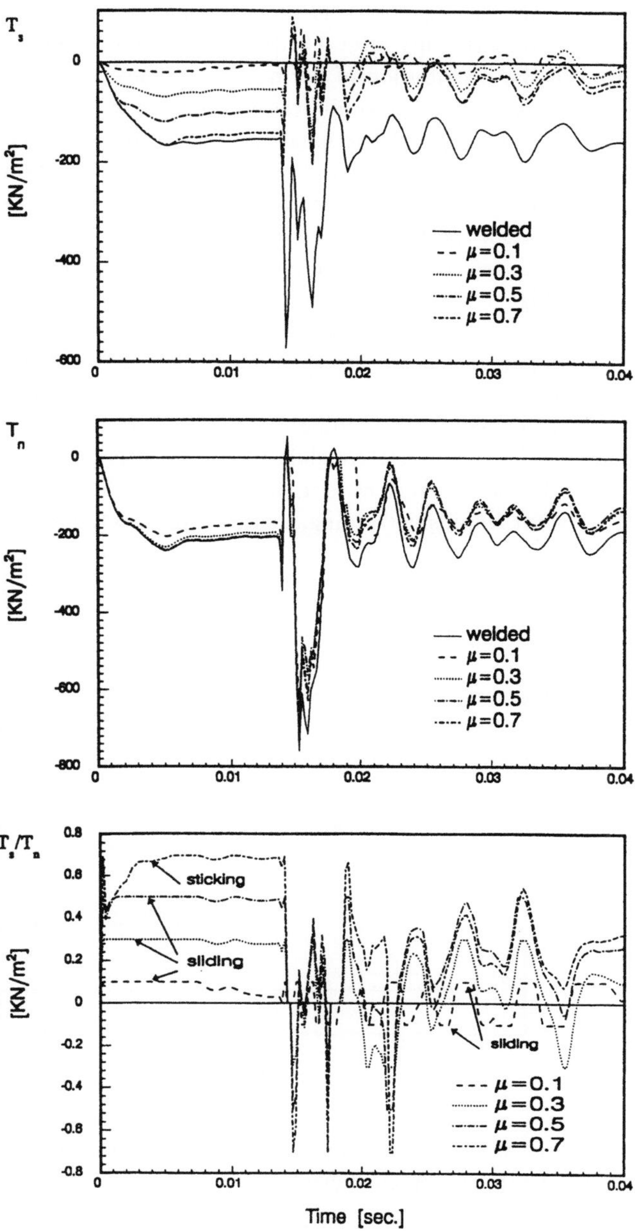

Fig. 6.16: Time history of the tangential traction, the normal traction, and their ratio at the point A: influence of different Coulomb friction coefficients μ. ("welded" means no debonding and no sliding)

This foundation is subjected to a transient vertical impulse with a duration of only two time steps, i.e. of $5.516 \cdot 10^{-4}$ secs, acting along 2 meters on the central part of the foundation's upper surface (see fig. 6.14). The impulse action with a constant intensity $p(t) = p_0 = -500$ KN/m starts only after 0.01348 secs $(= 60\Delta t)$. This time is needed in order that to reach the distribution of normal compressive stress resultants along the interface elements reaches a value equilibrating the dead weight of the foundation. Now, the implications of raising the frictional coefficient are analyzed. First, in fig. 6.15, the reduction of the sliding areas by taking higher Coulomb friction coefficients is very impressive. In fig. 6.16 there is depicted the time history of the tangential traction, and, also, the ratio between the tangential and the normal tractions at the point A, 0.005 m from the left interface corner of the foundation. This figure shows clearly the expected effects: the higher the frictional resistance, the more the "oscillations" are damped. Thus, only for the small Coulomb friction coefficient $\mu = 0.1$ both the friction limits indicating sliding are reached more than once. Finally, the time history of the relative slip (relative tangential displacement) $g_s = u_s^f + u_s^s$ at the point A is plotted in fig. 6.17. Obviously, as expected, the sliding due to the dead weight (the first 0.0135 secs$= 60\Delta t$) produces the larger relative shift (constant after some time steps), the smaller the Coulomb friction coefficient is assumed to be. The subsequent transient excitation, i.e. the short impulsive loading, causes a different, partially unexpected behavior.

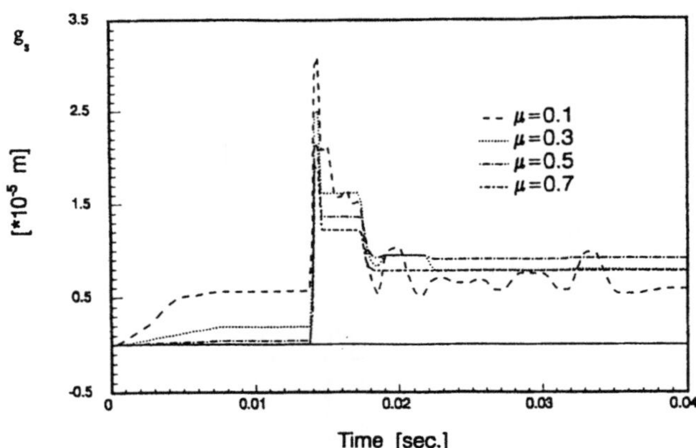

Fig. 6.17: Time history of the relative slip of the foundation at the point A: influence of the Coulomb friction coefficient μ

In the case of the small friction coefficient $\mu = 0.1$, the frequent changes from bonded to sliding contact produce also frequently alternating slips which finally tend towards the "static" dead weight slip. The higher the frictional resistance, the lower is the number of slip changes after the impulse action. In the case $\mu = 0.7$, for instance, only two different constant slip levels are reached. The final slip of the ($\mu = 0.3$)-case is smaller than that of the ($\mu = 0.5$)-solution, and, moreover, identical with the final value related to $\mu = 0.7$. The ($\mu = 0.3$)-solution reaches this rather unexpected value by the "last" sliding which does not happen in the higher frictional resistance case $\mu = 0.5$. This may be the reason for this ending with a smaller slip level, i.e. for this more "relaxed" ending.

As a second application we study the influence of incident waves on a flexible massive strip foundation of height $H = 1.0m$ and width $2B = 6.25m$. The foundation is resting on the free surface of a supporting homogeneous soil medium with mass density $\rho = 2000kg/m^3$ and Poisson's ratio $\nu = 0.25$. Under conditions of plane strain the velocities of the p-wave and s-wave are $c_1 = 346.4m/s$ and $c_2 = 200.0m/s$. The foundation boundary is discretized by 40 elements of equal length 0.39 m. In addition to the 16 interface elements, the soil is modelled by 8 elements at each of the free soil areas (see fig. 6.18).

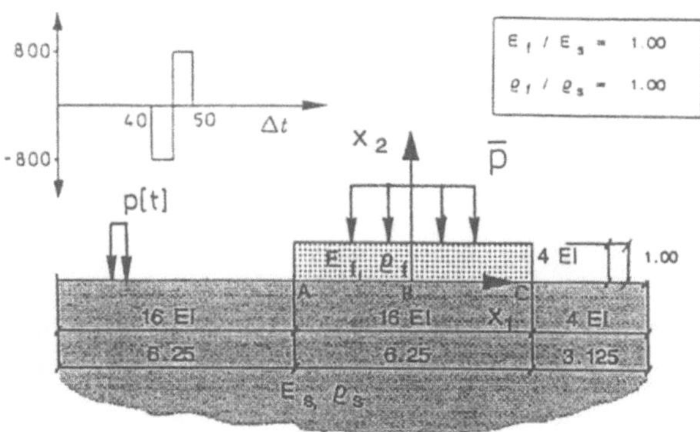

Fig. 6.18: Geometry and discretization of a flexible foundation on elastic halfspace, and location of the transient surface impulses

Along eight central elements the top of the foundation is considered to be subjected to a vertical impulse of constant intensity $\bar{p} = -256.0KN/m$. Additionally, this foundation will be excited by incident waves: after 0.05 secs ($= 40\Delta t$) two successive transient rectangular impulses start acting on a strip of $0.39m$(one element) width in a distance of about $7.60m$ from the foundation center (see fig. 6.18), the first compressive with an intensity of $800KN/m$ and duration of $5\Delta t$, the second tensile with equal intensity and duration.

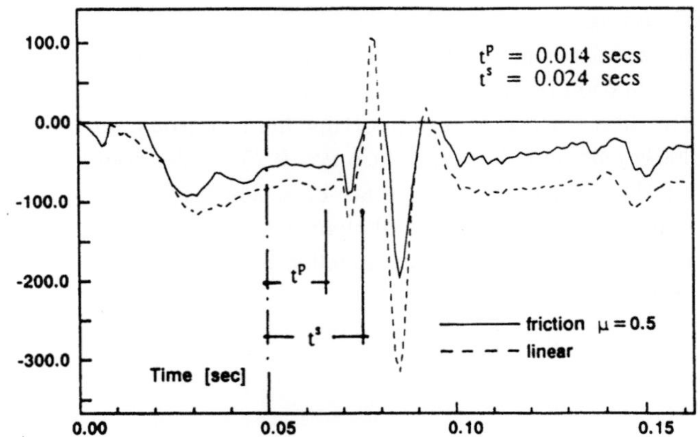

Fig. 6.19: Time history of normal traction at point A:
Influence of the incident wave ("linear" means without debonding and sliding, i.e. "clamped")

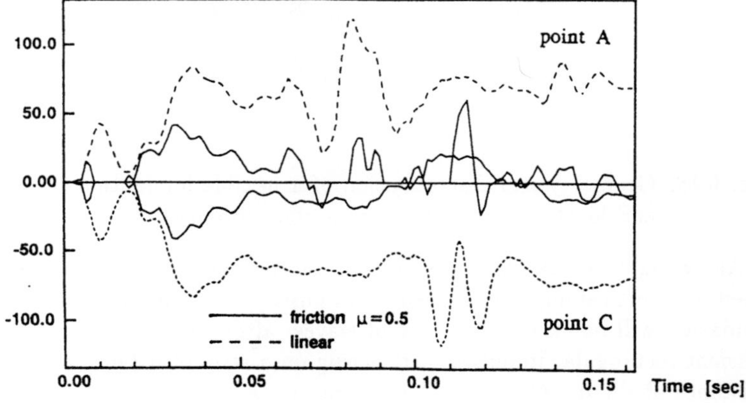

Fig. 6.20: Time history of shear traction at the points A and C:
Influence of the incident wave and frictional sliding

Besides the response to the "suddenly acting" dead load and the "static" loading $\bar{p}$, the plottings of the stresses show clearly the effect of the incident waves. Near the left corner of the foundation, at point A, both the traction components, the normal traction (fig. 6.19) and the shear traction (fig. 6.20) are influenced so much that briefly after the wave's arrival uplifting starts. Moreover, frictional sliding causes relaxation, i.e. the tractions are lower in the case of unilateral frictional contact than in the linear model without debonding and sliding friction; this is especially apparent from the shear traction at point C (fig. 6.20).

Chapter 7
Boundary Integral Formulations for the Signorini-Fichera Inequality Problem

7.1 Primal, Dual and Mixed Formulations of the B.V.P.

In this Chapter we introduce the basic method for the derivation of boundary integral formulations for inequality contact problems. Here the method is illustrated with respect to the Signorini-Fichera problem. In this Chapter we have avoided the mathematical study of the questions of existence of solution etc., which are treated in the next Chapter. Let us notice, in advance, that the procedure followed in all the forthcoming Chapters with respect to linear elastic bodies holds also for beams, plates and shells, or generally speaking for all structures permitting a formulation of the equilibrium problem in terms of Lagrangian functions.

We consider the equilibrium of a three-dimensional linear elastic body, assuming a geometrically linear theory (small deformations).

Let Ω be an open bounded subset of the three-dimensional Euclidean space $\mathbb{R}^3$ with a boundary Γ assumed to be "regular" enough (a Lipschitzian boundary, i.e. $C^{0,1}$-regularity, is sufficient). Ω is occupied by a linear elastic body in its undeformed state and the points $x \in \bar{\Omega}$ are referred to an orthogonal Cartesian coordinate system $Ox_1x_2x_3$. Γ is decomposed into three open mutually disjoint parts Γ_1, Γ_2 and Γ_3. We demand that on Γ_1 (resp. Γ_2) the displacements (resp. the tractions) are given and that on Γ_3 the Signorini boundary conditions (1.80) hold. We denote, as usual, by $n = \{n_i\}$ the outward unit normal vector to Γ and by $S = \{S_i\} = \{\sigma_{ij}n_j\}$ the traction vector on the boundary, where $\sigma = \{\sigma_{ij}\}$ is the stress tensor. Further, S_N (resp. S_T) is the normal (resp. the tangential) component of S with respect to Γ; u_N and u_T are the corresponding components of the displacement u. Let $\varepsilon = \{\varepsilon_{ij}\}$ be the strain tensor and $C = \{C_{ijhk}\}, i, j, h, k = 1, 2, 3$, the Hooke's tensor of elasticity obeying the well-known symmetry and ellipticity conditions (1.151a) and (1.151b).

On Γ_1 we assume that

$$u_i = \bar{U}_i, \qquad \bar{U}_i = \bar{U}_i(x), \tag{7.1}$$

and on Γ_2

$$S_i = \bar{T}_i, \qquad \bar{T}_i = \bar{T}_i(x). \tag{7.2}$$

The Signorini boundary conditions (1.80) on Γ_3 are completed with the condition in the tangential direction

$$S_{T_i} = C_{T_i}, \quad C_{T_i} = C_{T_i}(x) \text{ on } \Gamma_3 \tag{7.3}$$

where C_T is a prescribed tangential force distribution. Instead of the law (1.80) we can also consider the more general case depicted in fig. 1.5b which corresponds to the unilateral contact with a deformable support lying at distance h from the body. Note that the more realistic unilateral contact with a deformable support (e.g. Winkler springs) leads to analogous boundary integral expressions as the ones corresponding to fig. 1.6. Indeed, it is sufficient to enlarge the body Ω by fictitious linear-elastic or nonlinear elastic (monotone) springs of zero length along Γ_3.

Further we have

$$\sigma_{ij,j} + \bar{p}_i = 0 \quad \text{in } \Omega \tag{7.4}$$

$$\varepsilon_{ij} = \varepsilon_{ij}(u) = \frac{1}{2}(u_{i,j} + u_{j,i}) \text{ in } \Omega \tag{7.5}$$

$$\sigma_{ij} = C_{ijhk}\varepsilon_{hk} \quad \text{in } \Omega \tag{7.6}$$

where $\bar{p} = \{\bar{p}_i\}$ representes the volume force vector and the comma as usual denotes the partial derivation.

We denote further by $\tilde{V}$ the vector space of the displacement v_i and let V be a set of the kinematically admissible displacement fields

$$V = \{v|v = \{v_i\}, v_i \in \tilde{V}, v_i = \bar{U}_i, \quad i = 1,2,3, \text{ on } \Gamma_1\}, \tag{7.7}$$

without taking into account the constraints on Γ_3. Because of the Signorini boundary conditions on Γ_3 the kinematically admissible set of the displacements takes the form

$$K = \{v|v = \{v_i\}, \quad v \in V, \quad v_N \leq 0 \text{ on } \Gamma_3\}. \tag{7.8}$$

We denote by $(\bar{p}, v)$ the work done by the force $\bar{p} = \{\bar{p}_i\}$ for the displacement $v = \{v_i\}$ on Ω and by $[\bar{T}, v]_{\tilde{\Gamma}}$ the corresponding work on $\tilde{\Gamma} \subset \Gamma$ (i.e. $\int_{\Omega} \bar{p}_i v_i d\Omega$) etc. Further let

$$a(u,v) = (C\varepsilon(u), \varepsilon(v)) = \int_{\Omega} C_{ijhk}\varepsilon_{ij}(u)\varepsilon_{hk}(v)d\Omega \tag{7.9}$$

be the bilinear form of elasticity and let Π be the potential energy

$$\Pi(v) = \frac{1}{2}\alpha(v,v) - (\bar{p}, v) - [C_T, v_T]_{\Gamma_3} - [\bar{T}, v]_{\Gamma_2}. \tag{7.10}$$

It is well-known [Fich63;64;72, Duv72, Pan85], that the problem

$$\Pi(u) = \min\{\Pi(v)|v \in K\} \tag{7.11}$$

characterizes completely the position of equilibrium i.e. if (7.11) holds we have equilibrium and conversely. Problem (7.11) has one and only one solution for $v_i \in H^1(\Omega)$ $C_{ijhk} \in L^\infty(\Omega)$, $\bar{p}_i \in L^2(\Omega)$, $C_{T_i} \in L^2(\Gamma_3)$, $\bar{T}_i \in L^2(\Gamma_2)$, $\bar{U}_i \in H^{1/2}(\Gamma_1)$, which equivalently, satisfies the variational inequality

$$u \in K, \quad \alpha(u, v - u) - l(v - u) \geq 0 \quad \forall v \in K \tag{7.12}$$

with $l(v) = (\bar{p}, v) + [\bar{T}, v]_{\Gamma_2} + [C_T, v_T]_{\Gamma_3}$. Relations (7.11) or (7.12) are the primal formulations of the B.V.P. of Signorini-Fichera. In order to derive the mixed formulation we introduce the convex subset (which is also closed in the mentioned functional framework as we shall see in the next Chapter)

$$L = \{\mu_N|\mu_N \leq 0 \text{ on } \Gamma_3\} . \tag{7.13}$$

Let u_0 be a kinematically admissible displacement field such that $u_{0i} = \bar{U}_i$ on Γ_1, and let us perform the translation transformation

$$\bar{u} = u - u_0, \quad \bar{v} = v - u_0 \tag{7.14}$$

where

$$\bar{u}, \bar{v} \in V_0 = \{v|v = \{v_i\}, v_i \in \tilde{V}, v_i = 0 \text{ on } \Gamma_1\}. \tag{7.15}$$

Then (7.12) becomes: Find $u = \bar{u} + u_0 \in K$ such that

$$\alpha(\bar{u}, \bar{v} - \bar{u}) - l(\bar{v} - \bar{u}) + a(u_0, \bar{v} - \bar{u}) \geq 0 \quad \forall v = \bar{v} + u_0 \in K. \tag{7.16}$$

and (7.11) reads

$$\tilde{\Pi}(\bar{u}) = \min\{\tilde{\Pi}(\bar{v})|\bar{v} \in \tilde{K}\}, \tag{7.17}$$

where

$$\tilde{\Pi}(\bar{v}) = \frac{1}{2}a(\bar{v}, \bar{v}) - l(\bar{v}) + a(u_0, \bar{v}) = \Pi(v) - \Pi(u_0) \tag{7.18}$$

and

$$\tilde{K} = \{\bar{v}|\bar{v}_N + u_{0N} \leq 0 \text{ on } \Gamma_3\}. \tag{7.19}$$

The displacement field u_0 may represent any other initial strain field (temperature distributions, given dislocations etc). Further we denote by $\bar{l}$ the functional

$$\bar{l}(\bar{v}) = l(\bar{v}) - a(u_0, \bar{v}). \tag{7.20}$$

Through the new variable μ_N which is the Lagrange multiplier of the problem we shall introduce the boundary condition (1.80) on Γ_3: As it is easy to verify

$$\inf_{\tilde{K}} \tilde{\Pi}(\bar{v}) = \inf_{V_0} \left(\tilde{\Pi}(\bar{v}) + I_{\tilde{K}}(\bar{v}) \right)$$

where

$$I_{\tilde{K}}(\bar{v}) = \begin{cases} 0 \text{ if } \bar{v}_N + u_{0N} \leq 0 \text{ on } \Gamma_3 \\ \infty \text{ otherwise.} \end{cases} \tag{7.21}$$

But

$$I_{\tilde{K}}(\bar{v}) = \sup_{\mu_N \leq 0} \left[-[\mu_N, \bar{v}_N + u_{0N}]_{\Gamma_3} \right] \tag{7.22}$$

and thus

$$\inf_{\tilde{K}} \tilde{\Pi}(\bar{v}) = \inf_{\bar{v} \in V_0} \sup_{\mu_N \in L} \left\{ \tilde{\Pi}(\bar{v}) - [\mu_N, \bar{v}_N + u_{0N}]_{\Gamma_3} \right\}. \tag{7.23}$$

The Lagrangian of the problem is a real-valued function $\mathcal{L}$ on $V_0 \times L$ defined by the relation

$$\mathcal{L}(\bar{v}, \mu_N) = \frac{1}{2}a(\bar{v}, \bar{v}) - [\mu_N, \bar{v}_N + u_{0N}]_{\Gamma_3} - [C_T, \bar{v}_T]_{\Gamma_3} - [\bar{T}, \bar{v}]_{\Gamma_2} - (\bar{p}, \bar{v}) + a(u_0, \bar{v}). \tag{7.24}$$

Thus the mixed variational formulation of the Signorini-Fichera problem reads now (cf. [Eke, Has82;84, Hl]): Find the saddle point $\{\bar{w}, \lambda_N\} \in V_0 \times L$ of $\mathcal{L}$ on $V_0 \times L$, i.e.,

$$\mathcal{L}(\bar{w}, \mu_N) \leq \mathcal{L}(\bar{w}, \lambda_N) \leq \mathcal{L}(\bar{v}, \lambda_N) \quad \forall \bar{v} \in V_0, \quad \mu_N \in L. \tag{7.25}$$

It can be proved, as in [Eke] p.57, that problem (7.25) has a unique solution $\{\bar{w}, \lambda_N\} \in V_0 \times L$ in the above functional framework such that $\bar{w} = \bar{u} \in \tilde{K}$ and $\lambda_N = S_N(\bar{u})$ on Γ_3. The dual formulation with respect to the stresses results from (7.25) if we consider the statically admissible set of the stresses

$$\Lambda = \{\sigma|\sigma = \{\sigma_{ij}\}, \sigma_{ij} = \sigma_{ji}, \ \sigma_{ij,j} + \bar{p}_i = 0 \text{ in } \Omega, S_N \in L, \tag{7.26}$$
$$S_{T_i} = C_{T_i} \text{ on } \Gamma_3, S_i = \bar{T}_i \text{ on } \Gamma_2\}.$$

As shown, e.g. in [Pan85], the stress field at the position of equilibrium σ is completely characterized by the minimum of the complementary energy

$$\Pi^c(\sigma) = \frac{1}{2}A(\sigma, \sigma) - [\bar{U}, \sigma]_{\Gamma_1} \text{ over } \Lambda \text{ i.e.,} \tag{7.27}$$

$$\Pi^c(\sigma) = \min\{\Pi^c(\tau)|\tau \in \Lambda\}. \tag{7.28}$$

Here $\frac{1}{2}A(\sigma, \sigma) = (c\sigma, \sigma)$ where $c = \{c_{ijhk}\}$ is the inverse tensor to C.

7.2 Integral Formulation with Respect to the Tractions of the Contact Area

If $\{\bar{w}, \lambda\} \in V_0 \times L$ is the solution of the problem (7.25), then we can write it equivalently in the form

$$\mathcal{L}(\bar{w}, \lambda_N) = \inf_{V_0} \sup_{L} \mathcal{L}(\bar{v}, \mu_N) = \sup_{L} \inf_{V_0} \mathcal{L}(\bar{v}, \mu_N). \tag{7.29}$$

Let us set

$$\inf_{V_0} \mathcal{L}(\bar{v}, \mu_N) = \tilde{\Pi}_1(\mu_N) \tag{7.30}$$

assuming for the moment that $\mu_N \in L$ is given. Then (7.30) is equivalent to the following equality problem: find $\tilde{u} = \tilde{u}(\mu_N) \in V_0$ such that

$$a(\tilde{u}, \bar{v}) - [\mu_N, \bar{v}_N]_{\Gamma_3} - (\bar{p}, \bar{v}) - (\bar{T}, \bar{v})_{\Gamma_2} - [C_T, \bar{v}_T]_{\Gamma_3} + a(u_0, \bar{v}) = 0, \qquad \forall \bar{v} \in V_0. \tag{7.31}$$

We observe that (7.31) is the expression of the "principle" of virtual work for a fictive structure resulting from the initial one by ignoring the unilateral constraints on Γ_3 and by adding the corresponding reactions μ_N. The position of equilibrium of this structure is characterized by the minimization problem (7.30) for the potential energy. Now the solution $\tilde{u}$ of (7.31) can be considered, because of the linearity of (7.31), as the sum of $\tilde{u}_1 \in V_0$ and $\tilde{u}_2 \in V_0$ which are solutions of the two following equality or bilateral problems:

$$a(\tilde{u}_1, \bar{v}) - (\bar{p}, \bar{v}) - [\bar{T}, \bar{v}]_{\Gamma_2} - [C_T, \bar{v}_T]_{\Gamma_3} + a(u_0, \bar{v}) = 0, \qquad \forall \bar{v} \in V_0 \tag{7.32}$$

$$a(\tilde{u}_2, \bar{v}) - [\mu_N, \bar{v}_N]_{\Gamma_3} = 0, \qquad \forall \bar{v} \in V_0, \tag{7.33}$$

respectively. Both expressions correspond to the equilibrium configuration of two bilateral structures resulting from the initial one by ignoring the unilateral boundary condition Γ_3 and assuming zero loading on the appropriate parts of the boundary.

Thus in the case of (7.32) the structure is loaded by the forces $\bar{p}$ in Ω and C_T tangentially to Γ_3, $\bar{T}$ on Γ_2, and zero is the normal loading on Γ_3. Moreover the initial displacement field u_0 is superposed. In the case of (7.33) the structure is subjected to normal forces μ_N on Γ_3 and we assume zero forces in Ω, on Γ_2 and on Γ_3 in the tangential direction. Since we deal now with bilateral structures the solutions $\tilde{u}_1$ and $\tilde{u}_2$ are uniquely determined, as it is well known from the classical elasticity. For these two classical bilateral problems we can write the solution in terms of Green's operator G. This operator is the same for the two problems due to the "same" boundary conditions holding in both cases (cf. fig. 7.1). Thus we write in both cases the solution of the problem in the form:

$$\tilde{u}_1 = G(\bar{l}), \quad \tilde{u}_2 = G(\mu_N), \quad \tilde{u} = \tilde{u}_1 + \tilde{u}_2, \quad \bar{l} = \{\bar{p}, \bar{T}, C_T, u_0\}. \tag{7.34}$$

Note that the unknown force distribution μ_N must be determined.

Thus from (7.30), (7.32) and (7.33) we obtain by setting $\bar{v} = \tilde{u}_1$ in (7.32) and $\bar{v} = \tilde{u}_2$ in (7.33) etc. that

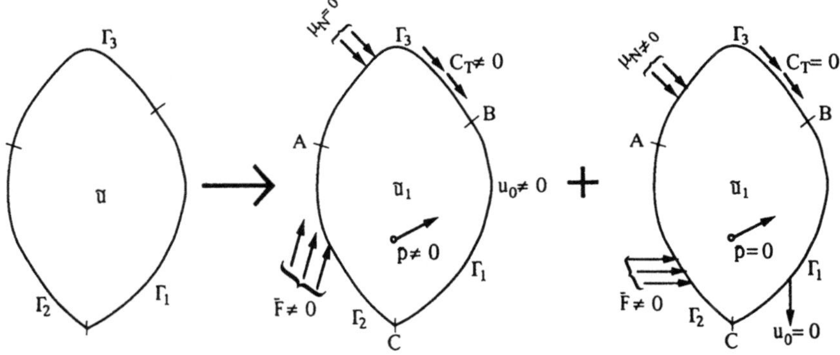

Fig. 7.1: On the meaning of $\tilde{u}_1$ and $\tilde{u}_2$ ($AB \equiv \Gamma_3$, $AC \equiv \Gamma_2$, $BC \equiv \Gamma_1$)

$$\tilde{\Pi}_1(\mu_N) = \mathcal{L}(\tilde{u}, \mu_N) \;=\; \frac{1}{2}a(\tilde{u}, \tilde{u}) - [\mu_N, \tilde{u}_N + u_{0N}]_{\Gamma_3} - [C_T, \tilde{u}_T]_{\Gamma_3} - (\bar{p}, \tilde{u}) \tag{7.35}$$
$$- [\bar{T}, \tilde{u}]_{\Gamma_2} + a(u_0, \tilde{u})$$
$$= -[\mu_N, \tilde{u}_{N_1}]_{\Gamma_3} - \frac{1}{2}[\mu_N, \tilde{u}_{N_2}]_{\Gamma_3} - \frac{1}{2}(\bar{p}, \tilde{u}_1) - \frac{1}{2}[\bar{T}, \tilde{u}_1]_{\Gamma_2}$$
$$- \frac{1}{2}[C_T, \tilde{u}_{T_1}]_{\Gamma_3} - [\mu_N, u_{0N}]_{\Gamma_3} + \frac{1}{2}a(u_0, \tilde{u}_1)$$
$$= -[\mu_N, [G(\bar{l})]_N]_{\Gamma_3} - \frac{1}{2}[\mu_N, [G(\mu_N)]_N]_{\Gamma_3} + \frac{1}{2}a(u_0, G(\bar{l}))$$
$$- \frac{1}{2}(\bar{p}, G(\bar{l})) - \frac{1}{2}[\bar{T}, G(\bar{l})]_{\Gamma_2} - \frac{1}{2}[C_T, G(\bar{l})]_{\Gamma_3} - [\mu_N, u_{0N}]_{\Gamma_3}.$$

Further we denote by β the bilinear form

$$\beta(\mu_N, \nu_N) = [\mu_N, [G(\nu_N)]_N]_{\Gamma_3} \tag{7.36}$$

and by γ the linear form

$$\gamma(\mu_N) = -\big[\mu_N, [G(\bar{l})]_N\big]_{\Gamma_3} - [\mu_N, u_{0N}]_{\Gamma_3}. \tag{7.37}$$

Thus

$$\begin{aligned}
\tilde{\Pi}_1(\mu_N) \;=\; & -\frac{1}{2}\beta(\mu_N,\mu_N) + \gamma(\mu_N) - \frac{1}{2}(\bar{p}, G(\bar{l})) - \frac{1}{2}[\bar{T}, G(\bar{l})]_{\Gamma_2} \\
& -\frac{1}{2}[C_T, G(\bar{l})]_{\Gamma_3} + \frac{1}{2}a(u_0, G(\bar{l})).
\end{aligned} \tag{7.38}$$

From the equations (7.29), (7.30) and (7.38) we obtain the following minimization problem with respect to the unknown boundary tractions μ_N:

$$\min\left\{ \Pi_1(\mu_N) = \frac{1}{2}\beta(\mu_N,\mu_N) - \gamma(\mu_N) \,\big|\, \mu_N \in L \right\}. \tag{7.39}$$

The term $-\frac{1}{2}(\bar{p}, G(\bar{l})) - \frac{1}{2}[\bar{T}, G(\bar{l})]_{\Gamma_2} - \frac{1}{2}[C_T, G(\bar{l})]_{\Gamma_3} + \frac{1}{2}a(u_0, G(\bar{l}))$ is not taken into account because it does not depend on μ_N. We denote obviously again by λ_N the solution of this problem.

From (7.30), (7.24) and (7.31) we obtain another expression for $\tilde{\Pi}_1$:

$$\begin{aligned}
\tilde{\Pi}_1(\mu_N) \;=\; & \inf_{V_0} \mathcal{L}(\tilde{v}, \mu_N) = \mathcal{L}(\tilde{u}, \mu_N) \\[4pt]
=\; & \frac{1}{2}a(\tilde{u}, \tilde{u}) - [\mu_N, \tilde{u}_N + u_{0N}]_{\Gamma_3} - [\bar{T}, \tilde{u}]_{\Gamma_2} - (\bar{p}, \tilde{u}) - [C_T, \tilde{u}_T]_{\Gamma_3} + a(u_0, \tilde{u}) \\[4pt]
=\; & \frac{1}{2}a(\tilde{u}, \tilde{u}) - a(\tilde{u}, \tilde{u}) - [\mu_N, u_{0N}]_{\Gamma_3} \\[4pt]
=\; & -\frac{1}{2}a(\tilde{u}, \tilde{u}) - [\mu_N, u_{0N}]_{\Gamma_3}.
\end{aligned} \tag{7.40}$$

For $\mu_N = \lambda_N$ it is obviously $\tilde{u} = \bar{w}$ and thus (7.40) implies that

$$\tilde{\Pi}_1(\lambda_N) = -\left[\frac{1}{2}a(\bar{w}, \bar{w}) + [\lambda_N, u_{0N}]_{\Gamma_3}\right]. \tag{7.41}$$

In the next Chapter we shall show that

$$\lambda_N = S_N(\bar{u}) \text{ on } \Gamma_3, \tag{7.42}$$

where $\bar{u} \in \tilde{K}$ is the solution of the primal problem (7.17). The proof uses the fact that the solution of (7.39) is unique.

It is interesting to note that the quadratic form $\beta(\mu_N, \mu_N)$ is symmetric. Indeed from (7.36) using the linearity of G and the reciprocal theorem of Betti on the theory of elasticity ([Sok] p.391) we can easily verify that (see also fig. 7.2).

$$\beta(\mu_N, \nu_N) = \beta(\nu_N, \mu_N). \tag{7.43}$$

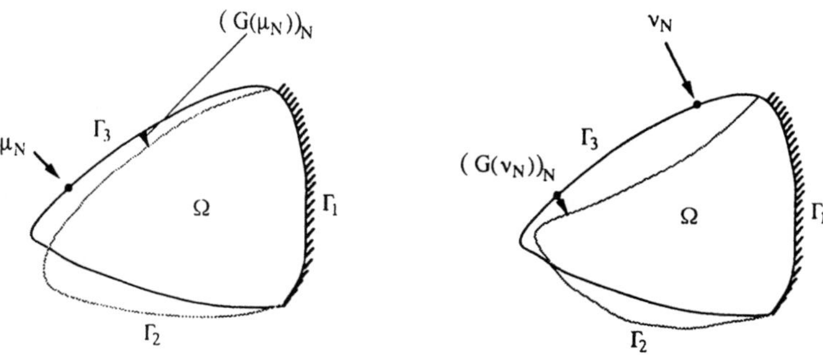

$$[\mu_N, [G(v_N)]_N]_{\Gamma_s} = [v_N, [G(\mu_N)]_N]_{\Gamma_s}$$

Fig. 7.2: The symmetry of $\beta(\cdot, \cdot)$

We give now two equivalent formulations of problem (7.39). The first is a variational inequality: find $\lambda_N \in L$ such as to satisfy

$$\beta(\lambda_N, \mu_N - \lambda_N) - \gamma(\mu_N - \lambda_N) \geq 0 \qquad \forall \mu_N \in L. \tag{7.44}$$

The second is a multivalued integral equation holding on the boundary part Γ_3 of the structure and reads

$$\gamma - \frac{1}{2}\mathrm{grad}\beta(\lambda_N, \lambda_N) \in \partial I_L(\lambda_N), \tag{7.45}$$

where $I_L(\lambda_N)$ is the indicator of the admissible set L, i.e.

$$I_L(\lambda_N) = \begin{cases} 0 & \text{if } \lambda_N \in L \\ \infty & \text{otherwise.} \end{cases} \tag{7.46}$$

Relation (7.44) is equivalent to (7.39) due to Prop. 1.2. Relation (7.45) is equivalent to (7.44) by the definition of the subdifferential ∂.

7.3 Integral Formulation with Respect to the Displacements of the Contact Area

Let us consider again the saddle-point problem (7.29). In the previous Section, we have considered $\sup_L \inf_{V_0} \mathcal{L}(\bar{v}, \mu_N)$ with μ_N as given in order to derive a variational formulation on the boundary with respect to the unknown unilateral contact tractions. Here we will work on $\inf_{V_0} \sup_L \mathcal{L}(\bar{v}, \mu_N)$, or on an equivalent saddle-point formulation, by considering $\bar{v}$ as given. We note that

$$\inf_{V_0} \sup_L \mathcal{L}(\bar{v}, \mu_N) = \sup_L \inf_{V_0} \mathcal{L}(\bar{v}, \mu_N) \tag{7.47}$$

$$= \sup_L \inf_{V_0} \left[\frac{1}{2}a(\bar{v}, \bar{v}) - [\mu_N, \bar{v} + u_{0N}]_{\Gamma_3} \right.$$

$$\left. -[C_T, \bar{v}_T]_{\Gamma_3} - [\bar{T}, \bar{v}]_{\Gamma_2} - (\bar{p}, \bar{v}) + a(u_0, \bar{v}) \right]$$

$$\begin{aligned}
&= \sup_{\mu_N \in L} \inf_{v \in V} \left[\frac{1}{2} a(v - u_0, v - u_0) - [\mu_N, v_N]_{\Gamma_3} - [C_T, v_T - u_{0T}]_{\Gamma_3} \right. \\
&\qquad\qquad \left. - [\bar{T}, v - u_0]_{\Gamma_2} - (\bar{p}, v - u_0) + a(u_0, v - u_0) \right] \\
&= \sup_{\mu_N \in L} \inf_{v \in V} \left[\frac{1}{2} a(v, v) + \frac{1}{2} a(u_0, u_0) - a(v, u_0) - [\mu_N, v_N]_{\Gamma_3} \right. \\
&\qquad\qquad - [C_T, v_T - u_{0T}]_{\Gamma_3} - [\bar{T}, v - u_0]_{\Gamma_2} - (\bar{p}, v - u_0) \\
&\qquad\qquad \left. + a(u_0, v) - \frac{1}{2} a(u_0, u_0) \right] \\
&= \inf_{v \in V} \sup_{\mu_N \in L} \left[\frac{1}{2} a(v, v) - [\mu_N, v_N]_{\Gamma_3} - [C_T, v_T]_{\Gamma_3} - [\bar{T}, v]_{\Gamma_2} \right. \\
&\qquad\qquad \left. - (\bar{p}, v) - \Pi(u_0) \right] = \inf_{v \in V} \sup_{\mu_N \in L} \tilde{\mathcal{L}}(v, \mu_N) - \Pi(u_0).
\end{aligned}$$

Here we have introduced the modified Lagrangian

$$\tilde{\mathcal{L}}(v, \mu_N) = \frac{1}{2} a(v, v) - [\mu_N, v_N]_{\Gamma_3} - [C_T, v_T]_{\Gamma_3} - [\bar{T}, v]_{\Gamma_2} - (\bar{p}, v). \tag{7.48}$$

We have (see Sect. 1.2.1) that

$$\sup_{\mu_N \leq 0} -[\mu_N, v_N]_{\Gamma_3} = \begin{cases} 0 & \text{if } v_N \leq 0 \\ \infty & \text{if } v_N > 0 \end{cases} = I_K(v) \tag{7.49}$$

and we can write

$$\inf_{v \in V} \sup_{\mu_N \in L} \tilde{\mathcal{L}}(v, \mu_N) = \inf_{v \in V} \left[\frac{1}{2} a(v, v) + I_K(v) - l(v) \right] = \inf_{v \in K} \Pi(v). \tag{7.50}$$

But

$$\frac{1}{2} a(v, v) = \frac{1}{2} (C\varepsilon(v), \varepsilon(v)) = \sup_{\tau \in \Sigma} \left((\tau, \varepsilon(v)) - \frac{1}{2} A(\tau, \tau) \right), \tag{7.51}$$

where Σ denotes the set of all symmetric stress-tensors. Indeed by simple derivation with respect to τ we find that the supremum is attained for

$$\varepsilon_{ij}(v) = c_{ijhk} \sigma_{hk} \quad \text{or} \quad \sigma_{ij} = C_{ijhk} \varepsilon_{hk}(v). \tag{7.51a}$$

which yields (7.51). Thus

$$\inf_{V} \sup_{L} \tilde{\mathcal{L}}(v, \mu_N) = \inf_{v \in K} \Pi(v) \tag{7.52}$$

$$= \inf_{v \in K} \sup_{\tau \in \Sigma} \left\{ (\tau, \varepsilon(v)) - (\bar{p}, v) - [\bar{T}, v]_{\Gamma_2} - [C_T, v_T]_{\Gamma_3} - \frac{1}{2} A(\tau, \tau) \right\}.$$

We note also that for every $v \in V$ the Green-Gauss theorem implies that

$$(\tau, \varepsilon(v)) = -(\tau_{ij,j}, v_i) + [T_N, v_N]_{\Gamma_3} + [T_T, v_T]_{\Gamma_3} + [T, v]_{\Gamma_2} + [T, \bar{U}]_{\Gamma_1}, \tag{7.53}$$

where T, T_N and T_T are the boundary tractions corresponding to τ. Note that (7.53) is the principle of virtual work for the free structure. From (7.52) and (7.53) we obtain that

$$\inf_{v \in K} \sup_{\tau \in \Sigma} \Bigg[-(\tau_{ij,j} + f_i, v_i) + [T - \bar{T}, v]_{\Gamma_2} + [T, \bar{U}]_{\Gamma_1} + [T_N, v_N]_{\Gamma_3}$$

$$+ [T_T - C_T, v_T]_{\Gamma_3} - \frac{1}{2} A(\tau, \tau) \Bigg] \qquad (7.54)$$

$$= \inf_{v \in K} \sup_{\tau \in \Sigma_1} \Bigg[[T, \bar{U}]_{\Gamma_1} + [T_N, v_N]_{\Gamma_3} - \frac{1}{2} A(\tau, \tau) \Bigg]$$

where

$$\Sigma_1 = \{\tau | \tau = \{\tau_{ij}\}, \tau_{ij} = \tau_{ji}, \tau_{ij,j} + \bar{p}_i = 0 \text{ in } \Omega, T_i = \bar{T}_i \text{ on } \Gamma_2, T_{T_i} = C_{T_i} \text{ on } \Gamma_3\}. \quad (7.55)$$

Indeed

$$\sup_{\tau \in \Sigma} \Big[-(\tau_{ij,j} + \bar{p}_i, v_i) + [T - \bar{T}, v]_{\Gamma_2} + [T_T - C_T, v]_{\Gamma_3} \Big]$$

$$= \begin{cases} 0 & \text{if } \tau_{ij,j} + \bar{p}_i = 0, \; T_i = \bar{T}_i, T_{T_i} = C_{T_i} \\ \infty & \text{otherwise.} \end{cases} \qquad (7.56)$$

Thus

$$\inf_{V} \sup_{L} \tilde{\mathcal{L}}(v, \mu_N) = \inf_{v \in K} \Pi(v) = \inf_{v \in K} \sup_{\tau \in \Sigma_1} L(v_N, \tau). \qquad (7.57)$$

where the new Lagrangian L is given by

$$L(v_N, \tau) = [T, \bar{U}]_{\Gamma_1} + [T_N, v_N]_{\Gamma_3} - \frac{1}{2} A(\tau, \tau). \qquad (7.58)$$

Note also that due to the duality between the primal problem (7.11) and the dual problem (7.27) the relation

$$\inf_{v \in K} \Pi(v) = \sup_{\tau \in \Lambda} -\Pi^c(\tau) = \sup_{\tau \in \Sigma_1} \left(-\frac{1}{2} A(\tau, \tau) + [T, \bar{U}]_{\Gamma_1} - I_L(\tau) \right) \qquad (7.59)$$

$$= \sup_{\tau \in \Sigma_1} \inf_{v \in K} \left(-\frac{1}{2} A(\tau, \tau) + [T, \bar{U}]_{\Gamma_1} + [T_N, v_N]_{\Gamma_3} \right)$$

$$= \sup_{\tau \in \Sigma_1} \inf_{v \in K} L(v_N, \tau)$$

holds. Here Λ is defined in (7.26). Moreover

$$I_L(T) = \begin{cases} 0 & \text{if } T_N \in L \\ \infty & \text{otherwise.} \end{cases} \qquad (7.60)$$

and

$$\inf_{v_N \leq 0} [T_N, v_N]_{\Gamma_3} = -I_L(T). \qquad (7.61)$$

Thus

$$\inf_{v \in K} \sup_{\tau \in \Sigma_1} L(v_N, \tau) = \sup_{\tau \in \Sigma_1} \inf_{v \in K} L(v_N, \tau) = L(u_N, \sigma). \tag{7.62}$$

The last equality in (7.62) can be verified by writing the previous relations for the actual solution $\{u, \sigma, S\}$ etc. (i.e. eq. (7.47), etc. without inf sup, inf etc.). A useful remark is that besides $\mathcal{L}$ and L another Lagrangian has been also derived; we mean the expression at the right-hand side of eq. (7.62). This Lagrangian can be also obtained if the saddle-point theory of [Eke] is applied.

From the above relations and the fact that the saddle-point problem (7.25) has a unique solution we can easily verify (cf. [Eke], p.57 and next Chapter) that the corresponding saddle-point problem with respect to L admits also a unique solution $\sigma \in \Sigma_1$ and $u_N = u_N(\sigma) \le 0$.

Let us introduce the notation

$$\sup_{\tau \in \Sigma_1} L(v_N, \tau) = \tilde{\Pi}_2(v_N) \quad \text{or} \quad \inf_{\tau \in \Sigma_1} -L(v_N, \tau) = -\tilde{\Pi}_2(v_N) \tag{7.63}$$

assuming that $v_N \le 0$ is prescribed. Then (7.63) corresponds to the variational inequality: find $\tilde{\sigma} = \tilde{\sigma}(v) \in \Sigma_1$ such that

$$A(\tilde{\sigma}, \tau - \tilde{\sigma}) - [\bar{U}, T - \tilde{S}]_{\Gamma_1} - [v_N, T_N - \tilde{S}_N]_{\Gamma_3} \ge 0, \quad \forall \tau \in \Sigma_1, \tag{7.64}$$

which expresses the principle of complementary virtual work for the structure on the assumption that the unilateral displacements on Γ_3 are assumed as given and are equal to v_N. Let us introduce a stress-field σ_0, which is statically admissible in the sense of Σ_1, i.e. it satisfies the equation of equilibrium and the static boundary conditions on Γ_2 and on Γ_3 in the tangential direction. Then we make the substitutions

$$\bar{\sigma} = \tilde{\sigma} - \sigma_0, \qquad \bar{\tau} = \tau - \sigma_0. \tag{7.65}$$

We verify easily that $\bar{\sigma}, \bar{\tau} \in \Sigma_0$ where

$$\Sigma_0 = \{\bar{\tau}|\bar{\tau}, = \{\bar{\tau}_{ij}\}, \bar{\tau}_{ij} = \bar{\tau}_{ji}, \bar{\tau}_{ij,j} = 0 \text{ in } \Omega, \ \bar{T}_i = 0 \text{ on } \Gamma_2, \bar{T}_{T_i} = 0 \text{ on } \Gamma_3\}. \tag{7.66}$$

Then (7.64) takes the new form: find $\bar{\sigma} = \bar{\sigma}(v) \in \Sigma_0$ such as to satisfy

$$A(\bar{\sigma} + \sigma_0, \bar{\tau} - \bar{\sigma}) - [\bar{U}, \bar{T} - \bar{S}]_{\Gamma_1} - [v_N, \bar{T}_N - \bar{S}_N]_{\Gamma_3} \ge 0, \quad \forall \bar{\tau} \in \Sigma_0. \tag{7.67}$$

Here $\bar{S}, \bar{T}$ are the boundary tractions corresponding to $\bar{\sigma}, \bar{\tau}$ respectively. But since Σ_0 is a linear space we may easily show (set $\bar{\tau} - \bar{\sigma} = \pm\varphi \in \Sigma_0$) that (7.67) is equivalent to the variational equality

$$A(\bar{\sigma}, \bar{\tau}) - [\bar{U}, \bar{T}]_{\Gamma_1} - [v_N, \bar{T}_N]_{\Gamma_3} + A(\sigma_0, \bar{\tau}) = 0, \quad \forall \bar{\tau} \in \Sigma_0. \tag{7.68}$$

Here

$$A(\sigma_0, \bar{\tau}) = \int_{\Omega} \varepsilon_{0ij}\bar{\tau}_{ij}d\Omega = -\int_{\Omega} \tilde{u}_{0i}\bar{\tau}_{ij,j}d\Omega + [\tilde{u}_{0N}, \bar{T}_N]_{\Gamma_3} + [\tilde{u}_{0T}, \bar{T}_T]_{\Gamma_3} \tag{7.69}$$

$$+[\tilde{u}_0, \bar{T}]_{\Gamma_2} + [\tilde{u}_0, \bar{T}]_{\Gamma_1} = [\tilde{u}_{0N}, \bar{T}_N]_{\Gamma_3} + [\tilde{u}_0, \bar{T}]_{\Gamma_1}, \quad \forall \bar{\tau} \in \Sigma_0$$

where $\varepsilon_0 = c\sigma_0$ and $\tilde{u}_0$ is a displacement field corresponding to σ_0. Note that we are free to assume that σ_0 in the unique solution of a bilateral problem having on Γ_1 and on Γ_3 zero displacements. In this case (7.69) takes the form

$$A(\sigma_0, \bar{\tau}) = 0, \qquad \forall \bar{\tau} \in \Sigma_0. \tag{7.70}$$

If σ_0 is such that the boundary condition (7.1) on Γ_1 is satisfied and on Γ_3, $\tilde{u}_{0N} = 0$, then

$$A(\sigma_0, \bar{\tau}) = [\bar{U}, \bar{T}]_{\Gamma_1}, \qquad \forall \bar{\tau} \in \Sigma_0. \tag{7.71}$$

Here we follow the general case of (7.69). The stress tensor $\bar{\sigma}$ in (7.68) can be written generally as the sum $\bar{\sigma}_1 + \bar{\sigma}_2$ where $\bar{\sigma}_1$ and $\bar{\sigma}_2$ are solutions of

$$A(\bar{\sigma}_1, \bar{\tau}) - [v_N, \bar{T}_N]_{\Gamma_3} = 0 \qquad \forall \bar{\tau} \in \Sigma_0, \tag{7.72}$$

$$A(\bar{\sigma}_2, \bar{\tau}) - [\bar{U} - \tilde{u}_0, \bar{T}]_{\Gamma_1} + [\tilde{u}_{0N}, \bar{T}_N]_{\Gamma_3} = 0 \qquad \forall \bar{\tau} \in \Sigma_0 \tag{7.73}$$

respectively. Both (7.72) and (7.73) are respectively expressions of the "principle" of complementary virtual work for fictive bilateral structures resulting from the previous one in the following way: In (7.72) (resp. (7.73)) we assume a structure Ω under the action of "given" displacements v_N (resp. $-\tilde{u}_{0N}$) on Γ_3, zero forces in Ω, on Γ_2 and tangentially on Γ_3, and given zero (resp. $\bar{U} - \tilde{u}_0$) displacements on Γ_1. Due to the fact that both variational equalities correspond to linear elastic structures subjected to classical boundary conditions, i.e. to bilateral linear elastic problems, $\bar{\sigma}_1$ and $\bar{\sigma}_2$ are uniquely determined as it results from mathematical elasticity [Sok]. From (7.72) and (7.73) we obtain that

$$\bar{\sigma}_1 = H(v_N), \quad \bar{\sigma}_2 = H(\bar{U}, \tilde{u}_0), \quad \bar{\sigma}_1 + \bar{\sigma}_2 = \bar{\sigma}, \quad \bar{\sigma} \in \Sigma_0, \tag{7.74}$$

where H denotes the "Green's operator" corresponding to the B.V.Ps (7.72) and (7.73). It is important to note that for both B.V.Ps, H is the same because for both problems the "same" boundary conditions hold (fig. 7.3).

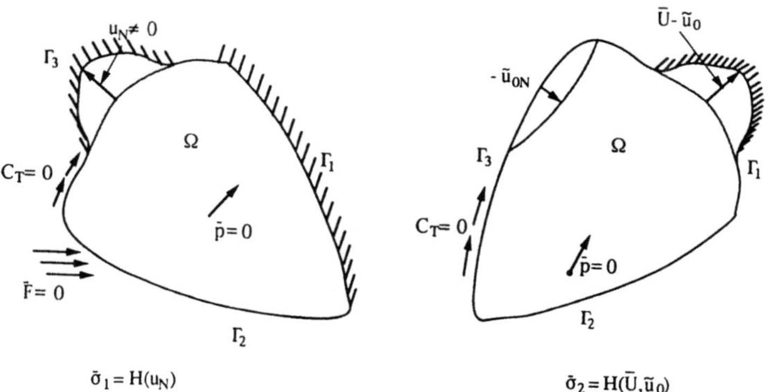

Fig. 7.3: On the meaning of $\bar{\sigma}_1$ and $\bar{\sigma}_2$

Using now (7.63) and (7.65), and combining them with (7.69) for $\bar{\tau} = \bar{\sigma}$, with (7.72) for $\bar{\tau} = \bar{\sigma}_1$, and with (7.73) written both for $\bar{\tau} = \bar{\sigma}$ and $\bar{\tau} = \bar{\sigma}_1$, we obtain that

$$
\begin{aligned}
\tilde{\Pi}_2(v_N) &= L(v_N, \tilde{\sigma}) = [\tilde{S}, \bar{U}]_{\Gamma_1} + [\tilde{S}_N, v_N]_{\Gamma_3} - \frac{1}{2}A(\tilde{\sigma}, \tilde{\sigma}) \qquad (7.75) \\
&= [\bar{S} + S_0, \bar{U}]_{\Gamma_1} + [\bar{S}_N + S_{0N}, v_N]_{\Gamma_3} - \frac{1}{2}A(\bar{\sigma}, \bar{\sigma}) - \frac{1}{2}A(\sigma_0, \sigma_0) - A(\bar{\sigma}, \sigma_0) \\
&= [\bar{S}, \bar{U}]_{\Gamma_1} + [\bar{S}_N, v_N]_{\Gamma_3} - \frac{1}{2}A(\bar{\sigma}, \bar{\sigma}) - \frac{1}{2}A(\sigma_0, \sigma_0) - [u_0, \bar{S}]_{\Gamma_1} \\
&\quad - [u_{0N}, \bar{S}_N]_{\Gamma_3} + [S_0, \bar{U}]_{\Gamma_1} + [S_{0N}, v_N]_{\Gamma_3} \\
&= [\bar{S}_2, \bar{U}]_{\Gamma_1} + [\bar{S}_{2N}, v_N]_{\Gamma_3} + \frac{1}{2}[\bar{S}_{1N}, v_N]_{\Gamma_3} - \frac{1}{2}[u_{0N}, \bar{S}_{2N}]_{\Gamma_3} - \frac{1}{2}[u_0, \bar{S}_2]_{\Gamma_1} \\
&\quad - \frac{1}{2}[\bar{U}, \bar{S}_2]_{\Gamma_1} + [S_0, \bar{U}]_{\Gamma_1} + [S_{0N}, v_N]_{\Gamma_3} - \frac{1}{2}A(\sigma_0, \sigma_0) \\
&= [\tilde{H}(\bar{U}, \tilde{u}_0), \bar{U}]_{\Gamma_1} + [[\tilde{H}(\bar{U}, u_0)]_N, v_N]_{\Gamma_3} + \frac{1}{2}[[\tilde{H}(v_N)]_N, v_N]_{\Gamma_3} \\
&\quad - \frac{1}{2}[u_{0N}, [\tilde{H}(\bar{U}, u_0)]_N]_{\Gamma_3} - \frac{1}{2}[u_0, \tilde{H}(\bar{U}, u_0)]_{\Gamma_1} \\
&\quad - \frac{1}{2}[\bar{U}, \tilde{H}(\bar{U}, u_0)]_{\Gamma_1} + [S_0, \bar{U}]_{\Gamma_1} + [S_{0N}, v_N]_{\Gamma_3} - \frac{1}{2}A(\sigma_0, \sigma_0).
\end{aligned}
$$

Here $\tilde{H}$ is defined by the relation

$$
\bar{S}_1 = \{\bar{S}_{1i}\} = \{\bar{\sigma}_{1ij}n_j\} \hat{=} H(v_N)\{n_j\} \hat{=} \{\tilde{H}(v_N)_i\} = \tilde{H}(v_N) \qquad (7.76)
$$

Let us introduce now the bilinear form

$$
\delta(v_N, w_N) = [[\tilde{H}(v_N)]_N, w_N]_{\Gamma_3} \qquad (7.77)
$$

and the linear form ($\tilde{H}$ is a linear operator)

$$
\zeta(w_N) = -[[\tilde{H}(U, u_0)]_N + S_{0N}, w_N]_{\Gamma_3}. \qquad (7.78)
$$

We denote also by $R(U, u_0, \sigma_0)$ the remaining constant terms in (7.75). Thus

$$
\begin{aligned}
\tilde{\Pi}_2(v_N) &= \sup_{\tau \in \Sigma_1} L(v_N, \tau) = L(v_N, \tilde{\sigma}) \qquad (7.79) \\
&= \frac{1}{2}\delta(v_N, v_N) - \zeta(v_N) + R(U, u_0, \sigma_0) = \Pi_2(v_N) + R(U, u_0, \sigma_0).
\end{aligned}
$$

Now (7.62) implies that we can formulate the following minimization problem with respect to the unknown boundary displacement v_N:

$$
\min\left\{\Pi_2(v_N) = \frac{1}{2}\delta(v_N, v_N) - \zeta(v_N) | v_N \leq 0\right\}. \qquad (7.80)
$$

Using the functional framework introduced in the next Chapter we will show that the minimum in (7.80) exists and is uniquely determined. Moreover if u_N is the solution of (7.80) then u_N corresponds to the stress field σ, i.e.

$$
u_N = u_N(\sigma), \qquad (7.81)
$$

where $\sigma \in \Lambda$ is the solution of the minimum complementary energy problem (7.28). Again we can verify by using Betti's theorem that δ is symmetric, i.e. that

$$\delta(v_N, w_N) = \delta(w_N, v_N). \tag{7.82}$$

We close this Section by giving the analogous formulation to (7.44) and (7.45). The first is a variational inequality: find $u_N \in K$ such that

$$\delta(u_N, v_N - u_N) - \zeta(v_N - u_N) \geq 0, \quad \forall v_N \in K. \tag{7.83}$$

The second is a multivalued boundary integral equation on the boundary part Γ_3 which reads

$$\zeta - \frac{1}{2}\mathrm{grad}\delta(u_N, u_N) \in \partial I_K(u_N). \tag{7.84}$$

Each one of the last two formulations are equivalent to the minimum problem (7.80).

7.4 The Numerical Treatment

As we have shown in the previous Sections two multivalued integral equations hold on the part of the boundary of the body subjected to the unilateral contact conditions, problem (7.45) with respect to the unknown reactions and problem (7.84) with respect to the unknown displacements. These integral formulations are equivalent to two minimum problems holding again on the boundary, the problems (7.39) and (7.80). Obviously the same method can be applied to discretized structures and minimum "principles" analogous to (7.39)) and (7.80) can be obtained. It should be noted here that in the framework of the discretized theory such minimum problems with respect to the unknown boundary displacements have been already formulated through the standard procedure of elimination of the internal degrees of freedom. For a continuous problem either the minimum problems (7.39) and (7.80), or the multivalued B.I.Es have to be discretized. This procedure has been discribed in [Has84] and uses an approximation of Green's operator obtained via a F.E. discretization. However in order to avoid mixing the F.E.M. with the use of boundary discretizations we proceed here as it follows.

A careful observation of the mechanical meaning of (7.39) and (7.80) and of the method for their derivation facilitates the procedure considerably and constitutes the advantage of the proposed method. Moreover it offers a method for the solution of the multivalued B.I.Es (7.45) and (7.84), or of their equivalent minimum problems, by using the B.E.M. for bilateral problems.

Let us consider first the minimum problem (7.39). In order to calculate the discrete form of Π_1 we proceed as is obvious from (7.32), (7.33), (7.36) and (7.37) as follows. We consider first the structure Ω_h obtained from the initial one by assuming only the kinematical constraints on Γ_1. Then we solve this structure for unit load $\mu_{N_1} = 1$ on the first node on Γ_3 using a classical B.E.M. routine and we obtain the corresponding normal displacements of all the m-nodes of Γ_3. These displacements constitute the first column of a matrix B. We repeat this procedure for the second node etc. and thus we form the whole matrix B. The normal displacements of the nodes of Γ_3 for the same structure under the given external actions constitute a vector $\boldsymbol{g}$. Then the solution of the discrete quadratic programming problem (Q.P.P.)

$$\min\left\{\frac{1}{2}\boldsymbol{\mu}^T\boldsymbol{B}\boldsymbol{\mu} - \boldsymbol{g}^T\boldsymbol{\mu}\,|\,\boldsymbol{\mu} \leq \boldsymbol{0}\right\}, \tag{7.85}$$

where $\boldsymbol{\mu} = \{\mu_{N1}, \ldots, \mu_{Nm}\}$ and m is the number of nodes of the contact area, supplies the unknown normal reactions on Γ_3. Once the optimum is obtained, the displacement and stress fields of the whole structure are calculated by back substitution.

We proceed analogously for the problem (7.80). We consider first a structure Ω_0 obtained from the initial one by assuming $\bar{U}_i = 0$ on Γ_1, $T_T = 0$ on Γ_3 and $\bar{T}_i = 0$ on Γ_2. Then we solve this structure by a classical B.E.M. code by imposing a unit normal displacement on the first node of Γ_3 and zero normal displacements on the other nodes of Γ_3. The solution of this kinematically overconstrained structure Ω_0' gives the corresponding normal reactions of all the m-nodes of Γ_3. They constitute the first column of a matrix $\boldsymbol{D}$. We repeat this procedure for the second node etc. and thus we form the whole matrix $\boldsymbol{D}$. The normal reactions of the nodes of Γ_3 for a structure Ω_0'' having $U_i - \tilde{u}_{0i}$ displacements on Γ_1, $T_T = 0$ on Γ_3, $\bar{T}_i = 0$ on Γ_2, and $\tilde{u}_{0N}$ normal displacements on Γ_3 together with the reactions S_{0N} which include the influence of the loading $C_T, \bar{T}, \bar{p}$ on the nodes of Γ_3, constitute a vector $\boldsymbol{z}$. Then the solution of the discrete Q.P.P.

$$\min\left\{\frac{1}{2}\boldsymbol{v}^T\boldsymbol{D}\boldsymbol{v} - \boldsymbol{z}^T\boldsymbol{v}\,|\,\boldsymbol{v} \leq \boldsymbol{0}\right\} \tag{7.86}$$

where $\boldsymbol{v} = \{v_{N1}, \ldots, v_{Nm}\}$, gives the unknown normal displacements on Γ_3. We should note at this point that matrices $\boldsymbol{B}$ and $\boldsymbol{D}$ should be symmetric due to Betti's theorem. However this symmetry may be lost because of the numerical method used for the unit load or unit displacement calculations. However in the numerical calculations of large problems this symmetry may be kept by considering instead of $\boldsymbol{B}$ or $\boldsymbol{D}$ the matrices $(\boldsymbol{B}+\boldsymbol{B}^T)/2$ or $(\boldsymbol{D}+\boldsymbol{D}^T)/2$, if one wants to apply a classical Q.P. algorithm. Note that if a displacement F.E. scheme is used for the calculation of the matrices $\boldsymbol{B}$, $\boldsymbol{D}$, etc. and the nodes of the F.E. discretization on Γ_3 coincide with the aforementioned m nodes of the contact area, then $\boldsymbol{B}$ and $\boldsymbol{D}$ are symmetric. If on the contrary, a B.E.M. code is used, then, generally, $\boldsymbol{B}$ and $\boldsymbol{D}$ are not symmetric. If we do not want to symmetrize the problem, then we note that the boundary variational inequalities (7.44) and (7.83), written in discrete forms ($\boldsymbol{B}$, $\boldsymbol{D}$, $\boldsymbol{g}$, $\boldsymbol{z}$, are obtained as before) are equivalent to the linear complementarity problems (L.C.Ps)

$$\boldsymbol{B}\boldsymbol{\mu} - \boldsymbol{g} \leq \boldsymbol{0}, \quad \boldsymbol{\mu} \leq \boldsymbol{0}, \quad \boldsymbol{\mu}^T(\boldsymbol{B}\boldsymbol{\mu} - \boldsymbol{g}) = 0 \tag{7.87}$$

and

$$\boldsymbol{D}\boldsymbol{v} - \boldsymbol{z} \leq \boldsymbol{0}, \quad \boldsymbol{v} \leq \boldsymbol{0}, \quad \boldsymbol{v}^T(\boldsymbol{D}\boldsymbol{v} - \boldsymbol{z}) = 0. \tag{7.88}$$

which can be treated by the corresponding general linear complementarity algorithms [Mur, Lem, Fah91].

It is to be noted that in the foregoing procedure we can have also as external "actions" initial strains. The boundary integral formulations (7.45) and (7.84) recall the classical duality between the "force" method and "displacement" method in the classical theory of elasticity. As we shall prove in the next Chapter, in the obtained minimum problems (7.39) and (7.80) the bilinear forms $\beta(\cdot,\cdot)$ and $\delta(\cdot,\cdot)$ are positive definite. Their discrete versions B and D are positive definite matrices as we can deduce from their derivation method. Thus the Q.P.Ps (7.85) and (7.86) are generally symmetric (recall previous remark) positive definite and have full matrices, but a small number of unknowns (i.e. the normal reactions or displacements on Γ_3) compared to the Q.P. formulation of the problem without elimination of the internal DOFs (see e.g. [Fre, Pan75;80, Dup]). It is also very important to note that the matrices D, B and the vectors z and g are obtained from the solution of bilateral structures under unit imposed displacements or forces by applying the B.E.M. as, for instance for infinite media. But also the F.E.M or any analytical approach available could be used.

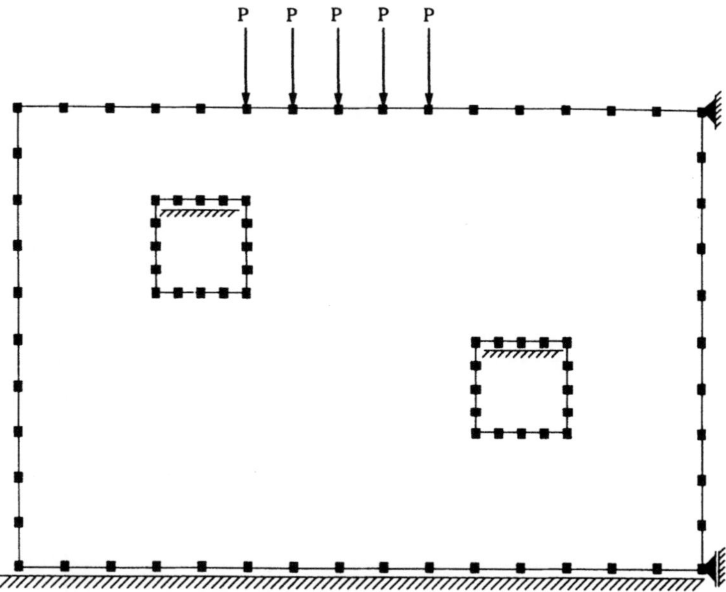

Fig. 7.4: A plane elastic problem in unilateral contact with a rigid support

As a first application we study the problem of fig. 7.4. It is a plane elasticity problem where the shaded areas of the interior and exteriors of the boundaries are in unilateral contact with a rigid support. In the tangential direction $C_T = 0$. The structure has a modulus of elasticity $E = 2.1 \times 10^6$ t/m^2, Poisson's ratio 0.33 and thickness $t = 0.1$ m. The length of the structure is 6m and the height 4m whereas the openings are squares having the dimensions 0.8m$\times$0.8m. The loading P=100 t and all the boundaries have an initial distance h=1 mm from the support. In fig. 7.5 the diagrams of the stresses σ_y

are depicted whereas in fig. 7.6 the diagram of the boundary reactions which indicates the contact and the noncontact areas.

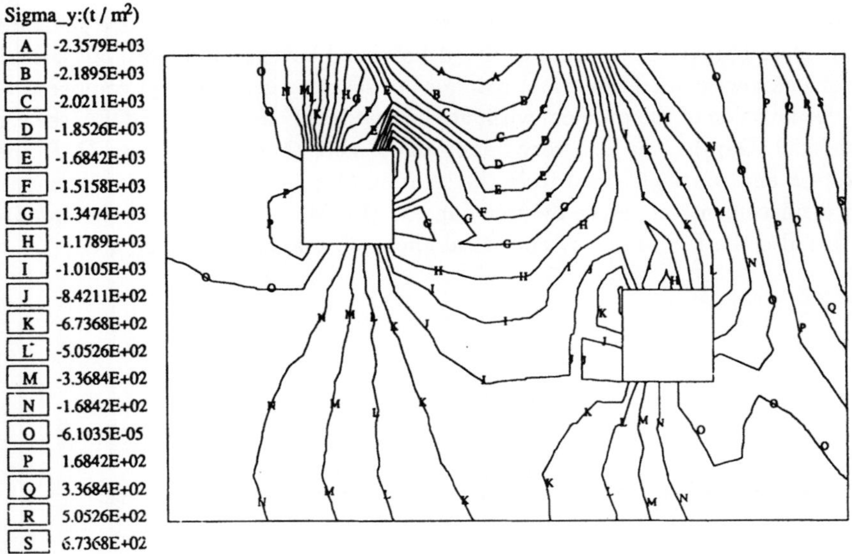

Fig. 7.5: The stresses σ_y

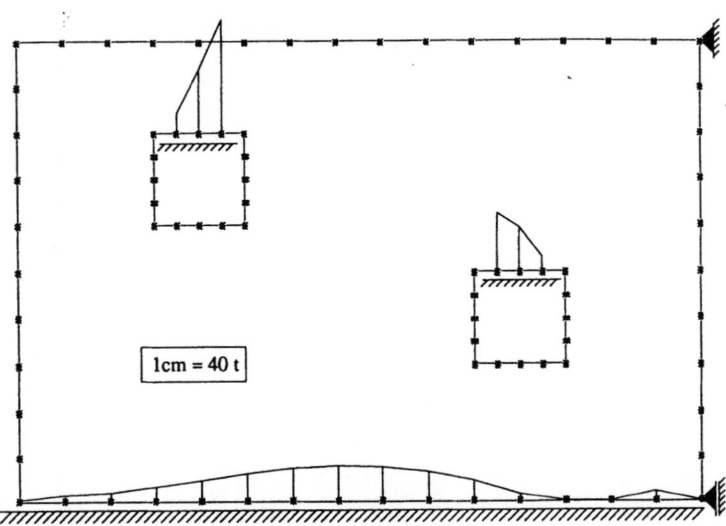

Fig. 7.6: The diagram of the reactions on the boundary

The calculation of the matrices $\boldsymbol{B}$ and $\boldsymbol{g}$ has been performed by direct B.E.M. scheme using linear interpolation boundary elements [Bre84a]. Then the arising Q.P.Ps has been solved by the algorithm of Hildreth and d' Esopo.

Chapter 8
Mathematical Study of the Boundary Integral Formulations of the Signorini-Fichera B.V.P.

8.1 The Signorini-Fichera B.V.P: The Multivalued B.I.E. with Respect to the Boundary Displacements

Let $\Omega \subset \mathbf{R}^3$ be an open bounded set with a Lipschitz boundary Γ, decomposed into three mutually disjoint parts Γ_1 Γ_2 and Γ_3, open in Γ and let mes $\Gamma_1 > 0$, mes $\Gamma_3 > 0$. Let $H^k(\Omega), k \geq 0$ integer, be the classical Sobolev space and let V_0 be the subspace of $\left[H^1(\Omega)\right]^3$ defined by

$$V_0 = \left\{ v | v = \{v_i\}, \ v_i \in H^1(\Omega), \gamma v = 0 \ \text{ on } \ \Gamma_1 \right\}, \tag{8.1}$$

where γv denotes the trace of v on Γ. The space of all traces of functions, belonging to V_0, is denoted by $\left[H^{1/2}(\tilde{\Gamma})\right]^3$ where $\tilde{\Gamma} = \Gamma_2 \cup \Gamma_3$. For the sake of simplicity the following notation will be used: We denote by $\boldsymbol{H}^{1/2}(\tilde{\Gamma})$ the space $\left[H^{1/2}(\tilde{\Gamma})\right]^3$, by $\boldsymbol{H}^{-1/2}(\tilde{\Gamma})$ the dual space to $\left[H^{1/2}(\tilde{\Gamma})\right]^3$ and by V_0' the dual space of V_0. The corresponding norms are defined in the usual way [Nec] and will be denoted by $|| \cdot ||_k$, and $|| \cdot ||_{1/2}$ whereas the dual norms are denoted by $|| \cdot ||_{V_0'}$, and $|| \cdot ||_{-1/2}$ respectively. If on Γ_1, $u_i = \bar{U}_i$, then this case will be transformed to the homogeneous one by the transformation $\bar{v} = v - v_0$ where $v_{0i}|_{\Gamma_1} = \bar{U}_i$ and $\bar{v} \in V_0$. Therefore this case will not be treated separately here. We assume now that Ω is occupied by a linear elastic body in its undeformed state. Then, the Signorini-Fichera B.V.P. is defined as in the previous Chapter by the eqs.(7.1) (with $\bar{U}_i = 0$), (7.2), (1.80), (7.3)$\div$(7.6).

Let, further, K be the closed, convex set of the admissible displacements

$$K = \left\{ v \in V_0 | v_N = v_i \, n_i \leq 0 \ \text{a.e on} \ \Gamma_3 \right\}. \tag{8.2}$$

We introduce the notations

$$\Sigma(\Omega) = \left\{ \tau | \tau = \{\tau_{ij}\}, \ \tau_{ij} = \tau_{ji} \ \text{a.e on} \ \Omega, \ \tau_{ij} \in L^2(\Omega), i, j = 1, 2, 3 \right\} \tag{8.3}$$

$$a(u, v) = \left(C\varepsilon(v), \varepsilon(u) \right) = \left(\sigma(v), \varepsilon(u) \right) = \int_\Omega \sigma_{ij}(v)\varepsilon_{ij}(u)d\Omega \tag{8.4}$$

$$l(v) = \int_\Omega \bar{p}_i v_i d\Omega + \int_{\Gamma_2} \bar{T}_i v_i d\Gamma + \int_{\Gamma_3} C_{T_i} v_{T_i} d\Gamma, \bar{p}_i \in L^2(\Omega), \bar{T}_i \in L^2(\Gamma_2), C_{T_i} \in L^2(\Gamma_3), \tag{8.5}$$

It is well-known [Fich63, Duv72, Pan85] that the potential energy of the body

$$\Pi(v) = \frac{1}{2}a(v, v) - l(v) \tag{8.6}$$

satisfies at the position of equilibrium $u \in K$ the minimum problem (primal problem) (7.11) over the kinematically admissible cone (8.2). Problem (7.11) has a unique solution. The dual formulation of (7.11) is the minimum problem of the complementary energy [Pan85]

$$\Pi^c(\tau) = (c\tau, \tau) = \int_\Omega c_{ijhk}\tau_{ij}\tau_{hk}d\Omega, \tag{8.7}$$

where c denotes the inverse tensor to $C = \{C_{ijhk}\}$, over the statically admissible set

$$\Lambda = \{\tau | \tau \in \Sigma(\Omega), \tau_{ij,j} + \bar{p}_i = 0 \quad \text{a.e on} \quad \Omega, T_i = \bar{T}_i \quad \text{a.e on} \quad \Gamma_2, \tag{8.8}$$

$$T_{T_i} = C_{T_i} \quad \text{a.e on} \quad \Gamma_3, T_N \leq 0 \quad \text{on} \quad \Gamma_3 \quad \text{in the sense of} \quad H^{-1/2}(\tilde{\Gamma})\}$$

Here T (resp. T_T) are the boundary tractions (resp. their tangential components) corresponding to the stress field $\tau = \{\tau_{ij}\}$. We can easily prove the following proposition

Proposition 8.1 The set

$$\tilde{K}^-(\Omega) = \{\tau | \tau \in \Sigma(\Omega), \ (\tau, \varepsilon(v)) \geq l(v) \quad \forall v \in K\} \tag{8.9}$$

is identified with Λ.

Proof: Applying the Green-Gauss theorem to the variational inequality of (8.9) for $v_i = \pm\varphi_i \in \mathcal{D}(\Omega)$ (set of infinitely differentiable function with compact support in Ω) we obtain the equations of equilibrium holding in the sense of distributions over Ω and since $\bar{p}_i \in L^2(\Omega)$ in the sense of $L^2(\Omega)$. Once again the application of Green-Gauss theorem together with the inequality in (8.9) yields the relation (cf. Sect. 1.3.2 and eq. (1.120))

$$\langle S_N, v_N\rangle_{1/2,\Gamma_3} + \langle S_T - C_T, v_T\rangle_{H_T,\Gamma_3} + \langle S - \bar{T}, v\rangle_{1/2,\Gamma_2} \geq 0 \quad \forall v \in K. \tag{8.10}$$

Here $\langle \cdot, \cdot\rangle_{1/2}$ denotes the duality pairing on $H^{1/2}(\tilde{\Gamma}) \times H^{-1/2}(\tilde{\Gamma})$ or on $\boldsymbol{H}^{1/2}(\tilde{\Gamma}) \times \boldsymbol{H}^{-1/2}(\tilde{\Gamma})$ and the subscripts Γ_3 and Γ_2 denote the restrictions of the functionals. For the definition of H_T see eq.(1.119). Note, for instance, that if $S_N \in L^2(\tilde{\Gamma})$ then $\langle S_N, v_N\rangle_{1/2,\Gamma_3} = \int_{\Gamma_3} S_N v_N d\Gamma$ etc. Now we take in (8.10) special variations: first we assume that $v = 0$ a.e. on Γ_3 which implies that $\langle S - \bar{T}, v\rangle_{1/2,\Gamma_2} \geq 0$ for every $v \in K$. Taking $v = \pm w \in \boldsymbol{H}^{1/2}(\tilde{\Gamma})$ implies that $S = \bar{T}$ on Γ_2 in the sense of $\boldsymbol{H}^{-1/2}(\tilde{\Gamma})$ and since $\bar{T}_i \in L^2(\Gamma_2)$ in the sense of $[L^2(\Gamma_2)]^3$. Analogously (8.10) implies that $S_{T_i} = C_{T_i}$ in the sense of $L^2(\Gamma_3)$. Thus (8.10) reduces to the inequality (1.126) which is a formulation of the Signorini boundary condition (1.80) on $H^{1/2}(\tilde{\Gamma}) \times H^{-1/2}(\tilde{\Gamma})$ q.e.d.

Now the dual problem takes the form: Find $\sigma \in \tilde{K}^-(\Omega)$ such that

$$\Pi^c(\sigma) = \inf\{\Pi^c(\tau)|\tau \in \tilde{K}^-(\Omega)\} \tag{8.11}$$

This problem has a unique solution $\sigma = \{\sigma_{ij}\}$ and

$$\sigma = C\varepsilon(u) \tag{8.12}$$

where u is the solution of the primal problem (7.11) (cf.e.g. [Pan85]).

Let now

$$V_{00} = \big\{ v \,|\, v = \{v_i\}, \ v_i \in H^1(\Omega), \ v_i = 0 \ \text{ on } \Gamma_1, \ i = 1,2,3, \ v_N = 0 \ \text{ on } \ \Gamma_3 \big\} \qquad (8.13)$$

and

$$\tilde{K}(\Omega) = \big\{ \tau \,|\, \tau \in \Sigma(\Omega), \ (\tau, \varepsilon(v)) = l(v) \quad \forall v \in V_{00} \big\}. \qquad (8.14)$$

Analogously to Lemma 8.1 we can prove that

$$\tilde{K}(\Omega) = \big\{ \tau \,|\, \tau \in \Sigma(\Omega), \tau_{ij,j} + \bar{p}_i = 0 \quad \text{a.e on} \ \ \Omega, \ T_i = \bar{T}_i \ \ \text{a.e on} \ \ \Gamma_2, \qquad (8.15)$$
$$T_{T_i} = C_{T_i} \ \ \text{a.e on} \ \ \Gamma_3 \big\}.$$

Thus $\tau \in \tilde{K}^-(\Omega)$ implies that

$$\tau \in \tilde{K}(\Omega), \quad \langle T_N, v_N \rangle_{1/2,\Gamma_3} \geq 0 \quad \forall v \in K \qquad (8.16)$$

and conversely. Thus the following Proposition holds:

Proposition 8.2 The function

$$\sup_{v \in K} \big\{ - \langle T_N, v_N \rangle_{1/2,\Gamma_3} \big\}, \quad \text{where} \ \ T_N = T_N(\tau) \ \ \text{and} \ \ \tau \in \tilde{K}(\Omega) \,, \qquad (8.17)$$

is the indicator function of $\tilde{K}^-(\Omega)$.

Thus we may write that

$$\inf_{\tau \in \tilde{K}^-(\Omega)} \Pi^c(\tau) = \inf_{\tau \in \tilde{K}(\Omega)} \sup_{v \in K} \big[\tilde{L}(\tau, v) \big] \qquad (8.18)$$

where

$$\tilde{L}(\tau, v) = \frac{1}{2} \big(c\tau, \tau \big) - \langle T_N(\tau), v_N \rangle_{1/2,\Gamma_3} \qquad (8.19)$$

is the Lagrangian defined on $\tilde{K}(\Omega) \times K$. Note that $\tilde{L}(\tau, v) = -L(v_N, \tau)$ defined in (7.58). According to [Has84, Eke] the saddle-point problem: "find $(\sigma^\star, u^\star) \in \tilde{K}(\Omega) \times K$ such that

$$\tilde{L}(\sigma^\star, v) \leq \tilde{L}(\sigma^\star, u^\star) \leq \tilde{L}(\tau, u^\star) \quad \forall \tau \in \tilde{K}(\Omega), \ \forall v \in K \text{ "} \qquad (8.20)$$

admits a unique solution $\{\sigma^\star, u^\star)\}$ such that

$$\sigma^\star = C\varepsilon(u^\star), \qquad (8.21)$$

$$u^\star = u, \qquad (8.22)$$

where $u \in K$ is the solution of the primal problem (7.11). We may now write that

$$\tilde{L}(\sigma^\star, u^\star) = \sup_{v \in K} \inf_{\tau \in \tilde{K}(\Omega)} \tilde{L}(\tau, v) \qquad (8.23)$$

and assuming that $v \in K$ is given, we denote by $\tilde{\Pi}_2$ the functional (cf. (7.63))

$$\tilde{\Pi}_2(v_N) = \inf\{ \tilde{L}(\tau, v) | \tau \in \tilde{K}(\Omega) \} \,. \qquad (8.24)$$

Further the explicit form of $\tilde{\Pi}_2$ will be derived. From (8.19) and (8.24) we see that $\inf\{\tilde{L}(\tau,v)|\tau \in \tilde{K}(\Omega)\}$ is the expression of the minimum of the complementary energy proposition for a classical bilateral problem. We denote by $\sigma = \sigma(v), v \in K$, the solution of this problem. Using classical arguments we obtain that

$$\sigma(v) = C\varepsilon(w)\,, \tag{8.25}$$

where the displacement field w solves the B.V.P. defined by the equation of equilibrium (7.4), by (7.5) and (7.6) and by the boundary conditions

$$w_i = 0 \quad \text{on} \quad \Gamma_1 \tag{8.26}$$

$$w_N = v_N \quad \text{on} \quad \Gamma_3 \tag{8.27}$$

$$T_T = C_T \quad \text{on} \quad \Gamma_3 \tag{8.28}$$

$$T_i = \bar{T}_i \quad \text{on} \quad \Gamma_2 \tag{8.29}$$

Now the following Proposition will be proved.

Proposition 8.3 We can write that

$$\tilde{K}(\Omega) = \tau_0 + K_{00}(\Omega), \tag{8.30}$$

where $\tau_0 \in \Sigma(\Omega)$ is such that

$$\big(\tau_0, \varepsilon(v)\big) = l(v) \quad \forall v \in V_{00} \tag{8.31}$$

and

$$K_{00}(\Omega) = \big\{\tau|\tau \in \Sigma(\Omega),\ \tau_{ij,j} = 0 \quad \text{a.e in}\ \ \Omega, \tag{8.32}$$
$$T_i = 0 \quad \text{a.e on}\ \ \Gamma_2, \quad T_T = 0 \quad \text{a.e on}\ \ \Gamma_3\big\}$$

Proof: Is obvious noting that (8.31) is equivalent to the nonhomogeneous equations of equilibrium and the nonhomogeneous boundary conditions on Γ_2 and Γ_3. This fact implies the result q.e.d.

Accordingly one can write that

$$\inf_{\tau \in \tilde{K}(\Omega)} \big\{\tilde{L}(\tau,v)\big\} = \inf_{\tau \in K_{00}(\Omega)} \tilde{L}(\tau + \tau_0, v) \tag{8.33}$$

where

$$\tilde{L}(\tau + \tau_0, \sigma) = \frac{1}{2}\big(c(\tau + \tau_0), \tau + \tau_0\big) - \langle T_N(\tau_0) + T_N(\tau), v_N\rangle_{1/2,\Gamma_3} \tag{8.34}$$

$$= \frac{1}{2}\big(c\tau, \tau\big) + \big(c\tau, \tau_0\big) + \frac{1}{2}\big(c\tau_0, \tau_0\big)$$

$$- \langle T_N(\tau_0), v_N\rangle_{1/2,\Gamma_3} - \langle T_N(\tau), v_N\rangle_{1/2,\Gamma_3}\,.$$

Let us now denote, by $\tilde{L}_1(\tau)$ the expression

$$\tilde{L}_1(\tau) = \frac{1}{2}\big(c\tau, \tau\big) + \big(c\tau, \tau_0\big) - \langle T_N(\tau), v_N\rangle_{1/2,\Gamma_3} \tag{8.35}$$

where $\tau \in K_{00}(\Omega)$, and $v \in K$ is considered as given. From (8.33) and (8.35) we are led to consider the problem

$$\tilde{L}_1(\bar{\tau}) = \inf \left\{ \tilde{L}_1(\tau) | \tau \in K_{00}(\Omega) \right\} . \tag{8.36}$$

Obviously if $\bar{\tau}$ is solution of (8.36) then $\bar{\tau} + \tau_0$ is solution of the first problem in (8.33). Note that (8.36) is equivalent to the expression

$$\left(c\bar{\tau}, \tau \right) + \left(c\tau_0, \tau \right) - \langle T_N(\tau), v_N \rangle_{1/2, \Gamma_3} = 0 \quad \forall \tau \in K_{00}(\Omega) . \tag{8.37}$$

But if u_0 is a displacement field corresponding to τ_0, i.e.

$$\tau_0 = C\varepsilon(u_0) , \tag{8.38}$$

then

$$\left(c\tau_0, \tau \right) = \int_{\Omega} \varepsilon_{ij}(u_0)\tau_{ij} d\Omega = \langle u_{0N}, T_N \rangle_{1/2, \Gamma_3} + \langle u_0, T \rangle_{1/2, \Gamma_1} \tag{8.39}$$

because $\tau \in K_{00}(\Omega)$. We can take $u_{0N} = 0$ on Γ_3 and $u_0 = 0$ on Γ_1 in order to find a displacement field u_0 corresponding to τ_0, and thus

$$\tilde{L}_1(\bar{\tau}) = \inf \left\{ \tilde{L}_1(\tau) | \tau \in K_{00}(\Omega) \right\} = \inf_{\tau \in K_{00}(\Omega)} \left\{ \frac{1}{2}(c\tau, \tau) - \langle T_N(\tau), v_N \rangle_{1/2, \Gamma_3} \right\} . \tag{8.40}$$

It can be easily verified that

$$\tilde{L}_1(\bar{\tau}) = -\frac{1}{2} \langle T_N(\bar{\tau}), v_N \rangle_{1/2, \Gamma_3} \tag{8.41}$$

and that

$$\bar{\tau} = C\varepsilon(\bar{u}) \tag{8.42}$$

where $\bar{u}$ is a displacement field solving the linear elasticity B.V.P. defined by $\bar{\tau} \in K_{00}(\Omega)$, by (7.5), by (7.6) and by the remaining boundary conditions

$$\bar{u}_i = 0 \quad \text{on} \quad \Gamma_1 \tag{8.43a}$$

$$\bar{u}_N = v_N \quad \text{on} \quad \Gamma_3 . \tag{8.43b}$$

Now we can define an operator $H: H^{1/2}(\tilde{\Gamma})\big|_{\Gamma_3} \to K_{00}(\Omega)$ such that

$$H(v_N) = \bar{\tau} \tag{8.44}$$

where $\bar{\tau} \in K_{00}(\Omega)$ is the solution of (8.40). From (8.41) and (8.44) we obtain that

$$\tilde{L}_1(\bar{\tau}) = -\frac{1}{2} \langle T_N(\bar{\tau}), v_N \rangle_{1/2, \Gamma_3} = -\frac{1}{2} \langle [\tilde{H}(v_N)]_N, v_N \rangle_{1/2, \Gamma_3} \tag{8.45}$$

where (cf.(7.76))

$$\left[\tilde{H}(v_N) \right]_N = H(v_N)_{ij} n_i n_j = \tilde{H}(v_N)_i n_i \tag{8.46}$$

Thus we get from (8.33) and from (8.45), by taking into account (8.35), that

$$\tilde{\Pi}_2(v_N) \;=\; \inf_{\tau \in \tilde{K}(\Omega)} \tilde{L}(\tau, v) \;=\; \inf_{\tau \in \tilde{K}_{oo}(\Omega)} \tilde{L}(\tau + \tau_0, v) = \tilde{L}(\bar{\tau} + \tau_0, v) \tag{8.47}$$

$$= \; \inf \left\{ \tilde{L}_1(\tau) - \langle T_N(\tau_0), v_N \rangle_{1/2,\Gamma_3} + \frac{1}{2}(c\tau_0, \tau_0) \right\}$$

$$= \; -\frac{1}{2}\langle \left[\tilde{H}(v_N)\right]_N, v_N \rangle_{1/2,\Gamma_3} - \langle T_N(\tau_0), v_N \rangle_{1/2,\Gamma_3} + \frac{1}{2}(c^{-1}\tau_0, \tau_0) \,.$$

Accordingly

$$\sup_{v \in K} \inf_{\tau \in \tilde{K}(\Omega)} \tilde{L}(\tau, v) = \sup_{v \in K} \tilde{\Pi}_2(v_N) \tag{8.48}$$

and therefore we are led to the problem

$$\inf \left\{ \frac{1}{2}\langle \left[\tilde{H}(v_N)\right]_N, v_N \rangle_{1/2,\Gamma_3} + \langle T_N(\tau_0), v_N \rangle_{1/2,\Gamma_3} \,|\, v \in K \right\} \tag{8.49}$$

which is a minimum problem on the boundary Γ. It is identical to the minimum problem (7.80) derived in Ch.7 and thus to the boundary integral equation (multivalued) (7.84).

We denote further by $\delta(\cdot, \cdot)$ the bilinear form on $H^{1/2}(\tilde{\Gamma}) \times H^{1/2}(\tilde{\Gamma})$, $\delta(v_N, v_N) = \langle [H(v_N)]_N, v_N \rangle_{1/2,\Gamma_3}$ and by $\zeta(\cdot)$ the linear form on $H^{1/2}(\tilde{\Gamma})$, $\zeta(v_N) = -\langle T_T(\tau_0), v_N \rangle_{1/2,\Gamma_3}$. Thus (8.49) can be written as

$$\inf \left\{ \Pi_2(v_N) | v_N \in H_-^{1/2}(\Gamma_3) \right\} \tag{8.50}$$

where $H_-^{1/2}(\Gamma_3) = \{v_N | v_N \in H^{1/2}(\tilde{\Gamma})|_{\Gamma_3}, \; v_N \leq 0 \text{ a.e on } \Gamma_3\}$, $H^{1/2}(\tilde{\Gamma})|_{\Gamma_3}$ denotes the restriction of the functions of $H^{1/2}(\tilde{\Gamma})$ to Γ_3 and

$$\Pi_2(v_N) = \frac{1}{2}\delta(v_N, v_N) - \zeta(v_N) \,. \tag{8.51}$$

Because of Betti's theorem δ is symmetric. We can prove now the following result.

Proposition 8.4 Problem (8.50) has exactly one solution.
Proof: We show first that

$$\delta(v_N, v_N) = \langle \left[H(v_N)\right]_N, v_N \rangle_{1/2,\Gamma_3} = |||v_N|||^2_{1/2,\Gamma_3} \tag{8.52}$$

where $||| \cdot |||_{1/2,\Gamma_3}$ denotes a norm of $H^{1/2}(\tilde{\Gamma})|_{\Gamma_3}$ equivalent to the classical $H^{1/2}$-norm. From this fact the existence and the uniqueness of the solution results immediately. We can endow V_0 in (8.1) with the norm

$$|||v||| = \left(C\varepsilon(v), \varepsilon(v)\right)^{1/2} \tag{8.53}$$

which by Korn's inequality is equivalent to the classical Sobolev space norm since $\mathrm{mes}\,\Gamma_1 > 0$. The space $H^{1/2}(\tilde{\Gamma})|_{\Gamma_3}$ of all traces $v_N = \{\gamma v\}_i n_i$ restricted to Γ_3, of functions $v \in V_0$, will be endowed with the norm

$$|||\varphi|||_{1/2,\Gamma_3} = \inf \left\{ |||v||| \,\big|\, v \in V_0, \; v_N = \varphi \text{ on } \Gamma_3 \right\}. \tag{8.54}$$

From this definition it results immediately that

$$|||\varphi|||_{1/2,\Gamma_3} = |||\bar{u}|||\tag{8.55}$$

where $\bar{u}$ is the unique solution of the B.V.P.

$$a(\bar{u}, v^\star) = \left(C\varepsilon(\bar{u}), \varepsilon(v^\star)\right) = 0 \quad \forall v^\star \in V_0\tag{8.56}$$

$$\bar{u}_N = \varphi \quad \text{on} \quad \Gamma_3.\tag{8.57}$$

Because of (8.45) and (8.37),(8.39) and (8.40) we can write that

$$\begin{aligned}
\tilde{L}_1(\bar{\tau}) &= -\frac{1}{2}\langle [\tilde{H}(v_N)]_N, v_N\rangle_{1/2,\Gamma_3} = -\frac{1}{2}\langle T_N(\bar{\tau}), v_N\rangle_{1/2,\Gamma_3}\\
&= -\frac{1}{2}(c\bar{\tau}, \bar{\tau}) = -\frac{1}{2}(\varepsilon(\bar{u}), \tau(\bar{u})) = -\frac{1}{2}|||\bar{u}|||^2 .
\end{aligned}\tag{8.58}$$

Here $\bar{u}$ is the solution of (8.56), (8.57) where φ is replaced by v_N. Thus

$$\langle [\tilde{H}(v_N)]_N, v_N\rangle_{1/2,\Gamma_3} = |||\bar{u}|||^2 = |||v_N|||_{1/2,\Gamma_3}^2 .\tag{8.59}$$

Note that the norm $|||v_N|||_{1/2,\Gamma_3}$ is equivalent to the classical $H^{1/2}$-norm.[3] The last relation (8.59) implies the existence and the uniqueness of the solution of (8.50) because of Prop. 1.16 q.e.d.

Proposition 8.5 The boundary minimum problem (8.50) can be equivalently written as the variational inequality problem: "Find $v_N^\star \in H_-^{1/2}(\Gamma_3)$ such that

$$\delta(v_N^\star, v_N - v_N^\star) - \zeta(v_N - v_N^\star) \geq 0 \quad \forall v_N \in H_-^{1/2}(\Gamma_3) \text{ "}\tag{8.60}$$

and as the multivalued boundary integral equation

$$\zeta - \frac{1}{2}\mathrm{grad}\delta(v_N^\star, v_N^\star) \in \partial I_{H_-^{1/2}(\Gamma_3)}(v_N^\star) .\tag{8.61}$$

Proof: the variational inequality (8.60) is obvious from the properties of δ because of Prop. 1.2. (8.61) results directly from the definitions of the subdifferential and of the indicator I of the cone $H_-^{1/2}(\Gamma_3)$ (see Ch.1) q.e.d.

We close this Section by noting that the solution $v_N^\star$ of (8.61) is equal to u_N, where u is solution of the primal problem. From the relation between primal problem (7.11) and dual problem (8.7), [Eke, Pan85], it results that (7.81) holds. The same can be obtained by using the uniqueness of the solution of (8.61) or equivalently of (8.50). Indeed let $\sigma \in \Lambda$ be the solution of the dual problem and let u_N correspond to σ, i.e. let (7.81) hold. Then (u_N, σ) satisfies the saddle point problem (8.20) and is the unique [Eke] solution of it. Thus, following the same method, as previously we can show that u_N is a solution of (8.50). The uniqueness of the solution of (8.50) implies the result.

---

[3]cf. [Hl] p.5 and [Nec].

8.2 The Signorini-Fichera B.V.P: The Multivalued B.I.E. with Respect to the Boundary Tractions

Here we summarize first some results, of Ch.7 in the rigorous mathematical setting of this Chapter. Let

$$L = H_-^{-\frac{1}{2}}(\Gamma_3) = \{\mu_N \in H^{-\frac{1}{2}}(\tilde{\Gamma})|_{\Gamma_3}|\langle\mu_N, v_N\rangle \geq 0 \quad \forall v \in K\}, \tag{8.62}$$

be a closed convex subset of $H^{-1/2}(\tilde{\Gamma})$ of Lagrange multipliers and let $\mathcal{L} : V_0 \times L \to \mathbb{R}$, be the Lagrange function (cf. (7.24) with $u_0 = 0$).

$$\mathcal{L}(v, \mu_N) = \frac{1}{2}\alpha(v, v) - \langle\mu_N, v_N\rangle_{1/2,\Gamma_3} - (l, v). \tag{8.63}$$

The mixed variational formulation of the Signorini-Fichera problem is the problem of finding a saddle-point $\{w, \lambda_N\} \in V_0 \times L$ of $\mathcal{L}$ on $V_0 \times L$ satisfying

$$\mathcal{L}(w, \mu_N) \leq \mathcal{L}(w, \lambda_N) \leq \mathcal{L}(v, \lambda_N) \quad \forall v \in V_0 \ \mu_N \in L \tag{8.64}$$

The following proposition holds (for the proof see e.g. [Eke]).

Proposition 8.6 There exists a unique solution (w, λ_N) of (8.64). Moreover,

$$w = u, \quad \lambda_N = S_N(u) \tag{8.65}$$

where $u \in K$ is the solution of the primal problem (7.11).

Let $\{w, \lambda_N\} \in V_0 \times L$ be the solution of (8.64). Then following the same procedure as in (7.29)÷(7.35) we arrive at the minimization problem (7.39) where β is the bilinear form on $\{H^{-1/2}(\tilde{\Gamma})|_{\Gamma_3} \times H^{-1/2}(\tilde{\Gamma})|_{\Gamma_3}\}$ and γ a linear form on $H^{-1/2}(\tilde{\Gamma})|_{\Gamma_3}$.
The following proposition holds.

Proposition 8.7 There exists a unique solution λ_N of (7.39). Moreover,

$$\lambda_N = S_N(u). \tag{8.66}$$

where $u \in K$ is the solution u of the primal problem.

Proof: Let u be the solution of the primal problem and let $S_N(u)\hat{=}\lambda_N$ be the corresponding normal reaction; let also $w = u$. Then $\lambda_N \in L$ and $\{w, \lambda_N\}$ is a solution of (8.64). A pair $\{w, \lambda_N\}$ which is a solution of (8.64) satisfies relations analogous to (7.29). Thus

$$\mathcal{L}(w, \lambda_N) = \inf_{v \in V_0} \mathcal{L}(v, \lambda_N) = \tilde{\Pi}_1(\lambda_N), \tag{8.67}$$

where $\tilde{\Pi}_1$ is given by (7.38). On the other hand,

$$\mathcal{L}(w, \lambda_N) = \sup_L \inf_{V_0} \mathcal{L}(v, \lambda_N) = \sup_L \tilde{\Pi}_1(\mu_N). \tag{8.68}$$

By comparing (8.67) with (8.68) and using the relation between $\tilde{\Pi}_1$ and Π_1 (cf. (7.38)) we see that $S_N(u)$, is a minimizer of Π_1 on L. The uniqueness and the existence of the minimizer of Π_1 over L, which results from the $H^{-1/2}(\tilde{\Gamma})\big|_{\Gamma_3}$-coerciveness of the form β, yields the result. Note that the coerciveness will be proved later q.e.d.

Let we recall now the relations (8.53), (8.54) and (8.55). Let $v_N = \gamma_N v$ where $\gamma_N \colon V_0 \to H^{1/2}(\tilde{\Gamma})$. By Σ we denote here the space of symmetric tensors $\sigma = \{\sigma_{ij}\}$ the divergence of which $\operatorname{div}\sigma = \{\sigma_{ij,j}\}$ is square integrable in Ω, i.e.

$$\Sigma = \left\{ \sigma \,\big|\, \sigma = \{\sigma_{ij}\}, \ \sigma_{ij} = \sigma_{ji}, \ \sigma_{ij} \in L^2(\Omega), \ \operatorname{div}\sigma \in [L^2(\Omega)]^3 \right\}. \tag{8.69}$$

We endow Σ with the norm

$$\|\sigma\|_\Sigma = \left[(c\sigma, \sigma) + (\operatorname{div}\sigma, \operatorname{div}\sigma) \right]^{1/2}. \tag{8.70}$$

Let we denote further with Σ_0 the closed subspace of Σ consisting of functions with zero divergence, with $S_T = 0$ on Γ_3 and with $\{S_i\} = 0$ on Γ_2 (i.e. $C_T = 0$ and $\bar{T}_i = 0$). Obviously Σ_0 is a Hilbert space with norm

$$\|\sigma\|_{\Sigma_0} = (c\sigma, \sigma)^{1/2}. \tag{8.71}$$

The following Green-Gauss formula holds (cf. [Pan85] p.33, and [Hün] as main source, see also [Has82]): There exists a unique linear continuous[4] mapping $S_N \in \mathcal{L}(\Sigma_0, H^{-1/2} (\tilde{\Gamma})\big|_{\Gamma_3})$ such that

$$\big(\tau, \varepsilon(v)\big) = \langle S_N(\tau), \gamma_N(v) \rangle_{1/2,\Gamma_3} \quad \forall \tau \in \Sigma_0 \ \ \forall v \in V_0. \tag{8.72}$$

Proposition 8.8 S_N maps Σ_0 onto $H^{-1/2}(\tilde{\Gamma})\big|_{\Gamma_3}$.

Proof: Let $\varphi_N^\star \in H^{-1/2}(\tilde{\Gamma})\big|_{\Gamma_3}$ be given. Then the problem: "Find $u \in V_0$ such that

$$a(u, v) = \langle \varphi_N^\star, \gamma_N v \rangle_{1/2,\Gamma_3} \quad \forall v \in V_0 \text{ "} \tag{8.73}$$

has due to the Lax-Milgram theorem a unique solution $u = u(\varphi_N^\star)$. Setting $\sigma = C\varepsilon(u)$ and using (8.72) and (8.4) we obtain that $\sigma \in \Sigma_0$ and $S_N(\sigma) = \varphi_N^\star$. q.e.d.

Now we shall derive certain useful expressions for $\||\varphi_N^\star\||_{-1/2,\Gamma_3}$. First the usual dual norm, namely

$$\||\varphi_N^\star\||_{-1/2,\Gamma_3} = \sup_{\substack{H^{1/2}(\tilde{\Gamma})|_{\Gamma_3} \\ \varphi_N \neq 0}} \frac{\langle \varphi_N^\star, \varphi_N \rangle_{1/2,\Gamma_3}}{\||\varphi_N\||_{1/2,\Gamma_3}} \tag{8.74}$$

is defined, where $\||\varphi_N\||_{1/2,\Gamma_3}$ is given by (8.54),(8.55).

Proposition 8.9 The following equalities hold.

$$\||\varphi_N^\star\||_{-1/2,\Gamma_3} = \|\sigma\|_{\Sigma_0} = \||u(\varphi_N^\star)\|| \quad \varphi_N^\star \in H^{-1/2}(\tilde{\Gamma})\big|_{\Gamma_3}. \tag{8.75}$$

Here $u(\varphi_N^\star)$ is the solution of (8.73) and $\sigma = C\varepsilon(u)$.

---

[4] $\mathcal{L}(X,Y)$ denotes the space of continuous linear operators form X into Y.

Proof: Let $\tau \in \Sigma_0$ satisfy $S_N(\tau) = \varphi_N^\star$. Then for any $\varphi_N \in H^{1/2}(\tilde{\Gamma})\big|_{\Gamma_3}$, eq.(8.72) implies that for $\gamma_N v = \varphi_N$

$$\langle \varphi_N^\star, \varphi_N \rangle_{1/2,\Gamma_3} = \langle S_N(\tau), \varphi_N \rangle_{1/2,\Gamma_3} = \big(\tau, \varepsilon(v)\big) = \big(c^{-1/2}\tau, c^{1/2}\varepsilon(v)\big) \quad \forall v \in V_0 \,. \tag{8.76}$$

Hence
$$\langle \varphi_N^\star, \varphi_N \rangle_{1/2,\Gamma_3} \leq \|\tau\|_{\Sigma_0} |\!|\!|v|\!|\!| \quad \forall v \in V_0 \;\; \gamma_N v = \varphi_N \tag{8.77}$$

so that (cf. (8.55))

$$|\!|\!|\varphi_N^\star|\!|\!|_{-1/2,\Gamma_3} \leq \|\tau\|_{\Sigma_0} \quad \forall \tau \in \Sigma_0 \;\; S_N(\tau) = \varphi_N^\star \,. \tag{8.78}$$

Let now $u = u(\varphi_N^\star)$ be the solution of (8.73) and let $\sigma = C\varepsilon(u) \in \Sigma_0$. Then

$$\|\sigma\|_{\Sigma_0}^2 = (c\sigma, \sigma) = \big(C\varepsilon(u), \varepsilon(u)\big) = |\!|\!|u(\varphi_N^\star)|\!|\!|^2 \tag{8.79}$$

Inserting $v = u(\varphi_N^\star)$ into (8.73) implies, due to (8.54), that

$$\begin{aligned}
\langle \varphi_N^\star, \gamma_N u \rangle_{1/2,\Gamma_3} &= a(u,u) = \big(C\varepsilon(u), \varepsilon(u)\big) = |\!|\!|u(\varphi_N^\star)|\!|\!|^2 \\
&= \|\sigma\|_{\Sigma_0} |\!|\!|u(\varphi_N^\star)|\!|\!| \geq \|\sigma\|_{\Sigma_0} |\!|\!|\gamma_N u|\!|\!|_{1/2,\Gamma_3}
\end{aligned} \tag{8.80}$$

From (8.80) we obtain that

$$|\!|\!|\varphi_N^\star|\!|\!|_{-1/2,\Gamma_3} \geq \|\sigma\|_{\Sigma_0}. \tag{8.81}$$

From (8.81) and (8.78),(8.79) we obtain (8.75) q.e.d.

Proposition 8.10 The following expression holds

$$|\!|\!|\varphi_N^\star|\!|\!|_{-1/2,\Gamma_3} = \sup_{\substack{V_0 \\ v \neq 0}} \frac{\langle \varphi_N^\star, \gamma_N v \rangle_{1/2,\Gamma_3}}{|\!|\!|v|\!|\!|} \tag{8.82}$$

Proof: From (8.74) it results by means of (8.54) that

$$\langle \varphi_N^\star, \gamma_N v \rangle_{1/2,\Gamma_3} \leq |\!|\!|\varphi_N^\star|\!|\!|_{-1/2,\Gamma_3} |\!|\!|\gamma_N v|\!|\!|_{1/2,\Gamma_3} \leq |\!|\!|\varphi_N^\star|\!|\!|_{-1/2,\Gamma_3} |\!|\!|v|\!|\!| \tag{8.83}$$

Let also $u = u(\varphi_N^\star)$ be the solution of (8.73). Then for $\sigma = C\varepsilon(u)$ we obtain using (8.75) that

$$\langle \varphi_N^\star, \gamma_N u \rangle_{1/2,\Gamma_3} = a(u,u) = |\!|\!|u(\varphi_N^\star)|\!|\!|^2 = |\!|\!|u(\varphi_N^\star)|\!|\!| \, |\!|\!|\varphi_N^\star|\!|\!|_{-1/2,\Gamma_3} \,. \tag{8.84}$$

But (8.83) implies that

$$\sup_{\substack{V_0 \\ v \neq 0}} \frac{\langle \varphi_N^\star, \gamma_N v \rangle_{1/2,\Gamma_3}}{|\!|\!|v|\!|\!|} \leq |\!|\!|\varphi_N^\star|\!|\!|_{-1/2,\Gamma_3} \tag{8.85}$$

which together with (8.84) implies (8.82) q.e.d.

Now we can give an equivalent form of Π_1 from which its coerciveness follows. We have from $(7.36)\div(7.39)$ that

$$\tilde{\Pi}_1(\mu_N) = \frac{1}{2}\beta(\mu_N,\mu_N) - \gamma(\mu_N) = \frac{1}{2}\langle\mu_N, [G(\mu_N)]_N\rangle_{1/2,\Gamma_3} - \gamma(\mu_N) \qquad (8.86)$$
$$= \frac{1}{2}\langle\mu_N, \gamma_N z\rangle_{1/2,\Gamma_3} - \gamma(\mu_N),$$

where $z \in V_0$ is the solution of the problem (cf. $(7.33),(7.34)$)

$$a(z,v) = \langle\mu_N, \gamma_N v\rangle_{1/2,\Gamma_3} \quad \forall v \in V_0 . \qquad (8.87)$$

Thus setting in (8.87) $v = z$ we have due to (8.75)

$$\Pi^1(\mu_N) = \frac{1}{2}\langle\mu_N, \gamma_N z\rangle_{1/2,\Gamma_3} - \gamma(\mu_N) = \frac{1}{2}a(z,z) - \gamma(\mu_N) \qquad (8.88)$$
$$= \frac{1}{2}|||z|||^2 - \gamma(\mu_N) = \frac{1}{2}|||\mu_N|||^2_{-1/2,\Gamma_3} - \gamma(\mu_N) .$$

This last relation implies the coerciveness of Π^1 and thus Prop. 1.1 implies the existence and uniqueness of the solution of the minimum problem (7.39) of Π^1 over L where L is given in (8.62). The equivalence of (7.39) to (7.44) and to the multivalued integral equation (7.45) implies the existence and uniqueness of the solution $\lambda_N \in L$ of this boundary integral formulation of the Signorini-Fichera problem.

Chapter 9
Boundary Integral Formulation of the Frictional Unilateral Contact B.V.P.

9.1 The Signorini Problem with Given Friction. Primal Problem, Mixed Problem and Approximation Results

In this Chapter we deal with the simplified friction problem which arises when the normal force S_N is assumed as given. Only at the end of this Chapter and in the numerical application do we consider the realistic Coulomb friction problem. In the normal direction we assume contact with a rigid support, i.e. we assume the Signorini boundary conditions. As the aim of the present Chapter is to give also some approximation results, we study a two dimensional plane elasticity problem, i.e. $\Omega \subset \mathbb{R}^2$ is an open bounded subset. The mathematical results of this Chapter are based mainly on [Has84] and on [Has82] and can be easily extended along the lines of the previous Chapter to three dimensional bodies only with the exception of the approximation results which need the assumption that $\Omega \subset \mathbb{R}^2$ and the other simplifying assumptions we are making. An extension of these latter results to $\Omega \subset \mathbb{R}^3$ is not impossible but quite cumbersome and in any case outside of the scopes of the present work.

Ω is an open bounded subset of $\mathbb{R}^2$ with a Lipschitz boundary Γ, decomposed into two nonempty parts Γ_1 and Γ_3, open in Γ, and Γ_3 is a straight-line segment. Let V_0 be the subspace of $H^1(\Omega)$ defined by

$$V_0 = \left\{ v \in H^1(\Omega), \gamma v = 0 \quad \text{on} \quad \Gamma_1 \right\}$$

where γv denotes the trace of v on Γ. The space of all traces of functions, belonging to V_0, is denoted by $H^{1/2}(\Gamma_3)$. The following notation is used in the case of vector-valued functions:

$$\boldsymbol{H}^k(\Omega) = [H^k(\Omega)]^2, \quad \boldsymbol{V} = V_0 \times V_0, \quad \boldsymbol{H}^{1/2}(\Gamma_3) = H^{1/2}(\Gamma_3) \times H^{1/2}(\Gamma_3) \quad \text{etc.}$$

The corresponding norms, defined in the classical way [Nec], are denoted by $\| \cdot \|_k$, $\| \cdot \|_{1/2}$. Symbols $\boldsymbol{V}'$, $\boldsymbol{H}^{-1/2}(\Gamma_3)$ indicate the dual spaces to $\boldsymbol{V}$ and $\boldsymbol{H}^{1/2}(\Gamma_3)$ with dual norms $\| \cdot \|_{\star}$, $\| \cdot \|_{-1/2}$, respectively.

Let $v \in \boldsymbol{H}^1(\Omega)$ and let $\varepsilon(v)$ denote the infinitesimal strain tensor defined in (7.5). The relation between the stress tensor τ and the strain tensor ε is given by the general Hooke's law (7.6).

In order to give the variational formulation, we introduce the closed, convex set of admissible displacements

$$K = \{v | v \in \boldsymbol{V}, v_N = v_i \cdot n_i \leq 0 \quad \text{on} \quad \Gamma_3\} \tag{9.1}$$

Let n, t denote the unit outward normal and the tangential vector to Γ respectively.

Finally, $\Pi\colon V \to \mathbb{R}$, the potential energy functional for the Signorini problem with friction is given by (cf. [Pan85])

$$\Pi(v) \;=\; \frac{1}{2}\int_\Omega C_{ijkl}\varepsilon_{ij}(v)\varepsilon_{kl}(v)d\Omega + \int_{\Gamma_3} g|v_T|d\Gamma - \int_\Omega \bar{p}_i v_i d\Omega \equiv \frac{1}{2}\big(C\varepsilon(v),\varepsilon(v)\big) \quad (9.2)$$
$$+(g,|v_T|)_{\Gamma_3} - (\bar{p},v)$$

where $g \geq 0$ almost everywhere on Γ_3, $g \in L^\infty(\Gamma_3)$, $\bar{p} \in L^2(\Omega)$. Symbols $(,)_{\Gamma_3}$, $(,)$ denote scalar products in $L^2(\Gamma_3)$, $L^2(\Omega)$ respectively.

The primal variational formulation of the Signorini problem with given friction reads: Find $u \in K$ such that

$$\Pi(u) = \inf\{\Pi(v)|v \in K\}. \tag{9.3}$$

By using Prop. 1.1 we obtain easily the following existence and uniqueness result.

Proposition 9.1 There exists a unique solution of (9.3).

In order to approximate (9.3), we apply the Galerkin method which consists of replacing (9.3) by a family of finite dimensional (or discrete) problem. The problem reads

$$\text{``find } u_h \in K_h \text{ such that } \Pi(u_h) \leq \Pi(v_h) \quad \forall v_h \in K_h\text{''}. \tag{9.4}$$

Here $\{K_h\}$, $h \in (0,1)$, is a system of finite-dimensional approximations of K. Let us mention that Π is a nondifferentiable functional. The Galerkin method, directly applied to (9.3) leads to a nondifferentiable optimization problem. However the application of the duality technique transforms the nondifferentiable problem into a differentiable one.

In order to obtain the mixed variational formulation of the B.V.P. considered we introduce (cf. [Has82, Hl]) the sets

$$L_1 = H_-^{-1/2}(\Gamma_3) = \{\mu_1 \in H^{-\frac{1}{2}}(\Gamma_3)|\langle\mu_1,v_N\rangle \geq 0 \quad \forall v \in K\}, \tag{9.5}$$

$$L_2 = \{\mu_2 \in L^2(\Gamma_3)|\,|\mu_2| \leq g \text{ a.e. on } \Gamma_3\}, \tag{9.6}$$

which are two closed convex subsets of $H^{-1/2}(\Gamma_3)$ of Lagrange multipliers and let $\mathcal{L} :$ $V \times L_1 \times L_2 \to \mathbb{R}$, be a Lagrange function defined by means of the relation

$$\mathcal{L}(v,\mu_1,\mu_2) = \frac{1}{2}\big(C\varepsilon(v),\varepsilon(v)\big) - \langle\mu_1,v_N\rangle - \langle\mu_2,v_T\rangle - (\bar{p},v) \tag{9.7}$$

where $\langle,\rangle$ is the duality pairing between $H^{-1/2}(\Gamma_3)$ and $H^{1/2}(\Gamma_3)$.

The mixed variational formulation of the Signorini problem with given friction is defined as the problem of finding a saddle-point $\{w,\lambda_1,\lambda_2\} \in V \times L_1 \times L_2$ of $\mathcal{L}$ on $V \times L_1 \times L_2$ satisfying

$$\mathcal{L}(w,\mu_1,\mu_2) \leq \mathcal{L}(w,\lambda_1,\lambda_2) \leq \mathcal{L}(v,\lambda_1,\lambda_2) \quad \forall v \in V, \ \{\mu_1,\mu_2\} \in L_1 \times L_2. \tag{9.8}$$

The relation between (9.3) and (9.8) is given by [Has84]

Proposition 9.2 There exists a unique solution $\{w, \lambda_1, \lambda_2\}$ of (9.8). Moreover,

$$w = u, \quad \lambda_1 = S_N(u), \quad \lambda_2 = S_T(u), \tag{9.9}$$

where $u \in K$ is the solution of (9.3) and $S_N(u)$, $S_T(u)$ are the corresponding normal and tangential components of the traction vector on Γ_3.

An aproximation scheme of the Signorini problem with given friction, based on the mixed variational formulation, is defined as follows: let $\{V_h\}$, $\{L_H\}$ where h, $H \in (0,1)$, be two systems of finite dimensional subspaces of V, $H^{-1/2}(\Gamma_3)$, respectively and let $\{L_{1H}\}$, $\{L_{2H}\}$ be families of closed convex subsets of L_H.

By the approximation of (9.8), we mean a saddle-point $\{w_h, \lambda_{1H}, \lambda_{2H}\}$ of $\mathcal{L}$ on $V_h \times L_{1H} \times L_{2H}$ such that

$$\mathcal{L}(w_h, \mu_{1H}, \mu_{2H}) \leq \mathcal{L}(w_h, \lambda_{1H}, \lambda_{2H}) \leq \mathcal{L}(v_h, \lambda_{1H}, \lambda_{2H}), \tag{9.10}$$
$$\forall v_h \in V_h, \quad \forall \{\mu_{1H}, \mu_{2H}\} \in L_{1H} \times L_{2H}.$$

or equivalently the problem

$$\text{find } \{w_h, \lambda_{1H}, \lambda_{2H}\} \in V_h \times L_{1H} \times L_{2H} \text{ such that} \tag{9.10a}$$
$$\big(C\varepsilon(w_h), \varepsilon(v_h)\big) = \langle \lambda_{1H}, v_{hN} \rangle + \langle \lambda_{2H}, v_{hT} \rangle + (\bar{p}, v_h), \quad \forall v_h \in V_h,$$
$$\langle \mu_{1H} - \lambda_{1H}, w_{hN} \rangle + \langle \mu_{2H} - \lambda_{2H}, w_{hT} \rangle \geq 0, \quad \forall \{\mu_{1H}, \mu_{2H}\} \in L_{1H} \times L_{2H}.$$

The approximation based on the mixed variational formulation has several positive aspects. The main advantage consists in the fact that $S_N(u)$ and $S_T(u)$ can be calculated independently of u. It is also possible to analyse mathematically the relation between $\lambda_{1H}, \lambda_{2H}$ and λ_1, λ_2. Numerical tests show that $\lambda_{1H}, \lambda_{2H}$ are closer to the exact values $S_N(u)$, $S_T(u)$ than the values calculated from u_h by differentiation.

In order to prove convergence, we make the choice of $V_h, \Lambda_{1H}, \Lambda_{2H}$, more precise [Has84]: We assume now that $\Omega \subset \mathbb{R}^2$, is a polygonal domain and let $\{\mathcal{T}_h\}$, $h \to 0_+$ be a regular family of triangulations of $\bar{\Omega}$ compatible with the decomposition of Γ into Γ_1 and Γ_3. With any $\mathcal{T}_h \in \{\mathcal{T}_h\}$ will be associated a finite-dimensional space V_h, containing all the piecewise linear functions:

$$V_h = \{U_h \in [C(\bar{\Omega})]^2 | \; U_h|_T \in [P_1(T)]^2 \;\; \forall T \in \mathcal{T}_h, U_h = 0 \text{ on } \Gamma_1\}, \tag{9.11}$$

where $P_1(T)$, denotes the space of linear polynomials defined on the triangle $T \in \mathcal{T}_h$. Let $\{\mathcal{T}_H\}, H \to 0_+$ be a family of partitions of the part Γ_3, the nodes of which will be denoted by $\alpha_{1(H)}, \ldots, \alpha_{m(H)}$. We consider regular families of $\{\mathcal{T}_H\}$, for $H \to 0_+$ in the following sense: There exists a positive constant β_0 such that

$$\min H_i / \max H_i \geq \beta_0,$$

where H_i is the length of $\alpha_i \alpha_{i+1}$. It is a special case, if the partition $\mathcal{T}_H$ of Γ_3 is generated by nodes of $\mathcal{T}_h$ lying on Γ_3. Then, we put $h = H$. Let now

$$
\begin{aligned}
L_H &= \{\mu_H \in L^2(\Gamma_3) | \mu_H|_{\alpha_i \alpha_{i+1}} \in P_0(\alpha_i \alpha_{i+1}), \quad \forall i = 1, \ldots, m(H)\}, \tag{9.12} \\
L_{1H} &= \{\mu_{1H} \in L_H | \mu_{1H} \leq 0 \text{ on } \Gamma_3\} \\
L_{2H} &= \{\mu_{2H} \in L_H | \; |\mu_{2H}|_{\alpha_i \alpha_{i+1}} \leq g_i \; \; \forall i = 1, \ldots, m(H)\},
\end{aligned}
$$

where $P_0(\alpha_i\alpha_{i+1})$ is the set of all constant functions defined on $\alpha_i\alpha_{i+1}$ and g_i is the average value of g on $\alpha_i\alpha_{i+1}$. The following convergence result was established in [Has82] where we refer for the proof.

Proposition 9.3 For any $v \in K$, make the hypothesis that a sequence $\{v_h\}$ can be determined

$$v_h \in K_{hH} = \{v_h \in \boldsymbol{V}_h, \langle \mu_{1H}, v_{hN} \rangle \geq 0 \quad \forall \mu_{1H} \in L_{1H}\},$$

such that $v_h \to v$ on $\boldsymbol{V}$ for $h \to 0_+$. Moreover assume that $h \to 0_+$ if and only if $H \to 0_+$. Then

$$\begin{cases} w_h \to w & \text{in } \boldsymbol{H}^1(\Omega), \\ \lambda_{iH} \to \lambda_i & \text{in } L^2(\Gamma_3) \text{ (weakly)} \quad i = 1, 2. \end{cases} \tag{9.13}$$

Note that the assumptions of Prop. 9.3 are satisfied, if there exists a finite number of points $\bar{\Gamma}_1 \cap \bar{\Gamma}_3$ [Has84]. Stronger results are obtained in [Has82], where also rates of convergence are proved based on some stronger assumptions.

Now we describe briefly how (9.8) can be treated numerically. Since (9.10) is a saddle-point formulation in a finite dimensional space, Uzawa's method will be applied ([Eke] p.187): Let $\{\lambda_{1H}^{(0)}, \lambda_{2H}^{(1)}\} \in L_{1H} \times L_{2H}$ be a starting point. From $\{\lambda_{1H}^{(k)}, \lambda_{2H}^{(k)}\} \in L_{1H} \times L_{2H}$ given, we define $w_h^{(k)} \in \boldsymbol{V}_h$ as the solution of the minimization problem

$$\mathcal{L}(w_h^{(k)}, \lambda_{1H}^{(k)}, \lambda_{2H}^{(k)}) \leq \mathcal{L}(v_h, \lambda_{1H}^{(k)}, \lambda_{2H}^{(k)}) \quad \forall v_h \in \boldsymbol{V}_h, \tag{9.14}$$

or equivalently of the equation

$$\big(C\varepsilon(w_h^{(k)}), \varepsilon(v_h)\big) = \langle \lambda_{1H}^{(k)}, v_{hN} \rangle + \langle \lambda_{2H}^{(k)}, v_{hT} \rangle + (\bar{p}, v_h) \quad \forall v_h \in \boldsymbol{V}_h. \tag{9.15}$$

Then we replace $\{\lambda_{1H}^{(k)}, \lambda_{2H}^{(k)}\}$ by the new values $\{\lambda_{1H}^{(k+1)}, \lambda_{2H}^{(k+1)}\}$ given by

$$\lambda_{1H}^{(k+1)} = P_1(\lambda_{1H}^{(k)} - \rho w_{hN}^{(k)}), \quad \lambda_{2H}^{(k+1)} = P_2(\lambda_{2H}^{(k)} - \rho w_{hT}^{(k)}), \tag{9.16}$$

where $\rho > 0$ and P_1, P_2 is the projection of $\boldsymbol{V}_h$ on L_{1H}, L_{2H} respectively. For an appropriate choice of ρ, Uzawa's method converges in the first component, i.e. $w_h^{(k)} \to w_h$, as $k \to \infty$, and if $\{\lambda_{1H}, \lambda_{2H}\} \in L_{1H} \times L_{2H}$ is unique, then

$$\lambda_{1H}^{(k)} \to \lambda_{1H}, \quad \lambda_{2H}^{(k)} \to \lambda_{2H}, \quad \text{as } k \to \infty.$$

as well. Looking at (9.15), we see that $w_h^{(k)}$ are approximate solutions of the combined linear elasticity problems, in which only the normal and tangential tractions along Γ_3 change. The solutions of the arising classical linear elasticity problems can be obtained by any classical B.E.M. or F.E.M. routine.

9.2 The Derivation of a Multivalued B.I.E. for the Signorini Problem with Given Friction.

In the mixed variational formulation, presented in the last section, normal and tangential tractions along Γ_3 play the role of Lagrange multipliers associated with the inequality constraint and the nondifferentiable term of Π. Elimination of the displacement field w leads to a minimum problem for $S_N(u)$, $S_T(u)$ only, which is equivalent to a multivalued B.I.E. The derivation of this formulation and the study of the corresponding finite dimensional approximation will be the subject of the present section.

Let $\{w, \lambda_1, \lambda_2\} \in V \times L_1 \times L_2$ be the solution of (9.8). Then

$$\mathcal{L}(w, \lambda_1, \lambda_2) = \inf_{V} \sup_{L_1 \times L_2} \mathcal{L}(v, \mu_1, \mu_2) = \sup_{L_1 \times L_2} \inf_{V} \mathcal{L}(v, \mu_1, \mu_2) . \tag{9.17}$$

Let us set

$$\tilde{\Pi}(\mu_1, \mu_2) = \inf_{V} \mathcal{L}(v, \mu_1, \mu_2). \tag{9.18}$$

We derive the explicit form of $\tilde{\Pi}$. For $\{\mu_1, \mu_2\} \in L_1 \times L_2$ fixed, (9.18) is equivalent to the following linear B.V.P.: Find $u = u(\mu_1, \mu_2) \in V$ such that

$$\big(C\varepsilon(u), \varepsilon(v)\big) = \langle \mu_1, v_N \rangle + \langle \mu_2, v_T \rangle + (\bar{p}, v), \quad \forall v \in V. \tag{9.19}$$

Due to the linearity of (9.19) one can split u into u_1 and u_2 and write $u = u_1 + u_2$ where $u_1, u_2 \in V$ are the unique solutions of the linear elliptic B.V.Ps

$$\big(C\varepsilon(u_1), \varepsilon(v)\big) = (\bar{p}, v) \quad \forall v \in V, \tag{9.20}$$

$$\big(C\varepsilon(u_2), \varepsilon(v)\big) = \langle \mu_1, v_N \rangle + \langle \mu_2, v_T \rangle \quad \forall v \in V. \tag{9.21}$$

In the sequel, we denote by $G: V' \to V$ the Green's operator, corresponding to V and to the bilinear form $\big(C\varepsilon(u), \varepsilon(v)\big)$ and instead of (9.20),(9.21) we write

$$u_1 = G(\bar{p}), \quad u_2 = G(\mu_1, \mu_2), \quad \{\mu_1, \mu_2\} \in L_1 \times L_2 . \tag{9.22}$$

Since $u \in V$ is a minimizer of the quadratic functional $\mathcal{L}(v, \mu_1, \mu_2)$, we have

$$\begin{aligned} \mathcal{L}(u, \mu_1, \mu_2) &= -\frac{1}{2}\langle \mu_1, u_N \rangle - \frac{1}{2}\langle \mu_2, u_T \rangle - \frac{1}{2}(\bar{p}, u) = -\frac{1}{2}\langle \mu_1, u_{2N} \rangle \\ &\quad -\frac{1}{2}\langle \mu_2, u_{2T} \rangle - \frac{1}{2}\langle \mu_1, u_{1N} \rangle - \frac{1}{2}\langle \mu_2, u_{1T} \rangle - \frac{1}{2}(\bar{p}, u_2) - \frac{1}{2}(\bar{p}, u_1) \end{aligned} \tag{9.23}$$

Setting $v = u_2$ in (9.20) and using the symmetry of $\big(C\varepsilon(u), \varepsilon(v)\big)$ and (9.21), implies

$$\langle \mu_1, u_{1N} \rangle + \langle \mu_2, u_{1T} \rangle = \big(C\varepsilon(u_2), \varepsilon(u_1)\big) = \big(C\varepsilon(u_1), \varepsilon(u_2)\big) = (\bar{p}, u_2) \tag{9.24}$$

Thus (9.23) becomes

$$\begin{aligned} \mathcal{L}(u, \mu_1, \mu_2) &= -\frac{1}{2}\langle \mu_1, u_{2N} \rangle - \frac{1}{2}\langle \mu_2, u_{2T} \rangle - \langle \mu_1, u_{1N} \rangle - \langle \mu_2, u_{1T} \rangle - \frac{1}{2}(\bar{p}, u_1) \tag{9.25} \\ &\equiv -\frac{1}{2}\langle \mu_1, G(\mu_1, \mu_2)n \rangle - \frac{1}{2}\langle \mu_2, G(\mu_1, \mu_2)t \rangle - \langle \mu_1, G(\bar{p})n \rangle - \langle \mu_2, G(\bar{p})t \rangle \\ &\quad - \frac{1}{2}(\bar{p}, u_1) . \end{aligned}$$

Let us denote by $\beta: [H^{-1/2}(\Gamma_3)]^2 \times [H^{-1/2}(\Gamma_3)]^2 \to \mathbb{R}$ the bilinear form

$$\beta(\mu, \nu) = \langle \mu_1, G(\nu_1, \nu_2)n \rangle + \langle \mu_2, G(\nu_1, \nu_2)t \rangle, \quad \mu = \{\mu_1, \mu_2\}, \nu = \{\nu_1, \nu_2\} \tag{9.26}$$

and let $l: [H^{-1/2}(\Gamma_3)]^2 \to \mathbb{R}$ be given by

$$l(\mu) = -\langle \mu_1, G(\bar{p})n \rangle - \langle \mu_2, G(\bar{p})t \rangle . \tag{9.27}$$

Thus, we can rewrite $\tilde{\Pi}$ as follows:

$$\tilde{\Pi}(\mu) = \tilde{\Pi}(\mu_1, \mu_2) = -\frac{1}{2}\beta(\mu, \mu) + l(\mu) - \frac{1}{2}(\bar{p}, u_1) \tag{9.28}$$

Finally, let us set

$$\tilde{\Pi}_1(\mu) = -\tilde{\Pi}(\mu) - \frac{1}{2}(\bar{p}, u_1) = \frac{1}{2}\beta(\mu, \mu) - l(\mu). \tag{9.29}$$

Now we can define the following minimum problem. It is the problem of finding $\lambda = \{\lambda_1, \lambda_2\} \in L_1 \times L_2$ such that

$$\tilde{\Pi}_1(\lambda) = \inf\{\tilde{\Pi}_1(\mu)|\mu = \{\mu_1, \mu_2\} \in L_1 \times L_2\} \tag{9.30}$$

Note that (9.30) is equivalent (prop. 1.2) to the variational inequality: "Find $\lambda = \{\lambda_1, \lambda_2\} \in L_1 \times L_2$ such that

$$\beta(\lambda, \mu - \lambda) - l(\mu - \lambda) \geq 0 \quad \forall \mu = \{\mu_1, \mu_2\} \in L_1 \times L_2 \text{ "} \tag{9.31}$$

and to the multivalued B.I.E.

$$l - \frac{1}{2}\mathrm{grad}\beta(\lambda, \lambda) \in \partial I_{L_1 \times L_2}(\lambda) \tag{9.32}$$

where I is the indicator of $L_1 \times L_2$.

Now the relation between the primal problem and (9.30) or equivalently (9.31) or (9.32) is given by the following proposition.

Proposition 9.4 There exists a unique solution $\{\lambda_1, \lambda_2\}$, of (9.30) or (9.32). Moreover,

$$\lambda_1 = S_N(u), \quad \lambda_2 = S_T(u), \tag{9.33}$$

where $u \in K$ is the solution u of the primal problem (9.3).

Proof: Let u be the solution of (9.3) and let $S_N(u) = \lambda_1$ and $S_T(u) = \lambda_2$ be the corresponding normal and tangential forces on Γ_3; let also $w = u$. Then $\{\lambda_1, \lambda_2\} \in L_1 \times L_2$ and the triple $\{w, \lambda_1, \lambda_2\}$ is a solution of (9.8) (Prop. 9.2), i.e.

$$\mathcal{L}(w, \mu_1, \mu_2) \leq \mathcal{L}(w, \lambda_1, \lambda_2) \leq \mathcal{L}(v, \lambda_1, \lambda_2), \forall v \in \mathbf{V}, \quad \{\mu_1, \mu_2\} \in L_1 \times L_2. \tag{9.34}$$

The second inequality yields

$$\mathcal{L}(w, \lambda_1, \lambda_2) = \inf_{v \in \mathbf{V}} \mathcal{L}(v, \lambda_1, \lambda_2) = \tilde{\Pi}(\lambda_1, \lambda_2), \tag{9.35}$$

where $\tilde{\Pi}$ is given by (9.28). On the other hand,

$$\mathcal{L}(w, \lambda_1, \lambda_2) = \sup_{L_1 \times L_2} \inf_{\mathbf{V}} \mathcal{L}(v, \mu_1, \mu_2) = \sup_{L_1 \times L_2} \tilde{\Pi}(\mu_1, \mu_2) . \tag{9.36}$$

By comparing (9.35) with (9.36) and using the relation between Π_1 and $\tilde{\Pi}_1$ we see that $\big(S_N(u), S_T(u)\big)$ is a minimizer of $\tilde{\Pi}_1$ on $L_1 \times L_2$. The $[H^{-1/2}(\Gamma_3)]^2$-ellipticity of the form β, which will be proved later, implies the uniqueness and the existence of the minimizer of $\tilde{\Pi}_1$ or $\tilde{\Pi}$ over $L_1 \times L_2$. The uniqueness implies the result q.e.d.

Another usefull expression for $\tilde{\Pi}$, is

$$\tilde{\Pi}(\mu_1, \mu_2) = \mathcal{L}(u, \mu_1, \mu_2) = \inf_V \mathcal{L}(v, \mu_1, \mu_2) = -\frac{1}{2}\big(C\varepsilon(u), \varepsilon(u)\big), \qquad (9.37)$$

where $u \in V$ is the solution of (9.19). Setting $\mu_1 = \lambda_1$, $\mu_2 = \lambda_2$ implies

$$\tilde{\Pi}(\lambda_1, \lambda_2) = -\frac{1}{2}\big(C\varepsilon(w), \varepsilon(w)\big), \qquad (9.38)$$

where w is the first component of the saddle-point of $\mathcal{L}$ on $V \times L_1 \times L_2$.

Let us give now some approximation results for the multivalued B.I.E. (9.32). We introduce, as before, two families $\{L_{1H}\}, \{L_{2H}\}$ of closed convex subsets. Each L_{1H}, L_{2H} will be imbedded into a finite-dimensional space L_H. The usual Galerkin method, consisting of replacing (9.32) or equivalently (9.30) by the corresponding finite dimensional minimum problem is not applicable, since the explicit form of G is known only in special cases. Therefore, some approximations of G must be used. Here we describe some of the possible approximations: Let A_h be a stiffness matrix, corresponding to the restriction of the bilinear form $\big(C\varepsilon(z), \varepsilon(v)\big)$ to a finite-dimensional subspace V_h of V. As an approximation G_h on G, we consider the inverse of A_h. Another approach would be the following: as an approximation of G we can consider an operator G_H giving the normal and tangential displacements due to μ_{1H} and μ_{2H} according to (9.22). Operator G_H results, for instance, by applying a B.E.M. scheme to the bilateral problem (9.20) and (9.21).

By the approximation of the multivalued B.I.E. (9.32) or equivalently of (9.30), we mean the problem of finding $\{\lambda_{1H}, \lambda_{2H}\} \in L_{1H} \times L_{2H}$ such that

$$\tilde{\Pi}_d(\lambda_{1H}, \lambda_{2H}) \leq \tilde{\Pi}_d(\mu_{1H}, \mu_{2H}) \quad \forall\{\mu_{1H}, \mu_{2H}\} \in L_{1H} \times L_{2H} \qquad (9.39)$$

where

$$\tilde{\Pi}_d(\mu_{1H}, \mu_{2H}) = \frac{1}{2}\langle \mu_{1H}, G_d(\mu_{1H}, \mu_{2H})n\rangle + \frac{1}{2}\langle \mu_{2H}, G_d(\mu_{1H}, \mu_{2H})t\rangle \qquad (9.40)$$
$$+ \langle \mu_{1H}, G_d(\bar{p}_d)n\rangle + \langle \mu_{2H}, G_d(\bar{p}_d)t\rangle \, .$$

Here G_d denotes the discretized operators G_h or G_H, $\bar{p}_d$ is the loading for the discrete problem, and $\tilde{\Pi}_d$ is the discretized energy functional.

The case $G_d = G_h$ has been fully examined in [Has84]. Here we shall study the case of the B.E.M. approximation of the unilateral problem, i.e. the case $G_d = G_H$. Replacing in all the previous relations μ_1 (resp. λ_1) and μ_2 (resp. λ_2) by μ_{1H} (resp. λ_{1H}) and μ_{2H} (resp. λ_{2H}) we obtain as in (9.20)$\div$(9.28), i.e. setting in (9.21) μ_{1H}, μ_{2H} etc., that

$$u_{1H} = G_H(\bar{p}_H), \quad u_{2H} = G_H(\mu_{1H}, \mu_{2H}), \quad \{\mu_{1H}, \mu_{2H}\} \in L_{1H} \times L_{2H} \, . \qquad (9.41)$$

Then we may write that

$$\tilde{\Pi}_H(\mu_H) \;=\; \tilde{\Pi}_H(\mu_{1H}, \mu_{2H}) = -\frac{1}{2}\beta_H(\mu_H, \mu_H) + l_H(\mu_H) - \frac{1}{2}(\bar{p}_H, u_{1H}) \tag{9.42}$$

$$\;=\; -\tilde{\Pi}_{1H}(\mu_H) - \frac{1}{2}(\bar{p}_H, u_{1H})$$

where $\beta_H \colon L_H \times L_H \to \mathbb{R}$ is given by

$$\beta_H(\mu_H, \nu_H) = \int_{\Gamma_3} \mu_{1H} G_H(\nu_{1H}, \nu_{2H}) n d\Gamma + \int_{\Gamma_3} \mu_{2H} G_H(\nu_{1H}, \nu_{2H}) t d\Gamma \tag{9.43}$$

and $l_H \colon L_H \to \mathbb{R}$ is given by

$$l_H(\mu_H) = -\int_{\Gamma_3} \mu_{1H} G_H(\bar{p}_H) n d\Gamma - \int_{\Gamma_3} \mu_{2H} G_H(\bar{p}_H) t d\Gamma. \tag{9.44}$$

Then the following proposition holds.

Proposition 9.5 Suppose that the B.E. scheme is such that $u_{1H}, u_{2H} \in \boldsymbol{V}$. Moreover let the condition

$$\langle \mu_{1H}, v_{HN} \rangle + \langle \mu_{2H}, v_{HT} \rangle = 0 \quad \forall v_H \in \boldsymbol{V} \iff \mu_{1H} = \mu_{2H} = 0 \tag{9.45}$$

hold. Then the minimization problem of $\tilde{\Pi}_{1H}$ over $L_{1H} \times L_{2H}$ admits a unique solution $\{\lambda_{1H}, \lambda_{2H}\}$.

Proof: From (9.43) we obtain using (9.41) and the Green-Gauss theorem that

$$\beta_H(\mu_H, \mu_H) \;=\; \langle \mu_{1H}, u_{2HN} \rangle + \langle \mu_{2H}, u_{2HT} \rangle \tag{9.46}$$

$$\;=\; (C\varepsilon(u_{2H}), \varepsilon(u_{2H})) \geq c\|u_{2H}\|^2 \quad c \text{ const} > 0.$$

This last inequality results from Korn's inequality and from (9.45). But (9.46) implies the existence and uniqueness of $\{\lambda_{1H}, \lambda_{2H}\}$ q.e.d.

Now, we make several observations concerning the norm of $(H^{-1/2}(\Gamma_3))^2$. Let $\boldsymbol{V}$ be endowed with the norm (8.53) which is denoted here as $\|\|v\|\|$. The set $\boldsymbol{H}^{1/2}(\Gamma_3)$ of all traces of functions, belonging to $\boldsymbol{V}$, will have the norm

$$\|\varphi\|_{1/2,\Gamma_3} = \inf\left\{ \|\|v\|\| \,\big|\, v \in \boldsymbol{V}, \ \gamma v = \varphi \ \varphi = (\varphi_1, \varphi_2) \in \boldsymbol{H}^{1/2}(\Gamma_3) \right\} \tag{9.47}$$

Here γv is the trace of $v \in \boldsymbol{V}$. It results that

$$\|\varphi\|_{1/2,\Gamma_3} = \|\|u\|\| \tag{9.48}$$

where $u \in \boldsymbol{H}^1(\Omega)$ is the unique solution of the B.V.P.

$$\begin{cases} (C\varepsilon(u), \varepsilon(v)) = 0 \quad \forall v \in \boldsymbol{V} \\ \gamma v = \varphi \quad \text{on} \quad \Gamma_3 \end{cases} \tag{9.48a}$$

Let $\delta\colon V \to H^{1/2}(\Gamma_3)$ be a mapping defined by the relation $\delta v = (v_N, v_T)$. Finally, by Σ, we denote the space of symmetric tensors, the divergence of which is square integrable in Ω:

$$\Sigma = \left\{\sigma \,\middle|\, \sigma = \{\sigma_{ij}\},\ \{\sigma_{ij}\} \in [L^2(\Omega)]^4,\ \sigma_{ij} = \sigma_{ji},\ \operatorname{div}\sigma = \{\sigma_{ij,j}\} \in [L^2(\Omega)]^2\right\}. \quad (9.49)$$

We equip Σ with the norm

$$\|\sigma\|_\Sigma = \{(c\sigma, \sigma) + (\operatorname{div}\sigma, \operatorname{div}\sigma)\}^{1/2}. \quad (9.50)$$

The closed subspace of Σ consisting of functions with zero divergence will be denoted by Σ_0. Clearly, Σ_0 is a Hilbert space with norm

$$\|\sigma\|_{\Sigma_0} = (c\sigma, \sigma)^{1/2}. \quad (9.50a)$$

The following Green-Gauss formulas hold for any $\tau \in \Sigma$ and $v \in V$:

(i) there exists a unique mapping $S = (S_1, S_2) \in \mathcal{L}(\Sigma, H^{-1/2}(\Gamma_3))$ such that

$$(\tau, \varepsilon(v)) + (\operatorname{div}\tau, v) = \langle S(\tau), \gamma v\rangle = \langle S_1(\tau), \gamma v_1\rangle + \langle S_2(\tau), \gamma v_2\rangle. \quad (9.51)$$

(ii) there exists a unique mapping $\tilde{S} = (S_N, S_T) \in \mathcal{L}(\Sigma, H^{-1/2}(\Gamma_3))$ such that

$$(\tau, \varepsilon(v)) + (\operatorname{div}\tau, v) = \langle \tilde{S}(\tau), \delta v\rangle = \langle S_N(\tau), v_N\rangle + \langle S_T(\tau), v_T\rangle. \quad (9.52)$$

Now we describe certain properties of S and $\tilde{S}$.

Proposition 9.6 S maps Σ_0 onto $H^{-1/2}(\Gamma_3)$.

Proof: Let $\varphi^\star \in H^{-1/2}(\Gamma_3)$ be given. Then the problem

$$\begin{cases} \text{find } u \in V \text{ such that} \\ (C\varepsilon(u), \varepsilon(v)) = \langle \varphi^\star, \gamma v\rangle, \quad \forall v \in V, \end{cases} \quad (9.53)$$

has a unique solution $u = u(\varphi^\star)$. Setting $\sigma = C\varepsilon(u)$ and using (i), implies immediately that $\sigma \in \Sigma_0$ and $S(\sigma) = \varphi^\star$ q.e.d.

The symbol $\|\varphi^\star)\|_{-1/2,\Gamma_3}$ denotes the usual dual norm, i.e.

$$\|\varphi^\star\|_{-1/2,\Gamma_3} = \sup_{\substack{H^{1/2}(\Gamma_3) \\ \varphi \neq 0}} \frac{\langle \varphi^\star, \varphi\rangle}{\|\varphi\|_{1/2,\Gamma_3}} \quad (9.54)$$

The following proposition holds:

Proposition 9.7 The equalities,

$$\|\varphi^\star\|_{-1/2,\Gamma_3} = \|\sigma\|_{\Sigma_0} = \||u(\varphi^\star)\|| \quad \forall \varphi^\star \in H^{-1/2}(\Gamma_3), \quad (9.55)$$

hold, where $u = u(\varphi^\star)$ is the solution of (9.53) and $\sigma = C\varepsilon(u)$.

Proof: Let $\tau \in \Sigma_0$ satisfy $S(\tau) = \varphi^\star$. Then, for any $\varphi \in H^{-1/2}(\Gamma_3)$, (i) yields

$$\langle \varphi^\star, \varphi\rangle = \langle S(\tau), \varphi\rangle = (\tau, \varepsilon(v)) = (C^{-1/2}\tau, C^{1/2}\varepsilon(v)) \quad \forall v \in V,\ \gamma v = \varphi. \quad (9.56)$$

Hence

$$\langle \varphi^\star, \varphi \rangle \leq ||\tau||_{\Sigma_0} |||v||| \quad \forall v \in \boldsymbol{V} \quad \gamma v = \varphi \tag{9.57}$$

so that

$$||\varphi^\star|||_{-1/2,\Gamma_3} \leq ||\tau||_{\Sigma_0} \quad \forall \tau \in \Sigma_0, \quad S(\tau) = \varphi^\star . \tag{9.58}$$

Let $u = u(\varphi^\star)$ be the solution of (9.53) and let $\sigma = C\varepsilon(u) \in \Sigma_0$. Then

$$||\sigma||_{\Sigma_0}^2 = (C^{-1}\sigma, \sigma) = \big(\varepsilon(u), C\varepsilon(u)\big) = |||u(\varphi^\star)|||^2 . \tag{9.59}$$

Inserting $v = u(\varphi^\star)$ into (9.53), using (9.59) and the definition of $|| \ ||_{1/2,\Gamma_3}$, yields

$$\begin{aligned}
\langle \varphi^\star, \gamma u \rangle &= \big(C\varepsilon(u), \varepsilon(u)\big) = |||u(\varphi^\star)|||^2 = ||\sigma||_{\Sigma_0} |||u||| \\
&\geq ||\sigma||_{\Sigma_0} ||\gamma u||_{1/2,\Gamma_3} .
\end{aligned} \tag{9.60}$$

Hence

$$||\varphi^\star||_{-1/2,\Gamma_3} \geq ||\sigma||_{\Sigma_0}. \tag{9.61}$$

This inequality and (9.58) yield the result.   q.e.d.

Proposition 9.8 The following expression holds:

$$||\varphi^\star||_{-1/2,\Gamma_3} = \sup_{\substack{\boldsymbol{V} \\ v \neq 0}} \frac{\langle \varphi^\star, \gamma v \rangle}{|||v|||}, \quad \forall \varphi^\star \in \boldsymbol{H}^{-1/2}(\Gamma_3) . \tag{9.62}$$

Proof: From the definition of $||\varphi^\star||_{-1/2,\Gamma_3}$ as the dual norm, it follows that

$$\langle \varphi^\star, \gamma v \rangle \leq ||\varphi^\star|||_{-1/2,\Gamma_3} ||\gamma v||_{1/2,\Gamma_3} \leq ||\varphi^\star||_{-1/2,\Gamma_3} |||v||| \tag{9.63}$$

and thus for $v \neq 0$

$$\sup_{\boldsymbol{V}} \frac{\langle \varphi^\star, \gamma v \rangle}{|||v|||} \leq ||\varphi^\star||_{-1/2,\Gamma_3} . \tag{9.64}$$

Let now $u = u(\varphi^\star)$ be the solution of (9.53). Then for $\sigma = C\varepsilon(u)$, we may write

$$\begin{aligned}
\langle \varphi^\star, \gamma u \rangle &= |||u(\varphi^\star)|||^2 = |||u(\varphi^\star)||| \cdot ||\sigma||_{\Sigma_0} \\
&= |||u(\varphi^\star)||| \cdot ||\varphi^\star||_{-1/2,\Gamma_3} ,
\end{aligned} \tag{9.65}$$

because of Prop. 9.7. Relations (9.64) and (9.65) imply the result.   q.e.d.

In a similar way one can prove the following Proposition.

Proposition 9.9 $\tilde{S}$ maps Σ_0 onto $\boldsymbol{H}^{-1/2}(\Gamma_3)$.

Let $\varphi^\star$ be an arbitrary element of $\boldsymbol{H}^{-1/2}(\Gamma_3)$. By virtue of Prop. 9.9, there exists $\tau \in \Sigma_0$ such that $\tilde{S}(\tau) = \varphi^\star$. Let us set $\mu^\star = \tilde{S}(\tau) \in \boldsymbol{H}^{-1/2}(\Gamma_3)$ and write $\mu^\star = \beta^\star \varphi^\star$ with $\beta^\star \colon \boldsymbol{H}^{-1/2}(\Gamma_3) \to \boldsymbol{H}^{-1/2}(\Gamma_3)$. It is readily seen that $\mu^\star$ does not depend on the choice of $\tau \in \Sigma_0$ where $\tilde{S}(\tau) = \varphi^\star$. By comparing (i), (ii), we obtain easily the following result:

Proposition 9.10 For any $\varphi^\star \in \boldsymbol{H}^{-1/2}(\Gamma_3)$,

$$||\varphi^\star||_{-1/2,\Gamma_3} = ||\beta^\star \varphi^\star||_{-1/2,\Gamma_3} . \tag{9.66}$$

Now by using previous results, one can give an equivalent form of $\tilde{\Pi}_1$ from which the coerciveness follows. We have due to (9.52) that

$$
\begin{aligned}
\tilde{\Pi}_1(\mu) &= \frac{1}{2}\beta(\mu,\mu) - l(\mu) = \frac{1}{2}\langle \mu_1, G(\mu_1,\mu_2)n\rangle + \frac{1}{2}\langle \mu_2, G(\mu_1,\mu_2)t\rangle - l(\mu) \qquad (9.67)\\
&= \frac{1}{2}\langle \mu_1, z_N\rangle + \frac{1}{2}\langle \mu_2, z_T\rangle - l(\mu) = \frac{1}{2}\langle \mu, \delta z\rangle - l(\mu)\\
&= \frac{1}{2}|||z(\mu)|||^2 - l(\mu), \quad \mu = \{\mu_1, \mu_2\}\,.
\end{aligned}
$$

Here $z \in V$ is the solution of

$$
\big(C\varepsilon(z), \varepsilon(v)\big) = \langle \mu, \delta v\rangle = \langle \beta^{\star -1}\mu, \gamma v\rangle, \quad \forall v \in V\,. \qquad (9.68)
$$

$\beta^{\star -1}: H^{-1/2}(\Gamma_3) \to H^{-1/2}(\Gamma_3)$ denotes the inverse of $\beta^\star$ and we have set in (9.68) $v = z$. From Prop. 9.7 and 9.10 we can write

$$
|||z(\mu)|||^2 = ||\beta^{\star -1}\mu||^2_{-1/2,\Gamma_3} = ||\mu||^2_{-1/2,\Gamma_3}\,. \qquad (9.69)
$$

Hence $\tilde{\Pi}_1$ can be written in the form

$$
\tilde{\Pi}_1(\mu) = \frac{1}{2}||\mu||^2_{-1/2,\Gamma_3} - l(\mu)\,. \qquad (9.70)
$$

A similar form can be written for $\tilde{\Pi}_{1H}$. Indeed, let $\mu = \{\mu_1, \mu_2\} \in H^{-1/2}(\Gamma_3)$ be given and let us introduce the norm (for $v_H \neq 0$)

$$
||\mu||_{-1/2,H} = \sup_{V} \frac{\langle \mu, \delta v_H\rangle}{|||v_H|||}\,. \qquad (9.71)
$$

It is easy to see that a necessary and sufficient condition for (9.71) to be a norm on $L_{1H} \times L_{2H}$ is that the condition (9.45) is satisfied. Next we shall consider only such B.E. schemes for which (9.45) is satisfied.

Let $\mu \in H^{-1/2}(\Gamma_3)$. Then

$$
||\mu||_{-1/2\,H} \leq |||z_H|||\,. \qquad (9.72)
$$

where $z_H \in V$ is the unique solution of

$$
\big(C\varepsilon(z_H), \varepsilon(v_H)\big) = \langle \mu, \delta v_H\rangle \quad \forall v_H \in V\,. \qquad (9.73)
$$

Inserting $v_H = z_H$ into (9.73) implies

$$
\langle \mu, \delta z_H\rangle = |||z_H|||^2\,, \qquad (9.73a)
$$

which, together with (9.71), (9.72) yields that

$$
||\mu||_{-1/2,H} = |||z_H|||\,. \qquad (9.74)
$$

Following the definition of $\tilde{\Pi}_{1H}$, we have, due to (9.43) and (9.73) that

$$\tilde{\Pi}_{1H}(\mu_H) = \frac{1}{2}\beta_H(\mu_H, \mu_H) - l_H(\mu_H) = \frac{1}{2}\|\mu_H\|^2_{-1/2,H} - l_H(\mu_H). \tag{9.75}$$

Proposition 9.11: Suppose that the B.E.M. scheme is such that as $H \to 0$

$$a(u_H, u_H) \to a(u, u) \tag{9.76}$$

Then

$$\tilde{\Pi}_H(\lambda_{1H}, \lambda_{2H}) \to \tilde{\Pi}(\lambda_1, \lambda_2). \tag{9.77}$$

Proof: (9.77) results immediately from (9.38) and the corresponding relation for $\tilde{\Pi}_H$ (cf. (9.73) and (9.43)) because of (9.76).   q.e.d.

For a more profound discussion of the approximation properties we need to make more precise the B.E.M. scheme applied for the discretization of the boundary. Note that with the same methods as in the previous Chapters 7 and 8 we can derive for the problem treated here a multivalued B.I.E. with respect to the unknown normal and tangential displacements on Γ_3. Since the results of the present Chapter are special cases of the results of Ch. 10 the reader is referred there in this context. Note finally that if in (9.31) λ_N is considered as fixed and given, i.e. $\lambda_N = C_N$, then we get the corresponding B.I.E. for the simple friction problem with given normal force.

9.3 On the Coulomb Friction Problem. Numerical Results

As in the previous Chapter the multivalued B.I.E. (9.32) gives rice, after the formulation of its discrete form, e.g., by a classical B.E.M. scheme, to a variational inequality with a full, positive definite matrix and with inequality subsidiary conditions induced by the sets L_{1H} and L_{2H}. The matrix is generally nonsymmetric because of the numerical scheme, although according to Betti's theorem it should be symmetric. Usually we symmetrize the problem by ignoring in the numerical calculations the fluctuations of the matrix elements causing the lack of symmetry, if they are small enough. This is done in order to obtain a Q.P.P. for which very efficient algorithms are available, otherwise we must apply an algorithm for a variational inequality [Glo].

Recall that the problem treated here is not the general unilateral contact problem with friction. Indeed this problem obeys the relation (1.92) and g depends on the unknown normal traction S_N. Thus in (9.2) the term $\int_{\Gamma_3} g|v_T|d\Gamma$, $g \in L^\infty(\Gamma_3)$, must be replaced by $\langle \mu(\tilde{S}_N), |v_T| \rangle$ where $\mu > 0$ is the friction coefficient and $\tilde{S}_N \in H_+^{-1/2}(\Gamma_3)$. Let $u = u(\tilde{S}_N)$ be the solution of the following Signorini problem with given friction equal to $\mu\tilde{S}_N$: Find $u = u(\tilde{S}_N) \in K$ such that

$$\big(C\varepsilon(u), \varepsilon(v-u)\big) + \langle \mu\tilde{S}_N, |v_T| - |u_T| \rangle \geq (\bar{p}, v-u) \quad \forall v \in K \tag{9.78}$$

Easily one can be prove that the normal traction $-S_N(u) \in H^{-1/2}(\Gamma_3)$. Thus we can define a mapping $\Phi: H_+^{-1/2}(\Gamma_3) \to H_+^{-1/2}(\Gamma_3)$ by setting

$$\Phi(\tilde{S}_N) = -S_N(u). \tag{9.79}$$

Then a weak solution of a Signorini problem with Coulomb friction is defined as the fixed point of Φ on $H_+^{-1/2}(\Gamma_3)$. The existence of this fixed point for small values of μ has been proved in [Neč80a, Ja83;84]. This procedure of the fixed point bas been previously introduced in a heuristic manner by the second author in [Pan75] and is described with respect to the more general B.V.P. defined by (1.100), (1.101) in Ch. 1.

Along the lines of the present Chapter multivalued B.I.Es can be derived for all other approaches to the unilateral frictional contact problem, e.g. for the quasivariational inequality approach [Co, Tel], the L.C.P. approach [Kla84;87;88, Li86;89, Kwa88;91, Fah92], the nonlocal friction approach [Mar, Duv80, Rab, De, Od], the differential inclusion approach (cf. e.g. [Mon] and the corresponding discrete mass problem [Mor88c]) and all other formulations and approaches to the friction problem ((e.g. [Ta, Li87;88;91, Jea85;87, Zh88;89, Kla90a,b,c, Spe82;85;87, Ra, Gol, Mit, An, Cam, Ode85, Cur86;88, Mar86, Pir82, Sa, Gas88a,b, Rao, Fei, Ka88a,b, Bis86;90, Kik]. Indeed in all the above problems, roughly speaking, the internal degrees of freedom can be eliminated, for the usual linear elasticity framework, thus giving rise to corresponding boundary formulations of the problem.

The following conclusions result from the numerical experience gained in the last twenty years of research of the group of the second author on the numerical methods for large scale Signorini problems with Coulomb friction:

  i)   The friction law should be treated directly with its complete vertical branch. A regularization of the friction law (cf. fig. 1.2) may lead to wrong results especially for large scale problems.

 ii)   The friction cone should not be linearized (i.e. replaced by a pyramid)

Here we have applied for the numerical treatment of the example the algorithm proposed in [Pan75], which has been named by Kalker [Ka88b;90], the PANA-algorithm. This algorithm reads: Suppose that $S_N^{(1)} = C_N^{(1)}$ where $C_N^{(1)}$ is given and then let us solve the simple friction problem without unilateral contact which yields a tangential traction field $S_T^{(1)}$. Then the simple Signorini-Fichera unilateral contact problem is solved for $S_T = S_T^{(1)}$. Its solution gives a new value of S_N, say $S_N^{(2)}$, which is again introduced into the simple friction problem and so on until the i-th iteration for which $|S_T^{(i+1)} - S_T^{(i)}|$ and $|S_N^{(i+1)} - S_N^{(i)}|$ become sufficiently small. Each of the subproblems is solved here by considering the multivalued B.I.E. (e.g. (7.45) for the Signorini-Fichera subproblem and (9.32) with λ_N given for the friction subproblem). In order to obtain the corresponding Q.P.P. the unit loading method as illustrated in Sect. 7.4 is used. It is combined with a classical (bilateral) B.E.M. for the determination of G_H. In fig. 9.1 we give the structure which has been studied. In fig. 9.2, 9.3 the stress diagramms are given, whereas in fig. 9.4 (resp. fig. 9.5) the diagrams of S_N and S_T (resp. $[u_N]$ and $[u_T]$) at the interfaces. The B.E.M. scheme applied is a direct classical linear interpolation scheme. To avoid the nondifferentiable term $\|[v_T]\|$ in the friction subproblem, the dual problem is considered, which is a classical Q.P.P.

The combination of the pure frictional with the pure unilateral contact problem follows here the algorithm described before. In order to treat large industrial applications (e.g. 2000-3000 contact nodes for the HP-750 workstation) some special procedures have been introduced by Bisbos [Bis90] which improve considerably the rate of convergence and the stability of the algorithm.

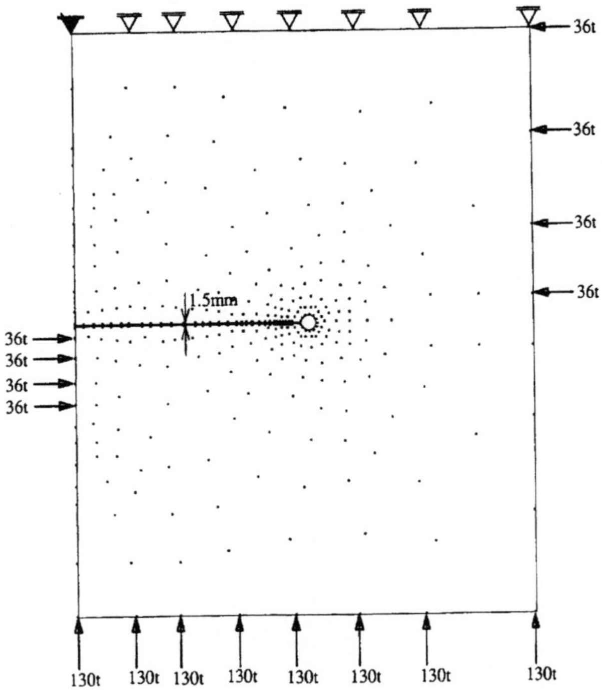

Fig. 9.1: The unilateral contact problem with friction. Boundary discretization and stress sampling points.
($E = 2.1 \times 10^7$ t/m^2, $\nu = 0.3$, $t = 0.1$ m, $\mu = 0.25$)

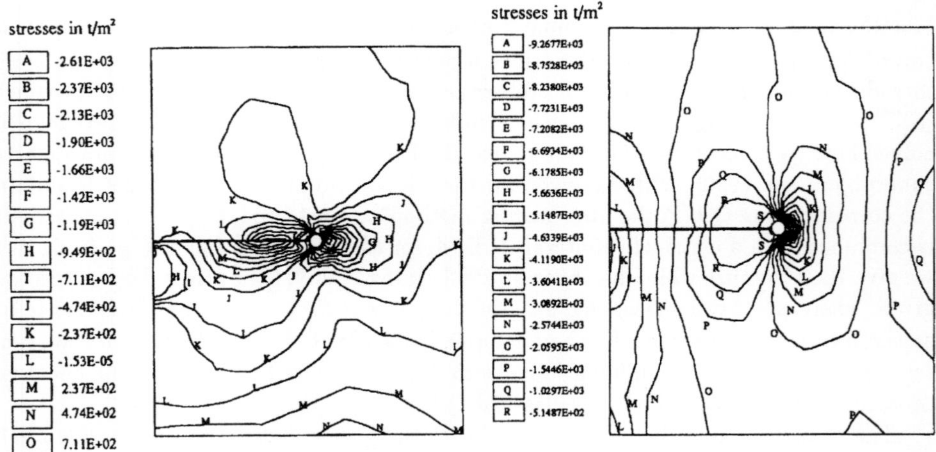

Fig. 9.2: The diagram of σ_x and σ_y

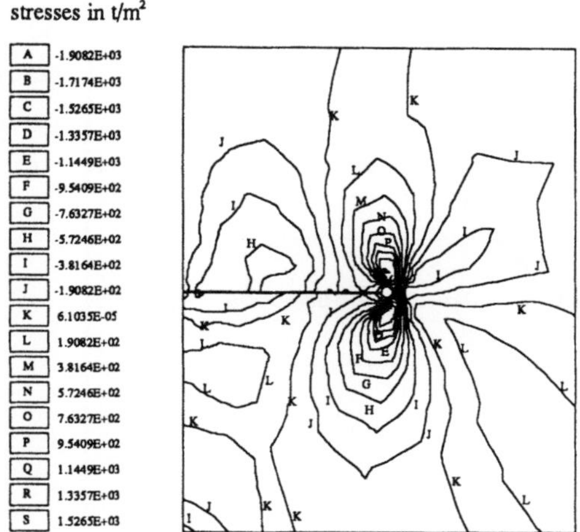

Fig. 9.3: The diagrams of σ_y and τ_{xy}

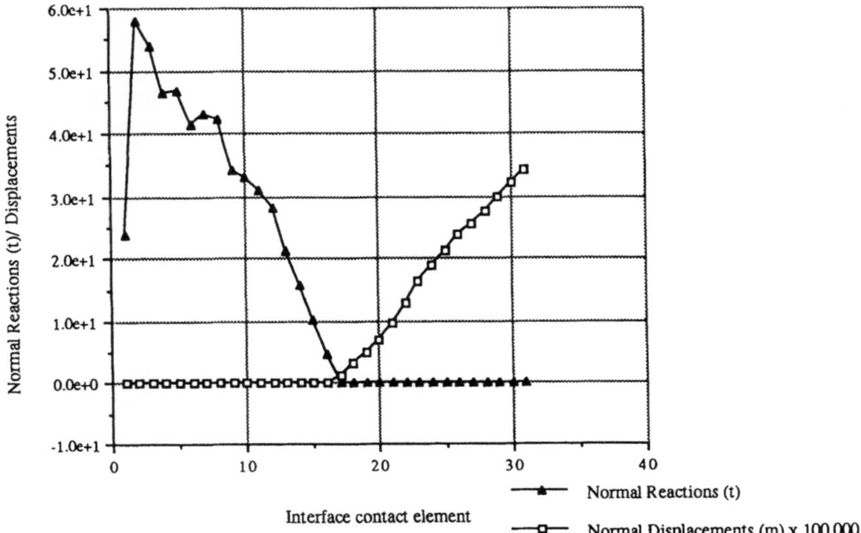

Fig. 9.4: The interface normal reactions S_N and relative displacements u_N

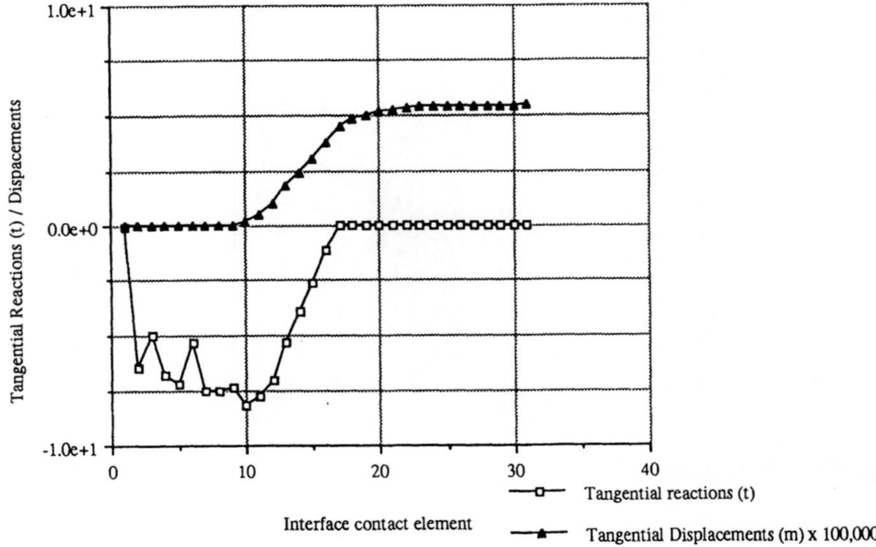

Fig. 9.5: The interface relative tangential slips $[u_T]$ and friction forces S_T

From the diagrams of S_N S_T and $[u_N]$ and $[u_T]$ one can determine the debonding and the contact regions, and within the contact region the adhesive and the sliding friction regions. As a qualibrative observation having to do with the example presented above one could note that the maximum stresses near the crack arresting hole, at the tip of the interface, have been reduced by a factor of approximately 3 solely due to the change of the mechanism of stress transfer that the taking into account of the unilateral contact and friction conditions brings in.

Chapter 10
Boundary Integral Formulations for the Monotone Multivalued Contact Boundary Conditions

10.1 Convex Problems. Primal, Dual and Mixed Problems

Let Ω be an open bounded[5] subset of the three-dimensional Euclidean space $\mathbb{R}^3$ with a boundary Γ (a Lipschitz boundary is sufficient). Ω is occupied by a linear elastic body in its undeformed state and we refer it to a Cartesian orthogonal coordinate system $Ox_1x_2x_3$. Γ is decomposed into three mutually disjoint parts Γ_1, Γ_2 and Γ_3 that are open subsets of Γ and mes $\Gamma_1 \neq 0$, mes $\Gamma_3 \neq 0$. It is assumed that on Γ_1 (resp. Γ_2) the displacements (resp. the tractions) are given and that on Γ_3 boundary conditions which are monotone and multivalued and which cause the inequality formulation of the problem hold. We call these conditions inequality boundary conditions. We denote as usual by $n = \{n_i\}$ the outward unit normal vector to Γ, by $S = \{S_i\} = \{\sigma_{ij}n_j\}$ the traction vector on the boundary, where $\sigma = \{\sigma_{ij}\}$ is the stress tensor and, by $u = \{u_i\}$ the corresponding displacement vector. Let $\varepsilon = \{\varepsilon_{ij}\}$ be the strain tensor (assumption of small strains) and $C = \{C_{ijhk}\}, i, j, h, k = 1, 2, 3$, Hooke's tensor of elasticity obeying the well-known symmetry and ellipticity conditions $(1.151a, b)$. On Γ_1 we assume that

$$u_i = \bar{U}_i, \qquad \bar{U}_i = \bar{U}_i(x), \tag{10.1}$$

and on Γ_2

$$S_i = \bar{T}_i, \qquad \bar{T}_i = \bar{T}_i(x). \tag{10.2}$$

The boundary conditions on Γ_3 are written in the general subdifferential form (cf. Sect. 1.3.1)

$$-S \in \partial j(u), \tag{10.3}$$

where j is a convex, lower semicontinuous, proper superpotential (see Sect. 1.2.1 and 1.3.1) and ∂ denotes the subdifferential. We denote, as usual, by S_N (resp. S_T) the normal (resp. the tangential) components of S with respect to Γ; u_N and u_T are the corresponding components of the displacement u. The method of the present Section remains also valid if (10.3) is replaced by

$$-S_N \in \partial j_N(u_N) \quad \text{and} \quad -S_T \in \partial j_T(u_T). \tag{10.4}$$

The equations of the B.V.P. read

$$\sigma_{ij,j} + \bar{p}_i = 0 \quad \text{in } \Omega, \tag{10.5}$$

---

[5]For Ω unbounded the same methodology is followed for the derivation of a boundary variational formulation of the problem. The classical conditions of decay at infinity of infinite linear elastic bodies guarantee the convergence of the "work" integrals.

$$\varepsilon_{ij} = \varepsilon_{ij}(u) = \frac{1}{2}(u_{i,j} + u_{j,i}) \text{ in } \Omega, \tag{10.6}$$

$$\sigma_{ij} = C_{ijhk}\varepsilon_{hk} \quad \text{in } \Omega, \tag{10.7}$$

where $\bar{p} = \{\bar{p}_i\}$ is again the volume force vector. Let $\tilde{V}$ be the vector space of the displacements v_i and let V be the kinematically admissible displacement set

$$V = \{v|v = \{v_i\}, v_i \in \tilde{V}, v_i = \bar{U}_i, \quad i = 1, 2, 3 \text{ on } \Gamma_1\}, \tag{10.8}$$

without the constraints on Γ_3. We use for the works the same notations as in Ch. 7. Moreover (7.9) holds also here for the bilinear form of elasticity. Let now Π be the potential energy

$$\Pi(v) = \frac{1}{2}a(v, v) + \Phi(v) - (\bar{p}, v) - [\bar{T}, v]_{\Gamma_2}, \tag{10.9}$$

where

$$\Phi(v) = \left\{ \begin{array}{l} \int\limits_{\Gamma_3} j(v)d\Gamma \text{ if } j(v) \in L^1(\Gamma) \\ \infty \text{ otherwise} \end{array} \right\}. \tag{10.10}$$

It is well-known (cf. e.g. [Pan85], Ch. 6) that the problem

$$\Pi(u) = \min\{\Pi(v)|v \in V\} \tag{10.11}$$

characterizes the position of equilibrium. Problem (10.11) has one and only one solution (for $\tilde{V} = H^1(\Omega)$, for $C_{ijhk} \in L^\infty(\Omega)$, $\bar{p}_i \in L^2(\Omega)$, $\bar{T}_i \in L^2(F)$, and on the assumption that there exists $u_{0i} \in H^1(\Omega)$ such that $u_{0i}|_{\Gamma_1} = \bar{U}_i$ in the sense of $H^{1/2}(\Gamma)$). The solution u equivalently, satisfies the variational inequality

$$u \in V, \quad a(u, v - u) + \Phi(v) - \Phi(u) - l(v - u) \geq 0 \quad \forall v \in V. \tag{10.12}$$

Relations (10.11) and (10.12) constitute the primal formulations of the B.V.P. with

$$l(v) = (\bar{p}, v) + [\bar{T}, v]_{\Gamma_2}. \tag{10.13}$$

In order to derive the mixed formulation a displacement field u_0 such that $u_{0i} = \bar{U}_i$ on Γ_1, is introduced, and let

$$\bar{u} = u - u_0, \quad \bar{v} = v - u_0 \tag{10.14}$$

where

$$\bar{u}, \bar{v} \in V_0 = \{v|v = \{v_i\}, v_i \in \tilde{V}, v_i = 0 \text{ on } \Gamma_1\}. \tag{10.15}$$

Then (10.12) becomes: Find $u = \bar{u} + u_0 \in V$ such that

$$a(\bar{u}, \bar{v} - \bar{u}) - l(\bar{v} - \bar{u}) + \Phi(\bar{v} + u_0) - \Phi(\bar{u} + u_0) + a(u_0, \bar{v} - \bar{u}) \geq 0 \quad \forall \bar{v} \in V_0 \tag{10.16}$$

and (10.11) becomes

$$\tilde{\Pi}(\bar{u}) = \min\{\tilde{\Pi}(\bar{v})|\bar{v} \in V_0\} \tag{10.17}$$

where

$$\tilde{\Pi}(\bar{v}) = \frac{1}{2}a(\bar{v}, \bar{v}) - l(\bar{v}) + \Phi(\bar{v} + u_0) + a(u_0, \bar{v}) = \tilde{\Pi}'(\bar{v}) + \Phi(\bar{v} + u_0). \tag{10.18}$$

More generally u_0 may be the displacement field corresponding to any prescribed initial strain field (temperature distributions, given dislocations etc.). Further let us denote by $\bar{l}$ the functional

$$\bar{l}(\bar{v}) = l(\bar{v}) - a(u_0, \bar{v}). \tag{10.19}$$

We consider now the Lagrange multiplier $\mu = \{\mu_i\}$ for the present problem and through it we will introduce the boundary condition (10.3) on Γ_3. We denote by L the admissible vector space of μ. In the functional framework considered previously L is the restriction of $\left[H^{-1/2}(\tilde{\Gamma})\right]^3$ to Γ_3; for the definition of $\tilde{\Gamma}$ we refer to Sect. 8.1. Moreover in the functional framework introduced the work expressions, e.g. $[\mu, \bar{v}]_{\Gamma_3}$, should be replaced by $\langle \mu, \bar{v} \rangle_{1/2, \Gamma_3}$ (cf. Sect. 8.1) and we observe that (cf. (1.29))

$$\Phi(\bar{v} + u_0) = \sup_{\mu \in L} \left([-\mu, \bar{v} + u_0)]_{\Gamma_3} - \Phi^c(-\mu) \right), \tag{10.20}$$

where Φ^c denotes the conjugate functional to Φ which has exactly the same properties as Φ (Prop. 1.10). Thus

$$\inf_{\bar{v} \in V_0} \tilde{\Pi}(\bar{v}) = \inf_{\bar{v} \in V_0} \sup_{\mu \in L} \{\tilde{\Pi}'(\bar{v}) - [\mu, \bar{v} + u_0]_{\Gamma_3} - \Phi^c(-\mu)\}. \tag{10.21}$$

The Lagrangian of the problem is a real-valued function $\mathcal{L}$ on $V_0 \times L$ defined by the relation

$$\mathcal{L}(\bar{v}, \mu) = \frac{1}{2}a(\bar{v}, \bar{v}) - [\mu, \bar{v} + u_0]_{\Gamma_3} - \Phi^c(-\mu) - [\bar{T}, \bar{v}]_{\Gamma_2} - (\bar{p}, \bar{v}) + a(u_0, \bar{v}). \tag{10.22}$$

Thus the mixed variational formulation of the problem reads now [Eke]: Find the saddle point $\{\bar{w}, \lambda\} \in V_0 \times L$ of $\mathcal{L}$ on $V_0 \times L$, i.e.,

$$\mathcal{L}(\bar{w}, \mu) \leq \mathcal{L}(\bar{w}, \lambda) \leq \mathcal{L}(\bar{v}, \lambda) \quad \forall \bar{v} \in V_0, \quad \mu \in L. \tag{10.23}$$

We can show in the previously mentioned functional framework by the methods given in [Eke] p. 57 that (10.23) has a unique solution $\{\bar{w}, \lambda\} \in V_0 \times L$ such that

$$\bar{w} = \bar{u} \in V_0 \tag{10.24}$$

and

$$\lambda = S(\bar{u}) \text{ on } \Gamma_3. \tag{10.25}$$

where $\bar{u}$ is the solution of the primal problem (10.17).

The dual formulation with respect to the stresses results easily from (10.23) if one considers the statically admissible set of the stresses

$$\Lambda = \{\sigma | \sigma = \{\sigma_{ij}\}, \sigma_{ij} = \sigma_{ji}, \sigma_{ij,j} + \bar{p}_i = 0 \text{ a.e. in } \Omega, S \in L \text{ on } \Gamma_3, S_i = \bar{T}_i \text{ a.e. on } \Gamma_2\}. \tag{10.26}$$

The stress field σ at the position of equilibrium is characterized by the minimum of the complementary energy [Pan85]

$$\Pi^c(\sigma) = \frac{1}{2}A(\sigma, \sigma) + \Phi^c(-S) - [\bar{U}, S]_{\Gamma_1} \text{ over } \Lambda \text{ i.e.,} \tag{10.27}$$

$$\Pi^c(\sigma) = \min\{\Pi^c(\tau)|\tau \in \Lambda\}. \tag{10.28}$$

Here $\frac{1}{2}A(\sigma,\sigma) = (c\sigma,\sigma)$, where $c = \{c_{ijhk}\}$ is the inverse tensor to C.

10.2 The Multivalued B.I.E. with Respect to the Boundary Tractions on Γ_3

The pair $\{\bar{w},\lambda\} \in V_0 \times L$ is the solution of problem (10.23). Then we can equivalently write (10.23) in the form

$$\mathcal{L}(\bar{w},\lambda) = \inf_{V_0}\sup_{L}\mathcal{L}(\bar{v},\mu) = \sup_{L}\inf_{V_0}\mathcal{L}(\bar{v},\mu). \tag{10.29}$$

We set

$$\inf_{V_0}\mathcal{L}(\bar{v},\mu) = \tilde{\Pi}_1(\mu) \tag{10.30}$$

assuming for the present that $\mu \in L$ is given. Then (10.30) is equivalent to the following bilateral problem: find $\tilde{u} = \tilde{u}(\mu) \in V_0$ such that

$$a(\tilde{u},\bar{v}) - [\mu,\bar{v}]_{\Gamma_3} - (\bar{p},\bar{v}) - (\bar{T},\bar{v})_{\Gamma_2} + a(u_0,\bar{v}) = 0, \qquad \forall \bar{v} \in V_0. \tag{10.31}$$

Note that (10.31) is the expression of the "principle" of virtual work for a structure which results from the initial unilateral one by ignoring the subdifferential constraints on Γ_3 and by adding the corresponding reactions $\mu = \{\mu_i\}$. This is a classical (bilateral) structure and therefore its position of equilibrium is characterized by the minimization problem (10.30) of the potential energy.

The linearity of (10.31) implies that the solution $\tilde{u}$ of it can be written as the sum of $\tilde{u}_1 \in V_0$ and $\tilde{u}_2 \in V_0$ which are solutions of the two following bilateral problems

$$a(\tilde{u}_1,\bar{v}) - (\bar{p},\bar{v}) - [\bar{T},\bar{v}]_{\Gamma_2} + a(u_0,\bar{v}) = 0, \qquad \forall \bar{v} \in V_0 \tag{10.32}$$

and

$$a(\tilde{u}_2,\bar{v}) - [\mu,\bar{v}]_{\Gamma_3} = 0, \qquad \forall v \in V_0, \tag{10.33}$$

respectively. Both variational equalities describe the equilibrium configuration of two bilateral structures resulting from the initial one by ignoring the subdifferential boundary conditions on Γ_3, and by assuming that certain boundary parts have zero loading: in the case of (10.32) the structure is loaded by the forces $\bar{p}$ in Ω and $\bar{T}$ on Γ_2, whereas on Γ_3 the loading is zero. Moreover the structure is subjected to an initial displacement field u_0. In the case of (10.33) the structure is loaded by a force $\mu = \{\mu_i\}$ on Γ_3 only; we assume zero forces in Ω and on Γ_2. The solutions $\tilde{u}_1$ and $\tilde{u}_2$ are uniquely determined, as it is well known from classical elasticity.

For the two bilateral structures, whose equilibrium is described by (10.32), (10.33), the solution can be written in terms of Green's operator G. This operator is the same for the two structures due to the same type of boundary conditions holding in both cases. Accordingly,

$$\tilde{u}_1 = G(\bar{l}), \quad \tilde{u}_2 = G(\mu), \quad \tilde{u} = \tilde{u}_1 + \tilde{u}_2, \quad \bar{l} = \{\bar{p},\bar{T},u_0\}. \tag{10.34}$$

The yet unknown force distribution $\mu \in L$ has to be determined. From (10.30), (10.32) and (10.33) we obtain that

$$
\begin{aligned}
\tilde{\Pi}_1(\mu) = \mathcal{L}(\tilde{u}, \mu) \;=\; & \frac{1}{2} a(\tilde{u}, \tilde{u}) - [\mu, \tilde{u} + u_0]_{\Gamma_3} - \Phi^c(-\mu) - (\bar{p}, \tilde{u}) \\
& - [\bar{T}, \tilde{u}]_{\Gamma_2} + a(u_0, \tilde{u}) \\
=\; & -[\mu, \tilde{u}_1]_{\Gamma_3} - \frac{1}{2}[\mu, \tilde{u}_2]_{\Gamma_3} - \frac{1}{2}(\bar{p}, \tilde{u}_1) - \frac{1}{2}[\bar{T}, \tilde{u}_1]_{\Gamma_2} \\
& - [\mu, u_0]_{\Gamma_3} + \frac{1}{2} a(u_0, \tilde{u}_1) - \Phi^c(-\mu) \\
=\; & -[\mu, G(\bar{l})]_{\Gamma_3} - \frac{1}{2}[\mu, G(\mu)]_{\Gamma_3} + \frac{1}{2} a(u_0, G(\bar{l})) - \frac{1}{2}(\bar{p}, G(\bar{l})) \\
& - \frac{1}{2}[\bar{T}, G(\bar{l})]_{\Gamma_2} - \Phi^c(-\mu) - [\mu, u_0]_{\Gamma_3}.
\end{aligned}
\tag{10.35}
$$

Further we denote by β the bilinear form

$$
\beta(\mu, \nu) = [\mu, G(\nu)]_{\Gamma_3}
\tag{10.36}
$$

and by γ the linear form

$$
\gamma(\mu) = -[\mu, G(\bar{l})]_{\Gamma_3} - [\mu, u_0]_{\Gamma_3}.
\tag{10.37}
$$

Thus

$$
\begin{aligned}
\tilde{\Pi}_1(\mu) \;=\; & -\frac{1}{2}\beta(\mu, \mu) + \gamma(\mu) - \Phi^c(-\mu) - \frac{1}{2}(\bar{p}, G(\bar{l})) - \frac{1}{2}[\bar{T}, G(\bar{l})]_{\Gamma_2} \\
& + \frac{1}{2} a(u_0, G(\bar{l})).
\end{aligned}
\tag{10.38}
$$

Eqs. (10.29), (10.30) and (10.38) imply that the following minimization problem holds with respect to the unknown boundary tractions μ:

$$
\inf \left\{ \Pi_1(\mu) = \frac{1}{2}\beta(\mu, \mu) + \Phi^c(-\mu) - \gamma(\mu) \,\middle|\, \mu \in L \right\}.
\tag{10.39}
$$

The term $-\frac{1}{2}(\bar{p}, G(\bar{l})) - \frac{1}{2}[\bar{T}, G(\bar{l})]_{\Gamma_2} + \frac{1}{2} a(u_0, G(\bar{l}))$ does not depend on μ and thus it is omitted. The solution of this problem is λ. Another expression for $\tilde{\Pi}_1$ can be obtained from (10.30), (10.22), and from (10.31) by setting $\bar{v} = \tilde{u}$ in it

$$
\begin{aligned}
\tilde{\Pi}_1(\mu) \;=\; & \inf_{V_0} \mathcal{L}(\bar{v}, \mu) = \mathcal{L}(\tilde{u}, \mu) \\
=\; & \frac{1}{2} a(\tilde{u}, \tilde{u}) - [\mu, \tilde{u} + u_0]_{\Gamma_3} - \Phi^c(-\mu) - [\bar{T}, \tilde{u}]_{\Gamma_3} - (\bar{p}, \tilde{u}) + a(u_0, \tilde{u}) \\
=\; & \frac{1}{2} a(\tilde{u}, \tilde{u}) - a(\tilde{u}, \tilde{u}) - [\mu, u_0]_{\Gamma_3} - \Phi^c(-\mu) \\
=\; & -\frac{1}{2} a(\tilde{u}, \tilde{u}) - [\mu, u_0]_{\Gamma_3} - \Phi^c(-\mu).
\end{aligned}
\tag{10.40}
$$

But for $\mu = \lambda$ we have $\tilde{u} = \bar{w}$ and therefore

$$
\tilde{\Pi}_1(\lambda) = - \left[\frac{1}{2} a(\bar{w}, \bar{w}) + [\lambda, u_0]_{\Gamma_3} + \Phi^c(-\lambda) \right].
\tag{10.41}
$$

We show now that

$$\lambda = S(\bar{u}) \text{ on } \Gamma_3, \tag{10.42}$$

where $\bar{u} \in V_0$ is the solution of the primal problem (10.11): Indeed let us choose $\lambda \in L$ such as to satisfy (10.42), where $\bar{u}$ is the solution of (10.17) and let $\bar{w} = \bar{u} \in V_0$. Then $\{\bar{u}, \lambda\}$ is the unique solution of the saddle point problem (10.23). The proof will be completed if we show that λ is a minimizer of $\tilde{\Pi}_1$ over L and that the solution of this minimization problem (10.39) is unique. Indeed (10.23) and (10.30) imply that

$$\mathcal{L}(\bar{w}, \lambda) = \inf_{\bar{v} \in V_0} \mathcal{L}(\bar{v}, \lambda) = \tilde{\Pi}_1(\lambda). \tag{10.43}$$

But

$$\mathcal{L}(\bar{w}, \lambda) = \sup_L \inf_{V_0} \mathcal{L}(\bar{v}, \mu) = \sup_L \tilde{\Pi}_1(\mu) \tag{10.44}$$

and thus

$$\tilde{\Pi}_1(\lambda) = \inf_L \tilde{\Pi}_1(\mu). \tag{10.45}$$

The uniqueness of the solution of (10.39) results easily from the strict convexity of $\beta(\cdot, \cdot)$. In this context we refer to the next proposition 10.1.

The quadratic form $\beta(\mu, \mu)$ is symmetric. Indeed from (10.36) by means of the linearity of G and the theorem of Betti we can verify that

$$\beta(\mu, \nu) = \beta(\nu, \mu). \tag{10.46}$$

We give further two equivalent formulations of problem (10.39). The first is a variational inequality: find $\lambda \in L$ such as to satisfy

$$\beta(\lambda, \mu - \lambda) + \Phi^c(-\mu) - \Phi^c(-\lambda) - \gamma(\mu - \lambda) \geq 0, \qquad \forall \mu \in L \tag{10.47}$$

and the second a multivalued integral equation on the boundary part Γ_3 of the structure which reads

$$\gamma - \frac{1}{2}\text{grad}\beta(\lambda, \lambda) \in \partial\Phi^c(-\lambda). \tag{10.48}$$

Relation (10.47) is equivalent to (10.39) by using a well-known result of the theory of variational inequalities (given in Prop. 1.2) and relation (10.48) is equivalent to (10.47) due to the definition of the subdifferential ∂. If instead of (10.3) relations (10.4) hold, then in (10.48) $\Phi^c(-\mu)$ is replaced by $\Phi_N{}^c(-\mu_N) + \Phi_T{}^c(-\mu_T)$, where Φ_N and Φ_T are defined from j_N and j_T respectively by relations analogous to (10.10).

Suppose, for instance, that the Signorini boundary conditions (1.80) and (7.3) hold on Γ_3 expressing the unilateral contact with a rigid support. In this case we may easily verify (see [Pan85]) that $\Phi_N{}^c(-\mu_N) = \{0$ if $\mu_N \leq 0$ in the sense of $H^{-1/2}(\tilde{\Gamma})\big|_{\Gamma_3}, \infty$ otherwise$\}$ and that $\Phi_T{}^c(-\mu_T) = \{0$ if $\mu_T = C_T$ a.e on Γ_3, ∞ otherwise$\}$. Therefore by applying to the problem (10.39) the classical procedure of convex analysis we may seek its solution over the admissible set $L_1 = \{\mu|\mu_N \leq 0$ in the sense of $H^{-1/2}(\tilde{\Gamma})\big|_{\Gamma_3}, \mu_T = C_T$ a.e on $\Gamma_3\}$ and omit $\Phi^c(-\mu)$ in the function $\Pi_1(\mu)$ which has to be minimized. If the Coulomb friction conditions hold on Γ_3 with given normal force

$S_N = C_N$ then $\Phi_T(u_T) = \int_{\Gamma_3} q|C_N||u_T|d\Gamma$, $\Phi_T{}^c(-\mu_T) = \{0$ if $|\mu_T| \leq q|C_N|$, a.e. on Γ_3, and ∞ otherwise$\}$ and $\Phi_N{}^c(-\mu_N) = \{0$ if $\mu_N = C_N, \infty$ otherwise$\}$, where q is the friction coefficient. In this case as well we seek the solution of problem (10.39) over the admissible set $L_2 = \{\mu||\mu_T| \leq q|C_N|$ and $\mu_N = C_N$ a.e. on $\Gamma_3\}$.

Let us close this Section with the following propositions.

Proposition 10.1 The multivalued B.I.E. (10.48) has a uniquely determined solution.

Proof: It is sufficient to prove that the solution of the equivalent minimum problem (10.39) exists and is unique. Indeed analogously to Ch. 9 we can show that β is defined on $\boldsymbol{H}^{-1/2}(\tilde{\Gamma})\big|_{\Gamma_3} \times \boldsymbol{H}^{-1/2}(\tilde{\Gamma})\big|_{\Gamma_3}$, where $\tilde{\Gamma} = \Gamma_2 \cup \Gamma_3$, and γ on $\boldsymbol{H}^{-1/2}(\tilde{\Gamma})\big|_{\Gamma_3}$ and that[6]

$$\beta(\mu,\mu) = \frac{1}{2}||\mu||^2_{-1/2,\Gamma_3} \tag{10.49}$$

Then due to the Hahn-Banach theorem there exist constants $c_1, c_2 \geq 0$ such that

$$\Phi^c(-\mu) \geq -c_1||\mu||_{-1/2,\Gamma_3} - c_2 \quad \forall \mu \in \boldsymbol{H}^{-1/2}(\tilde{\Gamma})\big|_{\Gamma_3} \tag{10.50}$$

Thus

$$\Pi_1(\mu) \geq \frac{1}{2}||\mu||^2_{-1/2,\Gamma_3} - c_1'||\mu||_{-1/2,\Gamma_3} - c_2 \quad c_1', c_2 \text{ const.} \tag{10.51}$$

which implies that $\Pi_1(\mu) \to \infty$ as $||\mu||^2_{-1/2,\Gamma_3} \to \infty$. Therefore according to Prop. 1.1 a solution of (10.39) exists. The solution is unique due to the strict convexity of Π_1 q.e.d.

10.3 The Multivalued B.I.E. with Respect to the Displacements u on Γ_3

We have considered in the previous section $\sup_{L}\inf_{V_0} \mathcal{L}(\bar{v}, \mu)$ with μ given. Here we will work on $\inf_{V_0}\sup_{L} \mathcal{L}(\bar{v}, \mu)$, or on an equivalent saddle point formulation by considering $\bar{v}$ as given. We have that

$$\inf_{V_0}\sup_{L} \mathcal{L}(\bar{v}, \mu) = \sup_{L}\inf_{V_0} \mathcal{L}(\bar{v}, \mu) \tag{10.52}$$

$$= \sup_{L}\inf_{V_0} \left[\frac{1}{2}a(\bar{v},\bar{v}) - [\mu,\bar{v} + u_0]_{\Gamma_3} - \Phi^c(-\mu) \right.$$
$$\left. -[\bar{T},\bar{v}]_{\Gamma_2} - (\bar{p},\bar{v}) + a(u_0,\bar{v})\right]$$

$$= \sup_{\mu \in L}\inf_{v \in V} \left[\frac{1}{2}a(v - u_0, v - u_0) - [\mu, v]_{\Gamma_3} - \Phi^c(-\mu) \right.$$
$$\left. -[\bar{T}, v - u_0]_{\Gamma_2} - (\bar{p}, v - u_0) + a(u_0, v - u_0)\right]$$

$$= \sup_{\mu \in L}\inf_{v \in V} \left[\frac{1}{2}a(v,v) + \frac{1}{2}a(u_0,u_0) - a(v,u_0) - [\mu, v]_{\Gamma_3} - \Phi^c(-\mu) \right.$$
$$\left. -[\bar{T}, v - u_0]_{\Gamma_2} - (\bar{p}, v - u_0) + a(u_0,v) - a(u_0,u_0)\right]$$

---

[6]Here $\boldsymbol{H}^{-1/2}(\tilde{\Gamma}) \doteq [H^{-1/2}(\tilde{\Gamma})]^3$ because $\Omega \subset \mathbb{R}^3$.

$$= \inf_{v\in V}\sup_{\mu\in L}\left[\frac{1}{2}a(v,v) - [\mu,v]_{\Gamma_3} - \Phi^c(-\mu) - [\bar{T},v]_{\Gamma_2}\right.$$

$$\left. -(\bar{p},v) - R(u_0)\right] = \inf_{v\in V}\sup_{\mu\in L}\tilde{\mathcal{F}}(v,\mu) - R(u_0).$$

Here

$$R(u_0) = \frac{1}{2}a(u_0,u_0) - [\bar{T},u_0]_{\Gamma_2} - (\bar{p},u_0) \tag{10.53}$$

and $\tilde{\mathcal{F}}$ is a modified Lagrangian

$$\tilde{\mathcal{F}}(v,\mu) = \frac{1}{2}a(v,v) - [\mu,v]_{\Gamma_3} - \Phi^c(-\mu) - [\bar{T},v]_{\Gamma_2} - (\bar{p},v). \tag{10.54}$$

But (cf. Sect. 1.2.1)

$$\sup_{\mu\in L}\{-[\mu,v]_{\Gamma_3} - \Phi^c(-\mu)\} = \Phi(v) \tag{10.55}$$

and thus

$$\inf_{v\in V}\sup_{\mu\in L}\tilde{\mathcal{F}}(v,\mu) = \inf_{v\in V}\left[\frac{1}{2}a(v,v) + \Phi(v) - l(v)\right] = \inf_{v\in V}\Pi(v). \tag{10.56}$$

We recall that

$$\frac{1}{2}a(v,v) = \frac{1}{2}(C\varepsilon(v),\varepsilon(v)) = \sup_{\tau\in\Sigma}\left\{(\tau,\varepsilon(v)) - \frac{1}{2}A(\tau,\tau)\right\} \tag{10.57}$$

where Σ denotes the set of all symmetric stress-tensors from $[L^2(\Omega)]^6$, as we may obtain by simple derivation with respect to τ. Indeed the supremum is attained for

$$\varepsilon_{ij}(v) = c_{ijhk}\sigma_{hk} \quad \text{or} \quad \sigma_{ij}(v) = C_{ijhk}\varepsilon_{hk}(v). \tag{10.58}$$

Accordingly we may write that

$$\inf_{v\in V}\sup_{\mu\in L}\tilde{\mathcal{F}}(v,\mu) = \inf_{v\in V}\Pi(v) \tag{10.59}$$

$$= \inf_{v\in V}\sup_{\tau\in\Sigma}\left\{(\tau,\varepsilon(v)) + \Phi(v) - (\bar{p},v) - [\bar{T},v]_{\Gamma_2} - \frac{1}{2}A(\tau,\tau)\right\}.$$

The Green-Gauss theorem for $v\in V$ implies that

$$(\tau,\varepsilon(v)) = -(\tau_{ij,j},v_i) + [T,v]_{\Gamma_3} + [\bar{T},v]_{\Gamma_2} + [T,\bar{U}]_{\Gamma_1}, \tag{10.60}$$

where T is the boundary tractions corresponding to τ. Relation (10.60) is the principle of virtual work for the unconstrained structure. From the last two relations we obtain that

$$\inf_{v\in V}\sup_{\tau\in\Sigma}\left[-(\tau_{ij,j} + \bar{p}_i,v_i) + [T - \bar{T},v]_{\Gamma_2} + [T,\bar{U}]_{\Gamma_1} + [T,v]_{\Gamma_3} + \Phi(v) - \frac{1}{2}A(\tau,\tau)\right] \tag{10.61}$$

$$= \inf_{v\in V}\sup_{\tau\in\Sigma_1}\left[[T,\bar{U}]_{\Gamma_1} + [T,v]_{\Gamma_3} + \Phi(v) - \frac{1}{2}A(\tau,\tau)\right]$$

where

$$\Sigma_1 = \{\tau | \tau = \{\tau_{ij}\}, \tau_{ij} = \tau_{ji} \in L^2(\Omega), \tau_{ij,j} + \bar{p}_i = 0 \text{ a.e. in } \Omega, \ T_i = \bar{T}_i \text{ a.e. on } \Gamma_2\} \tag{10.62}$$

because of the relation

$$\sup_{\tau \in \Sigma} \left[-(\tau_{ij,j} + \bar{p}_i, v_i) + [T - \bar{T}, v]_{\Gamma_2} \right] = \begin{cases} 0 & \text{if } \tau_{ij,j} + \bar{p}_i = 0 \text{ a.e. in } \Omega, \ T_i = \bar{T}_i \text{ a.e. on } \Gamma_2 \\ \infty & \text{otherwise.} \end{cases} \tag{10.62a}$$

Thus

$$\inf_{v \in V} \sup_{\mu \in L} \tilde{\mathcal{F}}(v, \mu) = \inf_{v \in V} \Pi(v) = \inf_{v \in V} \sup_{\tau \in \Sigma_1} \tilde{L}(v, \tau). \tag{10.63}$$

Here the new Lagrangian $\tilde{L}$ is given by

$$\tilde{L}(v, \tau) = [T, \bar{U}]_{\Gamma_1} + [T, v]_{\Gamma_3} + \Phi(v) - \frac{1}{2} A(\tau, \tau). \tag{10.64}$$

Due to the duality [Eke] between the primal (10.11) and the dual problem (10.28) the relation

$$\begin{aligned}
\inf_{v \in V} \Pi(v) &= \sup_{\tau \in \Sigma_1} -\Pi^c(\tau) = \sup_{\tau \in \Sigma_1} \left(-\frac{1}{2} A(\tau, \tau) + [T, \bar{U}]_{\Gamma_1} - \Phi^c(-T) \right) \\
&= \sup_{\tau \in \Sigma_1} \inf_{v \in V} \left(-\frac{1}{2} A(\tau, \tau) + [T, \bar{U}]_{\Gamma_1} + [T, v]_{\Gamma_3} + \Phi(v) \right) \\
&= \sup_{\tau \in \Sigma_1} \inf_{v \in V} \tilde{L}(v, \tau)
\end{aligned} \tag{10.65}$$

holds. Here

$$\inf_{v \in L} \{ [T, v]_{\Gamma_3} + \Phi(v) \} = -\Phi^c(-T) \tag{10.66}$$

by the definition of the conjugate function of Φ. Thus

$$\inf_{v \in V} \sup_{\tau \in \Sigma_1} \tilde{L}(v, \tau) = \sup_{\tau \in \Sigma_1} \inf_{v \in V} \tilde{L}(v, \tau) = \tilde{L}(u, \sigma). \tag{10.67}$$

The last equality in (10.67) is verified directly by writing the previous relations for the solution $\{u, \sigma, S\}$ of the problem (i.e. eqs. (10.59), (10.61) etc. without infsup, inf etc.). Within this procedure we have derived besides $\mathcal{L}$ and $\tilde{L}$ another Lagrangian, the expression of the right-hand side in eq. (10.59). The saddle-point theory of [Eke] leads to the same Lagrangian.

The foregoing relations and the fact that the saddle-point problem (10.23) admits a unique solution imply that the saddle-point problem (10.67) with respect to $\tilde{L}$ has also a unique solution $\sigma \in \Sigma_1$ and $u = u(\sigma)$. Let us denote further

$$\sup_{\tau \in \Sigma_1} \tilde{L}(v, \tau) = \tilde{\Pi}_2(v) \quad \text{or} \quad \inf_{\tau \in \Sigma_1} -\tilde{L}(v, \tau) = -\tilde{\Pi}_2(v) \tag{10.68}$$

assuming that v is given on Γ_3. The minimum problem (10.68) corresponds to the variational inequality: find $\tilde{\sigma} = \tilde{\sigma}(v) \in \Sigma_1$ such that

$$A(\tilde{\sigma}, \tau - \tilde{\sigma}) - [\bar{U}, T - \tilde{S}]_{\Gamma_1} - [v, T - \tilde{S}]_{\Gamma_3} \geq 0, \quad \forall \tau \in \Sigma_1. \tag{10.69}$$

Let us introduce a stress-field $\sigma_0 \in \Sigma_1$ and let us perform the substitutions

$$\bar{\sigma} = \tilde{\sigma} - \sigma_0, \qquad \bar{\tau} = \tau - \sigma_0, \tag{10.70}$$

where $\bar{\sigma}, \bar{\tau} \in \Sigma_0$ with

$$\Sigma_0 = \{\bar{\tau}|\bar{\tau} = \{\bar{\tau}_{ij}\}, \bar{\tau}_{ij} = \bar{\tau}_{ji} \in L^2(\Omega), \bar{\tau}_{ij,j} = 0 \text{ a.e. in } \Omega, \bar{T} = 0 \text{ a.e. on } \Gamma_2, i, j = 1, 2, 3\}. \tag{10.71}$$

Then (10.69) becomes: find $\bar{\sigma} = \bar{\sigma}(v) \in \Sigma_0$ such that

$$A(\bar{\sigma} + \sigma_0, \bar{\tau} - \bar{\sigma}) - [\bar{U}, \bar{T} - \bar{S}]_{\Gamma_1} - [v, \bar{T} - \bar{S}] \geq 0, \quad \forall \bar{\tau} \in \Sigma_0. \tag{10.72}$$

Here $\bar{S}, \bar{T}$ are the boundary tractions corresponding to $\bar{\sigma}, \bar{\tau}$ respectively[7]. But Σ_0 is a linear space and thus by setting $\bar{\tau} - \bar{\sigma} = \pm \tau_1 \in \Sigma_0$ we obtain that (10.72) is equivalent to the variational equality

$$A(\bar{\sigma}, \bar{\tau}_1) - [\bar{U}, \bar{T}_1]_{\Gamma_1} - [v, \bar{T}_1]_{\Gamma_3} + A(\sigma_0, \bar{\tau}_1) = 0, \quad \forall \bar{\tau}_1 \in \Sigma_0. \tag{10.73}$$

In the sequel we replace $\bar{\tau}_1$ by $\bar{\tau}$. Moreover

$$A(\sigma_0, \bar{\tau}) = \int_\Omega \varepsilon_{0ij} \bar{\tau}_{ij} d\Omega = -\int_\Omega \tilde{u}_{0i} \bar{\tau}_{ij,j} d\Omega + [\tilde{u}_0, \bar{T}]_{\Gamma_3} \tag{10.74}$$

$$+ [\tilde{u}_0, \bar{T}]_{\Gamma_2} + [\tilde{u}_0, \bar{T}]_{\Gamma_1} = [\tilde{u}_0, \bar{T}]_{\Gamma_3} + [\tilde{u}_0, \bar{T}]_{\Gamma_1}, \quad \forall \bar{\tau} \in \Sigma_0$$

where $\varepsilon_0 = c\sigma_0$ and $\tilde{u}_0$ is a displacement field corresponding to σ_0. We are free to determine $\sigma_0 \in \Sigma_1$ by assuming, e.g., that σ_0 is the unique solution of a bilateral problem having on Γ_1 and on Γ_3 zero displacements; in this case (10.74) takes the form

$$A(\sigma_0, \bar{\tau}) = 0, \qquad \forall \bar{\tau} \in \Sigma_0. \tag{10.75}$$

If σ_0 is such that (10.1) is satisfied on Γ_1 and $\tilde{u}_0 = 0$ on Γ_3 then

$$A(\sigma_0, \bar{\tau}) = [\bar{U}, \bar{T}]_{\Gamma_1}, \qquad \forall \bar{\tau} \in \Sigma_0. \tag{10.76}$$

Further we consider the general expression (10.74). We note that $\bar{\sigma}$ in (10.73) can be written as the sum $\bar{\sigma}_1 + \bar{\sigma}_2$ where $\bar{\sigma}_1$ and $\bar{\sigma}_2$ are solutions of the variational equalities

$$A(\bar{\sigma}_1, \bar{\tau}) - [v, \bar{T}]_{\Gamma_3} = 0, \qquad \forall \bar{\tau} \in \Sigma_0 \tag{10.77}$$

$$A(\bar{\sigma}_2, \bar{\tau}) - [U - \tilde{u}_0, \bar{T}]_{\Gamma_1} + [\tilde{u}_0, \bar{T}]_{\Gamma_3} = 0, \qquad \forall \bar{\tau} \in \Sigma_0 \tag{10.78}$$

respectively. Both (10.77) and (10.78) express respectively the "principle" of complementary virtual work for fictive bilateral structures resulting in the following way: for (10.77) (resp. (10.78)) we consider the structure Ω under the action of "given" displacements v (resp. $-\tilde{u}_0$) on Γ_3, zero forces in Ω and Γ_2, and zero (resp. $U - \tilde{u}_0$) displacements on Γ_1. The fact that both fictive structures are bilateral implies that $\bar{\sigma}_1$

---

and $\bar{\sigma}_2$ exist and are uniquely determined according to a well-known result of linear elasticity. From (10.77) and (10.78) we obtain that

$$\bar{\sigma}_1 = H(v), \quad \bar{\sigma}_2 = H(\bar{U}, \tilde{u}_0), \quad \bar{\sigma}_1 + \bar{\sigma}_2 = \bar{\sigma}, \quad \bar{\sigma} \in \Sigma_0 \tag{10.79}$$

where H denotes the "Green's" stress-displacement operator for the B.V.P. (10.77) and (10.78). Moreover let $\bar{S}_1 = \tilde{H}(v)$, $\bar{S}_2 = \tilde{H}(U, \tilde{u}_0)$, i.e. $\tilde{H}(v) = \{\tilde{H}(v)_i\} = \{[H(v)]_{ij} n_j\}$. Note that both fictive bilateral structures have the same H-operator. By means of (10.73), (10.74), (10.77), (10.78) the following relation results:

$$
\begin{aligned}
\tilde{\Pi}_2(v) \;=\; & \tilde{L}(v, \tilde{\sigma}) = [\tilde{S}, \bar{U}]_{\Gamma_1} + [\tilde{S}, v]_{\Gamma_3} + \Phi(v) - \frac{1}{2} A(\tilde{\sigma}, \tilde{\sigma}) \tag{10.80}\\[4pt]
=\; & [\tilde{S} + S_0, \bar{U}]_{\Gamma_1} + [\tilde{S} + S_0, v]_{\Gamma_3} - \frac{1}{2} A(\bar{\sigma}, \bar{\sigma}) - \frac{1}{2} A(\sigma_0, \sigma_0) - A(\bar{\sigma}, \sigma_0) + \Phi(v) \\[4pt]
=\; & [\tilde{S}, \bar{U}]_{\Gamma_1} + [\tilde{S}, v]_{\Gamma_3} - \frac{1}{2} A(\bar{\sigma}, \bar{\sigma}) - \frac{1}{2} A(\sigma_0, \sigma_0) - [\tilde{u}_0, \bar{S}]_{\Gamma_1} \\
& -[\tilde{u}_0, \bar{S}]_{\Gamma_3} + [S_0, \bar{U}]_{\Gamma_1} + [S_0, v]_{\Gamma_3} + \Phi(v) \\[4pt]
=\; & \frac{1}{2}[\bar{S}_2, \bar{U}]_{\Gamma_1} + [\bar{S}_2, v]_{\Gamma_3} + \frac{1}{2}[\bar{S}_1, v]_{\Gamma_3} - \frac{1}{2}[\tilde{u}_0, \bar{S}_2]_{\Gamma_3} - \frac{1}{2}[\tilde{u}_0, \bar{S}_2]_{\Gamma_1} \\
& +[S_0, \bar{U}]_{\Gamma_1} + [S_0, v]_{\Gamma_3} - \frac{1}{2} A(\sigma_0, \sigma_0) + \Phi(v) \\[4pt]
=\; & \frac{1}{2}[\tilde{H}(\bar{U}, \tilde{u}_0), \bar{U}]_{\Gamma_1} + [\tilde{H}(\bar{U}, \tilde{u}_0), v]_{\Gamma_3} + \frac{1}{2}[\tilde{H}(v), v]_{\Gamma_3} \\
& -\frac{1}{2}[u_0, \tilde{H}(\bar{U}, \tilde{u}_0)]_{\Gamma_3} - \frac{1}{2}[u_0, \tilde{H}(\bar{U}, \tilde{u}_0)]_{\Gamma_1} \\
& +[S_0, \bar{U}]_{\Gamma_1} + [S_0, v]_{\Gamma_3} - \frac{1}{2} A(\sigma_0, \sigma_0) + \Phi(v).
\end{aligned}
$$

Further the bilinear form

$$\delta(v, w) = [\tilde{H}(v), w]_{\Gamma_3} \tag{10.81}$$

and the linear form

$$\zeta(w) = [\tilde{H}(\bar{U}, \tilde{u}_0) + S_0, w]_{\Gamma_3} \tag{10.82}$$

are introduced. We denote by $\tilde{R}(U, \tilde{u}_0, \sigma_0)$ the remaining constant terms of (10.80). Thus

$$
\begin{aligned}
\tilde{\Pi}_2(v) \;=\; & \sup_{\tau \in \Sigma_1} \tilde{L}(v, \tau) = L(v, \tilde{\sigma}) \tag{10.83}\\[4pt]
=\; & \frac{1}{2}\delta(v, v) + \Phi(v) - \zeta(v) + \tilde{R}(U, \tilde{u}_0, \sigma_0).
\end{aligned}
$$

Relation (10.67) implies that the following minimization problem can be formulated with respect to the unknown boundary displacement v on Γ_3:

$$\inf \left\{ \Pi_2(v) = \frac{1}{2}\delta(v, v) + \Phi(v) - \zeta(v) \,|\, v \in N \right\}. \tag{10.84}$$

Here N is the space of traces $[H^{1/2}(\tilde{\Gamma})]^3\big|_{\Gamma_3}$. In the functional framework introduced previously we shall show further that (10.84) admits a unique solution. Now it can be shown that if u is the solution of (10.84) then u is the displacement field corresponding to the stress field σ, i.e. that

$$u = u(\sigma), \tag{10.85}$$

where $\sigma \in \Lambda$ is the solution of the minimum complementary energy problem (10.28): let us choose u as in (10.85), where σ is the solution of (10.28). Then $\{u, \sigma\} \in V \times \Sigma_1$ is the unique solution of the saddle-point problem (10.67) and thus $u(\sigma)$ is a solution of the minimum problem (10.84); the uniqueness of the solution of (10.84), which will be proved in Prop. 10.2, implies the result.

Again we can verify by using Betti's theorem that δ is symmetric, i.e. that

$$\delta(v, w) = \delta(w, v). \tag{10.86}$$

Finally we give the analogous formulations to (10.47) and (10.48). The first is a variational inequality: find $u \in N$ such that

$$\delta(u, v - u) + \Phi(v) - \Phi(u) - \zeta(v - u) \geq 0, \quad \forall v \in N. \tag{10.87}$$

The second is a multivalued boundary integral equation on the part Γ_3 of the boundary of the structure which reads

$$\zeta - \frac{1}{2}\mathrm{grad}\delta(u, u) \in \partial\Phi(u). \tag{10.88}$$

Each one of the last two formulations is equivalent to the minimum problem (10.84). If on Γ_3 conditions (1.80) and (7.3) hold then $\Phi(u) = \{0$ if $u_N \leq 0$ a.e. on Γ_3, ∞ otherwise$\}$. In this case we will seek the solution of (10.84) over the admissible cone $K = \{u | u \in N, u_N \leq 0$ a.e. on $\Gamma_3\}$ and we can set in the multivalued B.I.E. (10.88) according to (1.116) and to (1.122a) $\Phi(u) = \Phi_N(u_N) + \Phi_T(u_T) = -[C_T, u_T]_{\Gamma_3} + I_K(u_N)$. Analogously any other type of subdifferential conditions can be treated.

We shall close this Section by giving an existence and uniqueness result for the solution of (10.84).

Proposition 10.2 The multivalued B.I.E. (10.88) has a unique solution.

Proof: It is analogous to the proof of Prop. 8.4. Let us introduce in V_0 the equivalent norm (8.53). Then the space $H^{1/2}(\tilde{\Gamma})\big|_{\Gamma_3}$ will be endowed with the norm

$$||\varphi||_{1/2,\Gamma_3} = \inf\left\{|||v||| = \left(C\varepsilon(v), \varepsilon(v)\right)^{1/2} | v \in V_0 \ v = \varphi \text{ on } \Gamma_3\right\}. \tag{10.89}$$

From this definition it results that

$$|||\varphi|||_{1/2,\Gamma_3} = |||\bar{u}||| \tag{10.90}$$

where $\bar{u}$ is the unique solution of the B.V.P.

$$a(\bar{u}, v) = \left(C\varepsilon(\bar{u}), \varepsilon(v)\right) = 0 \quad \forall v \in V_0 \tag{10.91}$$

$$\bar{u} = \varphi \quad \text{on} \quad \Gamma_3. \tag{10.92}$$

This B.V.P. corresponds to an elastic body with $\bar{p}_i = 0$ in Ω, $\bar{T}_i = 0$ on Γ_2 and with $\bar{u} = 0$ on Γ_1 and $\bar{u} = \varphi$ on Γ_3 (cf. equivalently the relation (10.77)). Further we can write that

$$\delta(u, u) = [\tilde{H}(u), u]_{\Gamma_3} = \langle \tilde{H}(u), u\rangle_{1/2,\Gamma_3}. \tag{10.93}$$

But taking into account the definition of $\tilde{H}$ we note that $\tilde{H}(u)$ denotes the boundary tractions $S(u)$ on Γ_3 of the B.V.P. (10.77) or equivalently (10.91), (10.92). Thus applying the Green-Gauss theorem we may write that

$$\langle S(\bar{u}), \bar{u}\rangle_{1/2,\Gamma_3} = a(\bar{u}, \bar{u}) = |||\varphi|||^2_{1/2,\Gamma_3}. \tag{10.94}$$

Accordingly

$$\delta(u, u) = |||\gamma u|||^2_{1/2,\Gamma_3} \tag{10.95}$$

where γu denotes the trace of u on Γ. Now we proceed as in [Pan85] p.201: We have that

$$\Phi(v) \geq -(c_1|||\gamma v|||_{1/2,\Gamma_3} + c_2) \quad c_1, c_2 \text{ const } \geq 0 \tag{10.96}$$

due to the Hahn-Banach theorem. Due to (10.95), (10.96)

$$\Pi_2(v) \geq \frac{1}{2}|||\gamma v|||^2_{1/2,\Gamma_3} - c(|||\gamma v|||_{1/2,\Gamma_3} + 1) \tag{10.97}$$

which, due to Prop. 1.1, implies the existence and uniqueness of the solution, q.e.d.

10.4 Certain Semicoercive Multivalued B.I.Es. Existence Results

In this section we consider the case $\Gamma_1 = \emptyset$. Due to the lack of prescribed displacements we may have the appearence of rigid body displacements. Therefore the arising minimum problem of the boundary with respect to the displacements are semicoercive and for this reason we call the corresponding multivalued B.I.Es semicoercive.

Let us further denote by $\mathcal{R}$ the linear space of the rigid body displacements of the body Ω and by $\mathcal{R}_\Gamma$ the corresponding restriction of $\mathcal{R}$ to Γ_3. The classical norm on $[H^{1/2}(\Gamma_3)]$ is equivalent to the norm (cf. [Hl] p.5)

$$||v||_{1/2,\Gamma_3} = \inf\left\{||\tilde{v}||_{H^1(\Omega)} \big| \tilde{v} = v \text{ on } \Gamma_3\right\} \tag{10.98}$$

which, if $\operatorname{mes}\Gamma_1 \neq 0$, is equivalent to the norm (due to Korn's inequality)

$$|||v|||_{1/2,\Gamma_3} = \inf\left\{a(\tilde{v}, \tilde{v})^{1/2} \big| \tilde{v} = v \text{ on } \Gamma_3\right\}. \tag{10.99}$$

But for $\Gamma_1 = \emptyset$, $a(\cdot, \cdot)$ is only a seminorm on $H^1(\Omega)$. Note that if $\Gamma_1 = \emptyset$ we can easily show that the usual H^1-norm is equivalent to the norm $[a(v,v)^{1/2} + ||v||^2_{L^2(\Gamma_3)}]^{1/2}$ (apply the theorem of [Maz] p.27 noting that $||v||_{L^2(\Gamma_3)}$ is a continuous functional on $H^1(\Omega)$). Moreover $\mathcal{R}_\Gamma$ is the kernel of $|||v|||^2_{1/2,\Gamma_3}$. Using the aforementioned norm equivalence and (10.98) we obtain, by replacing in (10.98) $||\tilde{v}||_{H^1(\Omega)}$ by the equivalent norm, that $||||v||||_{1/2,\Gamma_3} = \left(|||v|||^2_{1/2,\Gamma_3} + |||v|||^2_{L^2(\Gamma_3)}\right)^{1/2}$ is a norm on $H^{1/2}(\Gamma_3)$ equivalent to the classical norm on this space. Moreover let Q be the orthogonal projector of $H^{1/2}(\Gamma_3)$ onto $\mathcal{R}_\Gamma$ for the $L^2(\Gamma_3)$-norm and let $\mathcal{R}_\Gamma^\perp$ be the orthogonal complement of $\mathcal{R}_\Gamma$. We write $v = \bar{v} + r$ where $\bar{v} \in \mathcal{R}_\Gamma^\perp$ and $r \in \mathcal{R}_\Gamma$. Then we have that

$$\||v\||_{1/2,\Gamma_3}^2 \;=\; \||v\||_{1/2,\Gamma_3}^2 + \|v\|_{L^2(\Gamma_3)}^2 = \||\bar{v}\||_{1/2,\Gamma_3}^2 + \|\bar{v}\|_{L^2(\Gamma_3)}^2 + \|r\|_{L^2(\Gamma_3)}^2 \qquad (10.100)$$
$$= \||\bar{v}\||_{1/2,\Gamma_3}^2 + \|r\|_{L^2(\Gamma_3)}^2$$

But $\||\bar{v}\||_{1/2,\Gamma_3}$ is equivalent to $\||\bar{v}\||_{1/2,\Gamma_3}$. Thus we have proved the following proposition by considering now that $v \in \left[H^{1/2}(\tilde{\Gamma})\right]^3\big|_{\Gamma_3}$ (cf. also [Pan93]).

Proposition 10.3 Suppose that $\Gamma_1 = \emptyset$. Then the expressions

$$\left(\||v\||_{1/2,\Gamma_3}^2 + \|v\|_{L^2(\Gamma_3)}^2\right)^{1/2} \qquad (10.101)$$

and

$$\left(\||v\||_{1/2,\Gamma_3}^2 + \|r\|_{L^2(\Gamma_3)}^2\right)^{1/2} \qquad (10.102)$$

are norms on $\left[H^{1/2}(\tilde{\Gamma})\right]^3\big|_{\Gamma_3}$ equivalent to the classical norm on this space.

The foregoing results permit us to apply Prop. 1.20 to the variational inequality (10.87), or equivalently to the multivalued B.I.E. (10.88). The following result holds:

Proposition 10.4 Suppose that $\Gamma_1 = \emptyset$. Then if (10.88) admits a solution then the relation

$$\zeta\big|_{\mathcal{R}_\Gamma} \in \overline{R(\partial\Phi_{u_0})} \quad \forall u_0 \in D(\Phi) \qquad (10.103)$$

holds. If at least one $u_0 \in D(\Phi)$ exists such that

$$\zeta\big|_{\mathcal{R}_\Gamma} \in \mathrm{rel\,int}\, R(\partial\Phi_{u_0}), \qquad (10.104)$$

then (10.88) has a solution.

Let us examine now some interesting special cases. We place ourselves in the functional framework of Ch. 8. We assume that $\Gamma_1 = \emptyset$, that on Γ_2 (7.2) holds and, that on Γ_3 (cf. eqs.(1.118) and (1.122a))

$$-S_N \in \partial j_N(u_N) \quad \text{and} \quad S_T = C_T. \qquad (10.105)$$

Here j_N is a convex, l.s.c., proper superpotential and $C_T = \{C_{T_i}\} \in L^2(\Omega)$. Then (10.87) takes the following form, as we can easily verify: Find $u_N \in H^{1/2}(\tilde{\Gamma})\big|_{\Gamma_3}$ such that

$$\delta(u_N, v_N - u_N) + \tilde{\zeta}(v_N - u_N) + \Phi(v_N) - \Phi(u_N) \geq 0 \quad \forall v_N \in H^{1/2}(\tilde{\Gamma})\big|_{\Gamma_3}. \qquad (10.106)$$

Here $\delta(\cdot,\cdot)$ is a bilinear symmetric form (see (8.52)) on $H^{1/2}(\tilde{\Gamma})\big|_{\Gamma_3} \times H^{1/2}(\tilde{\Gamma})\big|_{\Gamma_3}$, $\tilde{\zeta}(\cdot)$ a linear form on $H^{1/2}(\tilde{\Gamma})\big|_{\Gamma_3}$ and Φ is a convex, l.s.c., proper functional on $H^{1/2}(\tilde{\Gamma})\big|_{\Gamma_3}$ defined by the relation

$$\Phi(v_N) = \begin{cases} \displaystyle\int_{\Gamma_3} j_N(v_N)\,d\Gamma & \text{if } j_N(v_N) \in L^1(\Gamma_3) \\ \infty & \text{otherwise.} \end{cases} \qquad (10.107)$$

For this problem we can verify that a proposition analogous to Prop. 10.4 holds. If the Signorini boundary conditions hold then $\Phi(v_N) = I_K(v_N)$ where $K = \{v_N | v_N \in H^{1/2}(\tilde{\Gamma})|_{\Gamma_3}\ v_N \leq 0\}$. Then (10.106) takes the form of the Signorini-Fichera variational inequality (1.160). Thus we can apply, if $\Gamma_1 = \emptyset$, Prop. 1.17 and Prop. 1.18 which yield the following results (cf. e.g. [Pan85] p.122).

Proposition 10.5 Suppose that $\tilde{\mathcal{R}}_1 = \mathcal{R}_\Gamma \cap K$ and let $\tilde{\mathcal{R}}_2 = \{r | \pm r \in \tilde{\mathcal{R}}_1\}$. We assume that

$$\tilde{\zeta}(r) \leq 0 \quad \forall r \in \tilde{\mathcal{R}}_1 \tag{10.108}$$

and

$$\tilde{\zeta}(r) < 0 \quad \forall r \in \tilde{\mathcal{R}}_1 - \tilde{\mathcal{R}}_2 \tag{10.109}$$

Then the multivalued B.I.E. (10.88) (for Φ replaced by I_K) has a solution u_N. Any other solution is given by $u_N + r$, where

$$r \in \mathcal{R}_\Gamma, \quad \tilde{\zeta}(r) = 0 \quad u_N + r \in K. \tag{10.110}$$

Note that the relation $\tilde{\zeta}(r) < 0$ when transformed by the Green-Gauss theorem yields the classical sufficient condition for the semicoercive Signorini-Fichera problem (cf. e.g. [Pan85] p.121). Any other type of necessary or sufficient conditions for semicoercive variational inequalities can be adapted to the multivalued B.I.Es; for instance, the sufficient conditions given in [Pan85] (Prop. 6.2.4 and Prop. 6.2.5 p.203) which include as a special case the sufficient conditions for the friction problem. Thus the following two propositions hold:

Proposition 10.6 Assume that in (10.88) $\Phi = \Phi_1 + \Phi_2$ where Φ_1 and Φ_2 are convex l.s.c. and proper functionals such that $D(\Phi_1) \cap D(\Phi_2) \neq \emptyset$. Moreover assume that for $a > 1$ and for every $t > 0$

$$\Phi_1(0) = 0, \quad \Phi_1(tv) \leq t^a \Phi_1(v) \quad \forall v \in \left[H^{1/2}(\tilde{\Gamma})\right]^3\Big|_{\Gamma_3} \tag{10.111}$$

and that

$$\Phi_1(r) > 0 \quad \forall r \in \mathcal{R}_\Gamma \quad r \neq 0 \tag{10.112}$$

Then (10.88) has at least one solution.

Proposition 10.7 Assume that in (10.88)

$$\Phi(v) = \Phi_0(v) - \langle h, v \rangle, \tag{10.113}$$

where $h \in \left[H^{-1/2}(\tilde{\Gamma})\right]^3\Big|_{\Gamma_3}$ is given and Φ_0 is a convex, l.s.c. and proper functional such that

$$\Phi_0(0) = 0, \quad \text{and} \quad \Phi_0(tv) = t\Phi_0(v) \quad \forall t > 0 \ \forall v \in \left[H^{1/2}(\tilde{\Gamma})\right]^3\Big|_{\Gamma_3} \tag{10.114}$$

Then the condition

$$\Phi_0(r) \geq \tilde{\zeta}(r) + \langle h, r \rangle \quad \forall r \in \mathcal{R}_\Gamma \tag{10.115}$$

is a necessary condition, and

$$\Phi_0(r) > \tilde{\zeta}(r) + \langle h, r \rangle \quad \forall r \in \mathcal{R}_\Gamma \quad r \neq 0 \tag{10.116}$$

is a sufficient condition for the existence of a solution of the semicoercive B.I.E. (10.88).

The friction superpotential $\Phi(v_T) = \int_{\Gamma_3} \mu |C_N||v_T|$ fulfills the properties (10.113), (10.114) of the last proposition. Transforming the term $\tilde{\zeta}(r)$ by the Green-Gauss theorem [Pan93] yields the well-known necessary and sufficient conditions for the friction semicoercive variational inequality [Duv72].

Chapter 11
Elastodynamic Unilateral Problems. A B.I.E. Approach

11.1 The Time Discretization Scheme. Time-Difference Multivalued B.I.Es.

The variational inequalities corresponding to dynamic unilateral problem may be formulated as minimum problems, only after an appropriate discretization. In this way, a minimum problem results within each time step. Moreover in the unilateral contact problems the question of impact has to be carefully taken into account in a numerical method. In realistic problems we have to determine all the free boundaries as they vary with time and, in the case of contact, the time and the geometry of all the configurations in which the body is subjected to impact forces. It is important from the numerical point of view to formulate the problem only with respect to the "ambiguous" degrees of freedom. Taking as example the contact problem the term "ambiguous" means that it is not a priori known which parts of the body are in contact with the support or with the other parts of body; moreover, it is not a priori known which parts of the areas being in contact are in a state of sticking or sliding friction. In other classes of problems, e.g. in the plastic hinge problems in plates, other physical parameters like the extent of the plastic hinge along the boundary is not a priori known. From the numerical point of view, an algorithm for unilateral problems is considered as efficient if it can treat a large number of ambiguous points. Therefore the presented B.I.E. approach gains importance because it refers solely to the ambiguous degrees of freedom of an inequality problem.

Here the technique of the previous Sections is extended to dynamic variational inequalities: within each time step a parametric Lagrangian formulation of the initial variational inequality leads to time parametric minimum problems on the boundary which are equivalent to multivalued boundary integral equations within each time step, i.e. we get time-difference multivalued B.I.Es. Including of partial or total velocity reversal into our model permits the rational consideration of impact shocks. Numerical examples from the aseismic design and from the dynamic plasticity illustrate the theory.

This Chapter does not deal with nonmonotone unilateral contact problems leading to hemivariational inequalities where the lack of the convexity and the resulting nonuniqueness of the solution within each time step have left until now many unsolved problems. The present Chapter is tightly connected with the previous Chapter in the sense that it includes numerical applications related to the theory of Chapter 10.

We consider a three-dimensional linear elastic body; the method presented is general and also holds for plates, beams, etc., i.e. for all structures permitting a Langrangian formulation of the equilibrium problem.

Let Ω be an open bounded subset of the three-dimensional Euclidean space $\mathbb{R}^3$ with a boundary Γ. Ω is occupied by a linear elastic body in its undeformed case which is referred to an orthogonal Cartesian coordinate system $Ox_1x_2x_3$. Γ is decomposed as usual into three mutually disjoint parts Γ_1, Γ_2 and Γ_3. On Γ_1 (resp Γ_2) the displacements (resp. the tractions) are given and on Γ_3 boundary conditions giving rise to variational inequalities hold. We assume that the time t takes values in the time interval $[0, T]$.

On Γ_1

$$u_i = \bar{U}_i, \quad \bar{U}_i = \bar{U}_i(x, t), \tag{11.1}$$

and on Γ_2

$$S_i = \bar{T}_i, \quad T_i = \bar{T}_i(x, t). \tag{11.2}$$

The boundary condition on Γ_3 reads

$$-S \in \partial j(u, t) \text{ on } \Gamma_3 \times [0, T]. \tag{11.3}$$

Here j is a convex lower semicontinuous proper superpotential depending on t. Note that (11.3) might be replaced without affecting the method of the present Chapter by the two conditions

$$-S_N \in \partial j_N(u_N, t) \text{ and } -S_T \in \partial j_T(u_T, t) \text{ on } \Gamma_3 \times [0, T]. \tag{11.4}$$

The subdifferential is taken with respect to the displacement variable. The following equations of motion hold on the assumption of small displacements and small strains

$$\sigma_{ij,j} + \bar{p}_i(t) = \rho \ddot{u}_i + c\dot{u}_i \qquad \text{in } \Omega \times (0, T) \tag{11.5}$$

$$\epsilon_{ij} = \epsilon_{ij}(u) = 1/2(u_{i,j} + u_{j,i}) \qquad \text{in } \Omega \times (0, T) \tag{11.6}$$

$$\sigma_{ij} = C_{ijhk}\epsilon_{hk} \qquad \text{in } \Omega \times (0, T) \tag{11.7}$$

$$u_i = u_{i0}(x) \text{ at } t = 0 \tag{11.8}$$

$$\dot{u}_i = u_{i1}(x) \text{ at } t = 0 \tag{11.9}$$

where u_{i0} (resp. u_{i1}) denotes the initial displacements (resp. velocities), $\bar{p} = \{\bar{p}_i\}$ represents the volume force vector, the comma denotes the partial derivation, $\ddot{u}_i$ is the acceleration vector and ρ is the mass density. Let us assume that the damping term is proportional to the velocity $\dot{u}_i$ and let us denote the damping coefficient by $c > 0$. We apply the method of time discretization in order to reduce the problem into a static problem. The method of m-step linear difference operators is applied for the time discretization of the problem with respect to time. At time instant $t^{(p)}$ we obtain (cf., also [Fel, Mit83, Pan85])

$$\sum_{r=0}^{q} \alpha^{(r)} u^{(p-r)} = \Delta t \sum_{r=0}^{q} \beta^{(r)} \dot{u}^{(p-r)} \tag{11.10}$$

and

$$\sum_{r=0}^{q} \gamma^{(r)} u^{(p-r)} = \mu \Delta t^2 \sum_{r=0}^{q} \beta^{(r)} \ddot{u}^{(p-r)}, p \geq q, p > 1. \tag{11.11}$$

The coefficients $\alpha^{(r)}, \beta^{(r)}, \gamma^{(r)}$, and μ depend on the chosen finite-difference scheme. We assume that the time step size Δt remains constant and in order to have an implicit integration scheme that $\beta^{(0)}$ is nonzero. Thus the relations

$$a^{(0)}u^{(p)} - \Delta t\beta^{(0)}\dot{u}^{(p)} = -U\alpha + \Delta t\dot{U}\beta \tag{11.12}$$

$$\mu(\Delta t)^2\beta^{(0)}\ddot{u}^{(p)} = \gamma^{(0)}u^{(p)} + U\gamma - \mu(\Delta t)^2\ddot{U}\beta = \gamma^{(0)}u^{(p)} + \tilde{U}, \tag{11.13}$$

where

$$\alpha = [\alpha^{(1)}, \cdots, \alpha^{(q)}]^T, \quad \beta = [\beta^{(1)}, \cdots, \beta^{(q)}]^T$$
$$\gamma = [\gamma^{(1)}, \cdots, \gamma^{(q)}]^T, \text{ and } \quad U = [u^{(p-1)}, \cdots, u^{(p-q)}]$$

are obtained. Accordingly within the p-time interval $(t^{(p)}, t^{(p)} + \Delta t)$ we may write after applying the above time discretization to the equations of motion, that (we omit index p)

$$\sigma_{ij,j} + \bar{p}_i(t) = g_i(t) + Au_i \quad A > 0 \text{ in } \Omega \times (t, t + \Delta t). \tag{11.14}$$

Here A is a constant equal to $\gamma^{(0)}\rho/\mu(\Delta t)^2\beta^{(0)} + \alpha^{(0)}c/\Delta t\beta^{(0)} > 0$. The term g_i contains the terms of the previous time steps resulting from (11.12), (11.13). Note that thermal terms and terms due to dislocations may be included in the model provided we know their evolution with time. We denote further by $\tilde{f}_i$ the term

$$\bar{f}_i = \bar{p}_i - g_i$$

and we consider from now on only the behaviour of the structure in the time interval $(t, t + \Delta t)$. The above discretization process "replaces" roughly speaking the dynamic variational inequality formulation of the problem by a static variational formulation. In the absence of shocks it can be shown that as $\Delta t \to 0$ the solution of the dynamic problem is obtained (cf. eg. [Pan85]). In the realistic case of shocks, treated in the present Chapter, the mathematical problem of convergence is still open.

Let V be the set

$$V = \{v|v = \{v_i\}, \, v \in \tilde{V}, \, v_i = \bar{U}_i \quad i = 1, 2, 3 \text{ on } \Gamma_1\} \tag{11.15}$$

of the kinematically admissible displacements within the time interval $(t, t + \Delta t)$, where $\tilde{V}$ is the basic vector space for the displacements. We denote by $(\bar{f}, v)$ and by $[\bar{T}, v]_{\Gamma_2}$ etc. the corresponding work expressions as in Ch.7 and let

$$(u, v) = A\int_\Omega u_i v_i d\Omega \quad A > 0 \tag{11.16}$$

and

$$a(u, v) = (C\epsilon(u), \epsilon(v)) = \int_\Omega C_{ijhk}\epsilon_{ij}(u)\epsilon_{hk}(v)d\Omega \tag{11.17}$$

be the bilinear form of elasticity. Now let Π be the potential energy within the interval $(t, t + \Delta t)$

$$\Pi(v) = 1/2a(v, v) + 1/2(v, v) - (\bar{f}, v) + \Phi(v) - [\bar{T}, v]_{\Gamma_2} \tag{11.18}$$

where $\Phi(v) = \{\int_{\Gamma_3} j(v)d\Gamma$ if $j(v) \in L^1(\Gamma_3)$, ∞ otherwise $\}$.

We know, that [Pan85]

$$\Pi(u) = \min\{\Pi(v)|v \in V\} \text{ in } (t, t + \Delta t) \tag{11.19}$$

characterizes the position of equalibrium within each time interval. Problem (11.19) has one and only one solution for $\tilde{V} = [H^1(\Omega)]^3$, $C_{ijhk} \in L^\infty(\Omega)$, $\bar{f}_i \in L^2(\Omega)$, $\bar{T}_i \in L^2(\Gamma_2)$, $\bar{U}_i \in H^{1/2}(\Gamma_1)$ which satisfies equivalently the corresponding variational inequality. The procedure of Ch.10 is repeated within each time interval for the bilinear form $a(u, v) + (u, v) \hat{=} \bar{a}(u, v)$ and for the fictitious body forces $\bar{f}$. Obviously these changes do not affect the validity of all methods and transformations of Ch.10. Therefore here the results are the same as in Ch.10 with the aforementioned minor changes. Let us denote by $\tilde{\beta}$, $\tilde{\gamma}$, $\tilde{\delta}$ and $\tilde{\zeta}$ the bilinear and linear forms within the time interval which correspond to the bilinear and linear forms β, γ, δ and ζ of Ch.10.

Then we obtain the following minimization problem with respect to the unknown boundary tractions μ:

$$\min\{\Pi_1(\mu) = 1/2\tilde{\beta}(\mu, \mu) + \Phi^c(-\mu) - \tilde{\gamma}(\mu)|\mu \in L\} \quad L = \left[H^{-1/2}(\tilde{\Gamma})\right]^3\big|_{\Gamma_3} \tag{11.20}$$

We denote again by λ the solution of this problem. The form $\tilde{\beta}(\cdot, \cdot)$ is symmetric and coercive on $\boldsymbol{H}^{-1/2}(\tilde{\Gamma})\big|_{\Gamma_3} \times \boldsymbol{H}^{-1/2}(\tilde{\Gamma})\big|_{\Gamma_3}$. Then (11.20) is equivalent to the following variational inequality: Find $\lambda \in L$ such as to satisfy

$$\tilde{\beta}(\lambda, \mu - \lambda) + \Phi^c(-\mu) - \Phi^c(-\lambda) - \tilde{\gamma}(\mu - \lambda) \geq 0 \quad \forall \mu \in L. \tag{11.21}$$

Using the definition of the subdifferential, (11.21) is equivalent to the "multivalued" integral equation

$$\tilde{\gamma} - 1/2 \text{ grad } \tilde{\beta}(\lambda, \lambda) \in \partial\Phi^c(-\lambda) \text{ on } \Gamma_3 \tag{11.22}$$

which holds on Γ_3.

The minimum problem with respect to the unknown displacements u on Γ_3 reads in the time interval $(t, t + \Delta t)$

$$\min\{\Pi_2(v) = 1/2\tilde{\delta}(v, v) + \Phi(v) - \tilde{\zeta}(v)|v \in N\} \quad N = \left[H^{1/2}(\tilde{\Gamma})\right]^3\big|_{\Gamma_3}. \tag{11.23}$$

The bilinear form $\tilde{\delta}(\cdot, \cdot)$ is coercive and symmetric. Let us give now the two equivalent formulations of (11.23). The first is: Find $u \in \boldsymbol{H}^{1/2}(\tilde{\Gamma})\big|_{\Gamma_3}$ such as to satisfy within $(t, t + \Delta t)$ the variational inequality

$$\tilde{\delta}(u, v - u) + \Phi(v) - \Phi(u) - \tilde{\zeta}(v - u) \geq 0 \quad \forall v \in \boldsymbol{H}^{1/2}(\tilde{\Gamma})\big|_{\Gamma_3} \tag{11.24}$$

The second is: Find $u \in \boldsymbol{H}^{1/2}(\tilde{\Gamma})\big|_{\Gamma_3}$ solution of the multivalued B.I.E.

$$\tilde{\zeta} - 1/2 \text{ grad } \tilde{\delta}(u, u) \in \partial\Phi(u). \tag{11.25}$$

11.2 Numerical Applications

This section is based on two papers of the second author and his coworkers [Mit91; 93] and to [Zer]. The solution of the dynamic inequality problems considered, is formulated after an appropriate time discretization. Here we apply for the numerical solution, the weighted residual time discretization algorithm proposed by Zienckiewicz, Wood and Taylor [Zie80], which is a special case of the algorithm defined by eq. (11.10), (11.11). The algorithm is implicit and unconditionally stable. Note that explicit and conditionally stable algorithms are not applicable because the time step for them is to be chosen on the basis of formulas containing the frequencies of the system. But no frequencies can be defined in an inequality problem in the classical sense [Pan85].

The algorithm interpolates independently the displacement and velocity vectors, and therefore computation of acceleration terms is avoided. This is an advantage for the present problems, because a calculation of "initial" accelerations in the case of impact is not necessary.

As we have shown in the previous section two B.I.Es hold, within each time interval, on the boundary Γ_3 of the system. The two B.I.Es and their equivalent minimum problems are dual in the sense that the first has as unknowns the boundary forces on Γ_3, whereas the second has as unknowns the boundary displacements on Γ_3, with the remark that the work of the fictitious springs with constant A must be also considered, as well as the fictitious body forces $\bar{f}_i$ in order to take into account the influence of the time discretization scheme.

To calculate the discrete forms of Π_1 (resp. Π_2) we apply the same method as in Ch.7 which is based on unit force (resp. unit displacement) loadings of a bilateral structure and calculation of it by a classical B.E.M. or a F.E.M. Let us calculate the discrete form of Π_1. We discretize first the boundary of the elastic body under consideration by a B.E. or F.E. scheme. For this discretized system, the unilateral (inequality) constraints on the boundary Γ_3 refer to the m nodes of this boundary. We consider then the system Ω_0 obtained from the discretized one by assuming only the kinematical constraints on Γ_1. The resulting structural system is also appropriately modified by the fictitious springs of constant A introduced by the time discretization scheme. This discrete system is "solved" for a unit force corresponding to an "inequality" constrained degree of freedom, on the first node of Γ_3, and zero forces on the other nodes of Γ_3. In the case of interfaces we have pairs of nodes and pairs of unit forces corresponding to the nodes. The solution of the resulting underconstrained structure Ω_0, supplies the corresponding displacements in the directions of the "inequality" constrained degrees of freedom of the m nodes of Γ_3. They constitute the first column of a matrix $\tilde{B}$. This procedure is repeated for all the nodes m and thus the whole matrix $\tilde{B}$ of influence coefficients is calculated. Note at this point that if the unit force solutions are analytically given then an explicit form of the multivalued integral equation (11.22) on the boundary, would be obtained.

Within each time interval the displacements in the directions of the "inequality" degrees of freedom of the nodes, or in the case of interfaces of the node pairs of Γ_3, due to the external actions constitute a vector $\tilde{g}$. Then the discrete form of the minimum problem (11.20) reads (symmetrized problem)

$$\min\{\frac{1}{2}\boldsymbol{\mu}^T\tilde{\boldsymbol{B}}\boldsymbol{\mu} + \Phi^c(-\boldsymbol{\mu}) - \tilde{\boldsymbol{g}}^T\boldsymbol{\mu}|\boldsymbol{\mu} \in \mathbb{R}^m\}. \tag{11.26}$$

Here $\boldsymbol{\mu}$ is the vector of the unknown reactions on Γ_3, which are "inequality" constrained, i.e they must fulfill the inequality subsidiary conditions of the problem or are subjected to subdifferential boundary conditions.

The discrete form of the minimum problem of (11.23), can be written as:

$$\min\{\frac{1}{2}\boldsymbol{v}^T\tilde{\boldsymbol{D}}\boldsymbol{v} + \Phi(\boldsymbol{v}) - \tilde{\boldsymbol{z}}^T\boldsymbol{v}|\boldsymbol{v} \in \mathbb{R}^m\} \tag{11.27}$$

Here $\boldsymbol{v}$ is the vector of the unknown displacements on Γ_3, which are "inequality" constrained. The matrix $\tilde{\boldsymbol{D}}$ and the known vector $\tilde{\boldsymbol{z}}$ are obtained as follows: The discrete structural system Ω is solved by a bilateral B.E.M. by imposing a unit displacement corresponding to an "inequality" constrained degree of freedom on the first node of Γ_3 by zeroing the displacements of the other degrees of freedom of the same node and of the degrees of freedom of all the other nodes on Γ_3. The solution of the resulting overconstrained structure Ω_0', by the classical B.E.M. or the F.E.M., supplies the corresponding reactions of all the fictitious constraints of the m-nodes of Γ_3. They constitute the first column of a matrix $\tilde{\boldsymbol{D}}$. This procedure is repeated for all the nodes of Γ_3 and thus we obtain the whole matrix $\tilde{\boldsymbol{D}}$ of influence coefficients. Within each time interval the reactions of the nodes of all the fictitious constraints of the m nodes of Γ_3 due to the external actions constitute the vector $\tilde{\boldsymbol{z}}$. About the symmetry or not of $\tilde{\boldsymbol{B}}$ and $\tilde{\boldsymbol{D}}$ we refer to Ch.7. Both the discrete minimum problems are classical quadratic programming problems. $\tilde{\boldsymbol{B}}$ and $\tilde{\boldsymbol{D}}$ are full symmetric or symmetrized positive definite matrices but of relatively small size since only the unknowns on Γ_3 are involved.

Next two dynamic inequality contact problems of different nature are numerically studied. In both examples the displacement method is used for the solution. Within each time interval $(t, t + \Delta t)$ a modification of Hildreth d' Esopo's Q.P. algorithm is applied [Mit91]. This algorithm is applied for the solution of the quadratic programming problem (11.27) which has unknowns only the displacements at the Γ_3 boundary and which converges to the unique solution due to the positive definiteness of matrix $\tilde{\boldsymbol{D}}$.

As a first application we examine the problem of fig. 11.1. It is a main channel burried in a linear elastic homogeneous soil supported by a rigid bedrock on which a seismic excitation is given. A sinusoidal acceleration acts on the bedrock. Unilateral contact (Signorini) conditions with unprevented sliding are assumed to hold between the channel an the soil. No friction exists in this example.

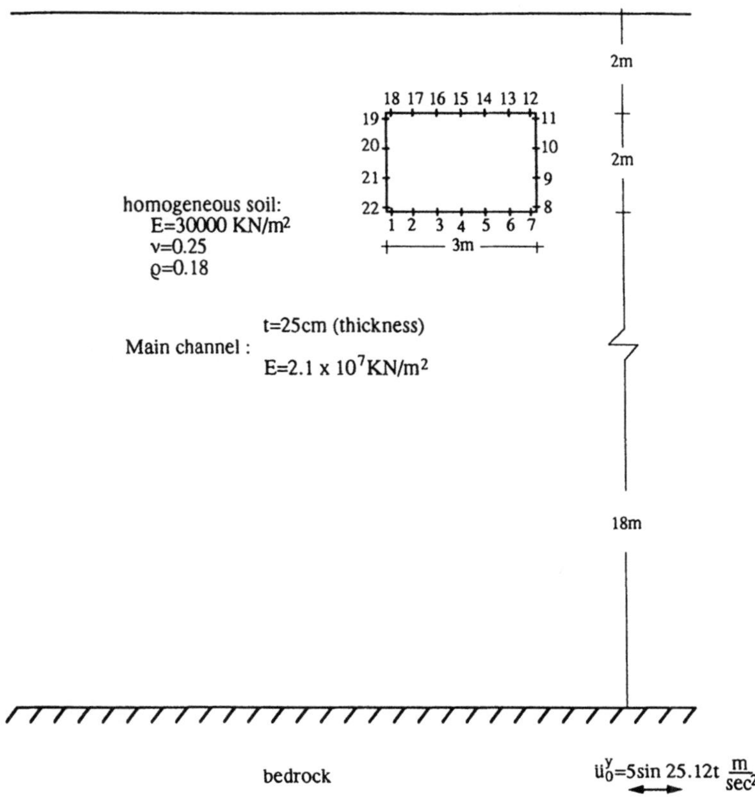

Fig. 11.1: The data of the first application

The vibration is considered to be undamped in the sense that c in eq. (11.5) is zero. When evaluating the dynamic response of the structure it is necessary to consider the collisions at the interface between channel and soil. The velocity towards the support of a point i before impact is $\dot{u}_i^-$. When the point has a contact with the support a part of the kinetic energy is lost due to impact. It is reasonable to accept a perfectly inelastic collision which dissipates the whole kinetic energy. Thus the velocity of the point i just after impact $(\dot{u}_i^+)$ is assumed to be zero. Thus although c was assumed to be zero an amount of damping is taken into account due to perfectly plastic collision.

For the time discretization of the equations of motion the algorithm of Zienkiewicz, Wood and Taylor with $a = \theta = 0.5$ (cf. [Tay]) is applied. At each time step the convergence of the Q.P. algorithm is very rapid. When a node i has a contact with the support the condition $\dot{u}_i^+ = 0$ is imposed as "initial" condition for the next time step.

The computer code can also take into accound any other type of velocity changes due to impact (e.g. velocity reversal in the case of elastic impact etc.). For the determination of the matrices of the discretized problem, i.e. of the Q.P.P., a classical direct B.E.M. code using boundary elements with linear interpolation was used. In fig. 11.2 the time history of the displacements of certain points at the channel-soil interface is depicted. In fig. 11.3 the displacements in the $y - y$ direction of the two corners of the channel are plotted. Finally in fig. 11.4 the displacements of the channel obtained by assuming everywhere nondebonding (i.e. bilateral contact) are compared with the same results when the possibility of debonding is not excluded (unilateral contact).

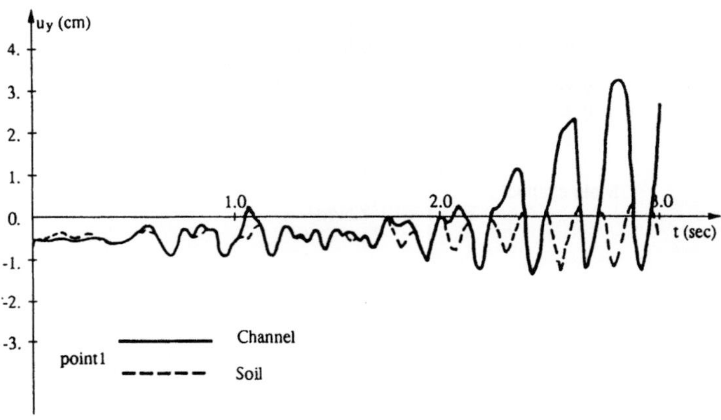

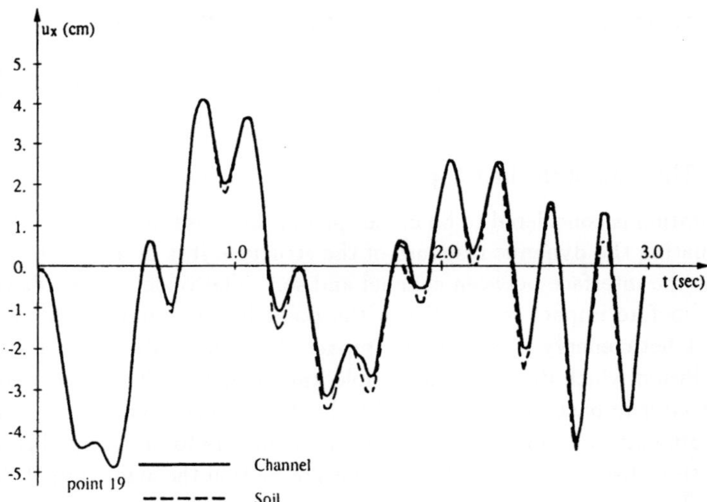

Fig. 11.2: The displacement evolution of points 1 and 19

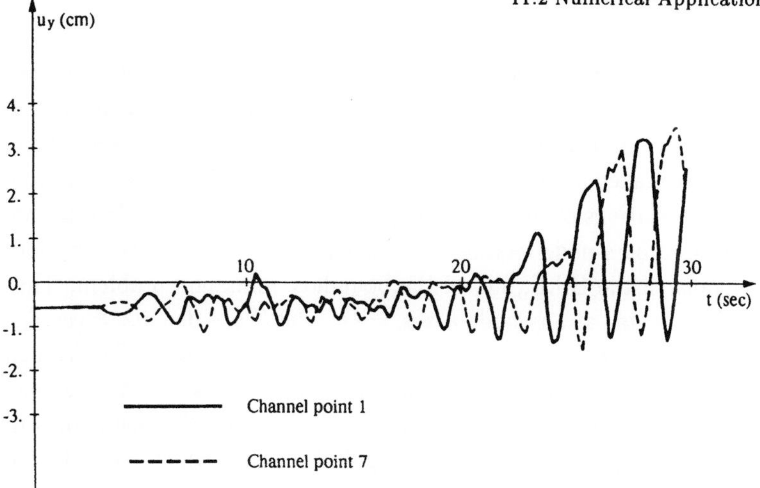

Fig. 11.3: The displacement evolution of points 1 and 7

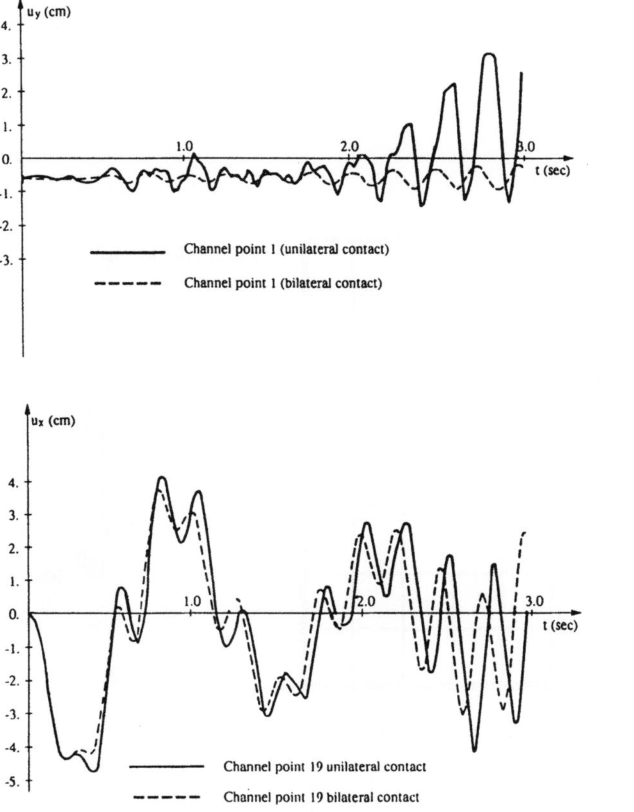

Fig. 11.4: Comparison of the results of the assumptions of unilateral and bilateral contact

As a second example we examine a rectangular Kirchhoff thin plate in bending with clamped rigid-plastic edges (fig. 11.5) [Mit93]. The displacements of any point in the direction of the coordinate axes are expressed as functions of the transverse displacements $w(x, y)$ which are taken as positive if they are directed in the z-direction (upwards), by the well-known relations of the Kirchhoff plate theory [Gir]. The plate as shown in fig. 11.5 is discretized by 4-node rectangular elements. This F.E.M. scheme offers here the discrete versions of the B.I.E., i.e. of the boundary minimum problem. The vector of nodal "displacements" at any node i is $u_i = \{w_i, \partial w_i/\partial y, -\partial w_i/\partial x\} \hat{=} \{w_i, \theta_{ix}, \theta_{iy}\}$ and the nodal actions corresponding to them are $\bar{p}_i = \{p_i, M_{ix}, M_{iy}\}$. The response of the plate, under a transverse uniform dynamic load, which varies with time as shown in fig. 11.6, is studied. Plastic hinges may be formed only along the boundaries during the loading process. In fig. 11.7 is shown the rigid-plastic law which gives the relationship between the edge moments M_T and the inelastic rotations θ_T, at the points of the boundary, and the loading and unloading paths.

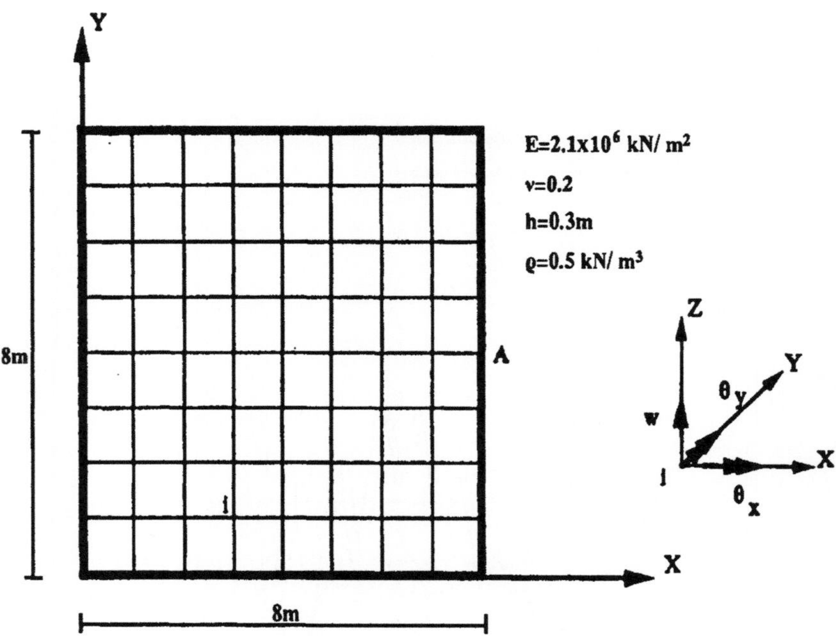

Fig. 11.5: The data of the second application (h=plate thickness, θ_x, θ_y = rotations)

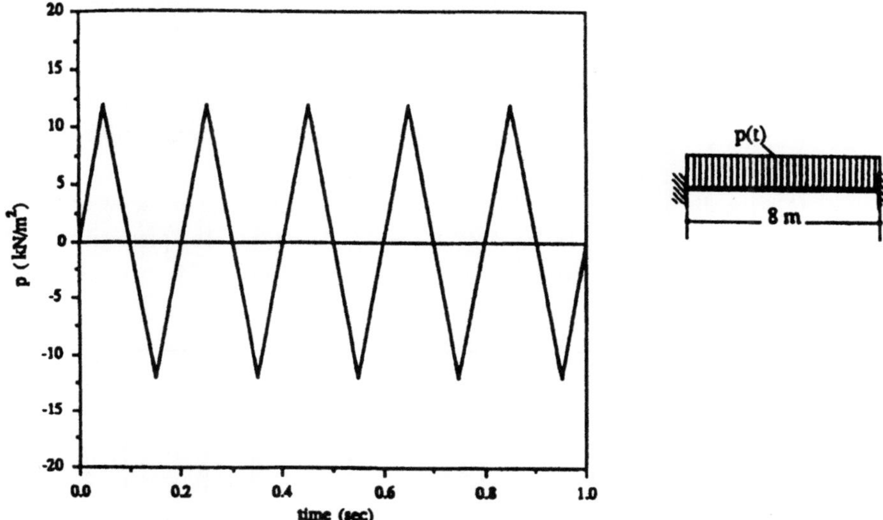

Fig. 11.6: The dynamic loading of the plate

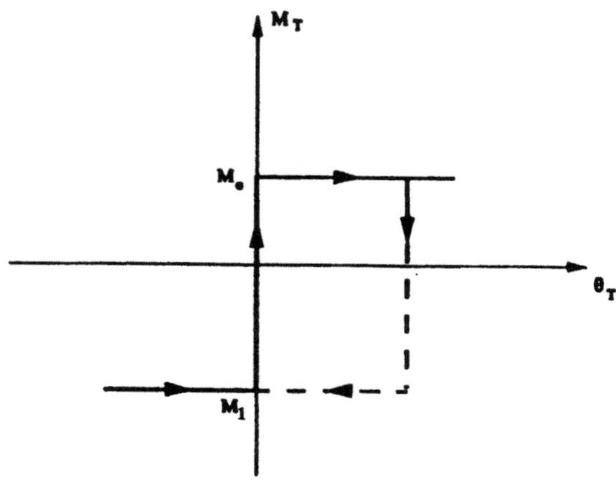

Fig. 11.7: The plastic hinge law of the plate boundary

For the numerical calculation we have considered the plastic hinge law (1.97) and we formulate the minimum problem (11.27) with respect to the rotations θ_T, within each time increment. After some minor transformations (splitting of θ_T into the positive and the negative part θ_{T+} and θ_{T-}) we may verify that the resulting minimum problem coincides with the minimum problem of de Donato and Maier [Don]. Thus we avoid the nonsmooth term $|\theta_T|$ in our minimization problem, something which would be also possible by passing to the dual problem; indeed the dual problem is a pure Q.P.P. with the inequality subsidiary condition $M_1 \leq M \leq M_0$. A serious disadvantage of the splitting of θ_T into θ_T^+ and θ_T^- is the doubling of the unknowns in the resulting minimum problem. This is the reason why in large scale problems involving the absolute value function in the energy, as it is in the friction problem, we avoid the procedure followed here and we prefer the passing to the dual problem. Certain numerical results are given further.

In fig. 11.8 the displacements w of the central node of the plate are shown for three different boundary conditions. In the first $M_0 = M_1 \to \infty$, i.e. the plasticity occurs, in the second $M_0 = M_1 = 15kNm$ in the third case $M_0 = 5kNm$, $M_1 = -15kNm$.

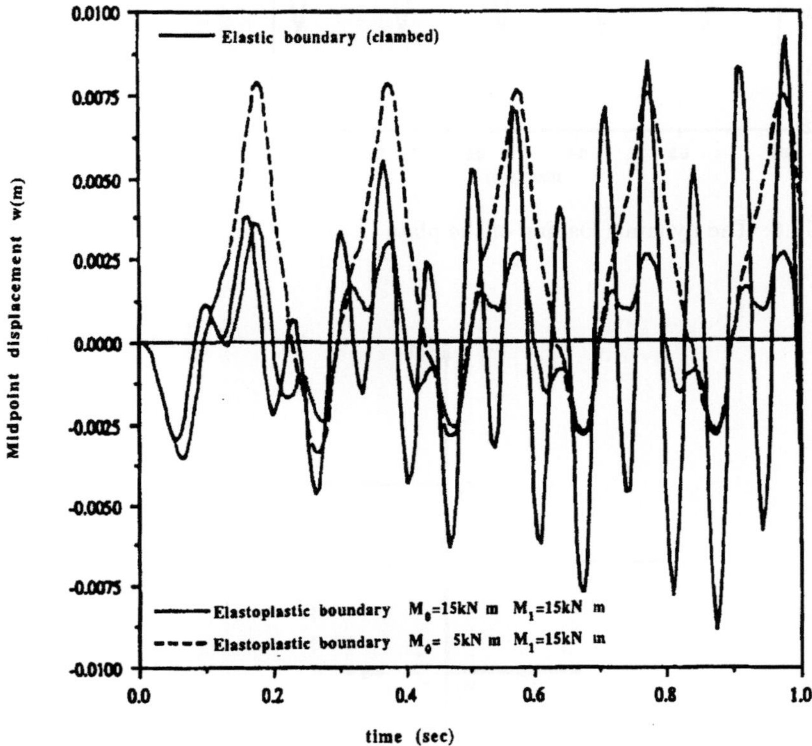

Fig. 11.8: Oscillations of the middle point of the plate for several types of boundary conditions

Finally in fig. 11.9 the bending moments of the point A of the plate boundary are shown.

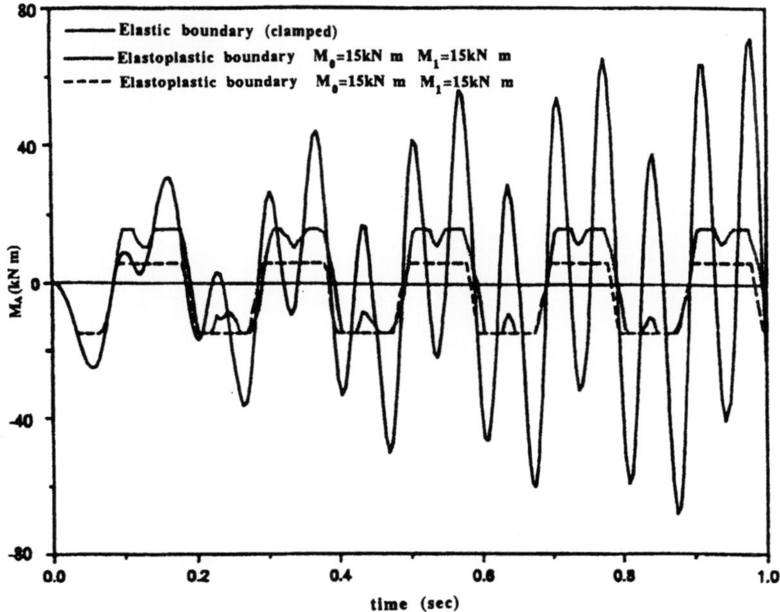

Fig. 11.9: The variation of M at the boundary point A of the plate.

Chapter 12
Nonconvex Unilateral Contact Problems.

12.1 A Boundary Integral Equation with Respect to the Boundary Tractions

We consider a three-dimensional linear elastic body subjected to nonmonotone multivalued boundary conditions which are obtained from nonconvex superpotentials (cf. Sect. 1.3.3). The procedure which we follow remains valid also for shells, plates, beams etc. Let Ω be an open bounded subset of the three-dimensional Euclidean space $\mathbb{R}^3$ with a Lipschitz boundary Γ. Ω is occupied by a linear elastic body in its undeformed state. We refer to a Cartesian orthogonal coordinate system $Ox_1x_2x_3$. Γ is decomposed into nonoverlapping parts Γ_1, Γ_2 and Γ_3 open in Γ, such that mes $\Gamma_1 \neq 0$, and mes $\Gamma_3 \neq 0$. It is assumed that on Γ_1 (resp. Γ_2) the displacements (resp. the tractions) are given and that on Γ_3 the boundary conditions causing the inequality formulation of the problem hold. With the same notation as in the two previous Chapters we assume that on Γ_1

$$u_i = \bar{U}_i, \qquad \bar{U}_i = \bar{U}_i(x), \tag{12.1}$$

and on Γ_2

$$S_i = \bar{T}_i, \qquad \bar{T}_i = \bar{T}_i(x). \tag{12.2}$$

The nonconvex superpotential boundary conditions have the general form

$$u \in \bar{\partial}\tilde{j}(-S) \tag{12.3}$$

or

$$-S \in \bar{\partial}j(u) \quad \text{on } \Gamma_3, \tag{12.3a}$$

where $\tilde{j}, j$ are nonconvex superpotentials which are locally Lipschitz. Here $\bar{\partial}$ denotes the generalized gradient. For the forms of $\tilde{j}, j$, and the corresponding mechanical problems we refer to Sect. 1.3.3, to [Pan85, Mor88a,b]. If S_N (resp. S_T) are the normal (resp. the tangential) components of S with respect to Γ and u_N and u_T are the corresponding components of the displacement u, the method presented here remains valid if (12.3) and (12.3a) are replaced by

$$u_N \in \bar{\partial}\tilde{j}_N(-S_N) \quad \text{and} \quad u_T \in \bar{\partial}\tilde{j}_T(-S_T) \text{ on } \Gamma_3 \tag{12.4}$$

or

$$-S_N \in \bar{\partial}j_N(u_N) \quad \text{and} \quad -S_T \in \bar{\partial}j_T(u_T) \text{ on } \Gamma_3. \tag{12.4a}$$

The equations of the B.V.P. read

$$\sigma_{ij,j} + \bar{p}_i = 0 \quad \text{in } \Omega, \tag{12.5}$$

$$\varepsilon_{ij} = \varepsilon_{ij}(u) = \frac{1}{2}(u_{i,j} + u_{j,i}) \text{ in } \Omega, \tag{12.6}$$

$$\sigma_{ij} = C_{ijhk}\varepsilon_{hk} \quad \text{in } \Omega, \tag{12.7}$$

where the comma denotes the partial derivation and $p = \{\bar{p}_i\}$ is the volume force vector. Let us denote by $\tilde{V}$ the linear space of the displacements v_i and by V the set of the kinematically admissible displacement fields

$$V = \{v|v = \{v_i\}, v \in \tilde{V}, v_i = U_i, i = 1, 2, 3 \text{ on } \Gamma_1\} \tag{12.8}$$

without taking yet into account the constraints on Γ_3. As in the previous Chapters the work of the force $\bar{p} = \{\bar{p}_i\}$ (resp $\bar{T} = \{\bar{T}_i\}$) for the displacement $v = \{v_i\}$ on Ω (resp. on Γ_2 is written as $(\bar{p}, v)$ (resp. as $[\bar{T}, v]_{\Gamma_2}$) etc. Note that if $\tilde{V} = \left[H^1(\Omega)\right]^3$ and $\bar{p}_i \in L^2(\Omega)$, $C_{ijhk} \in L^\infty(\Omega)$, $\bar{T}_i \in L^2(\Gamma)$ and $\bar{V}_i \in H^{1/2}(\Gamma)\big|_{\Gamma_1}$ then $[\bar{T}, v]_{\Gamma_2} = \langle\bar{T}, v\rangle_{1/2,\Gamma_2}$ etc. The bilinear form of elasticity is again denoted by $a(\cdot, \cdot)$ and the relation (7.9) holds. Moreover

$$l(v) = (\bar{p}, v) + \int_{\Gamma_2} \bar{T}_i v_i d\Gamma. \tag{12.9}$$

In order to make the problem homogeneous on Γ_1 we introduce a kinematically admissible displacement field u_0 such that $u_{0i} = \bar{U}_i$ on Γ_1, and let

$$\bar{u} = u - u_0, \qquad \bar{v} = v - u_0 \tag{12.10}$$

where

$$\bar{u}, \bar{v} \in V_0 = \{v|v = \{v_i\}, v \in V, v_i = 0 \text{ on } \Gamma_1\}. \tag{12.11}$$

We denote by L the admissible vector space of the tractions S on Γ_3 i.e. L is the restriction of $[H^{-1/2}(\tilde{\Gamma})]^3$ to Γ_3, where $\tilde{\Gamma} = \Gamma_1 \cup \Gamma_2$. Due to the lack of convexity of j or j_N, j_T it is not possible to apply the Lagrangian approach of the previous Chapters. Here we rely mainly on Betti's theorem of elasticity and we obtain multivalued B.I.Es on the boundary part Γ_3.

Now we assume that $S \in L$ is given on Γ_3 and is equal to $\mu = \{\mu_i\}$. Then the solution of the arising classical problem satisfies the following problem: Find $\bar{u} = \bar{u}(\mu) \in V_0$ such that

$$a(\bar{u}, \bar{v}) + a(u_0, \bar{v}) - [\mu, \bar{v}]_{\Gamma_3} - (\bar{p}, \bar{v}) - [\bar{T}, \bar{v}]_{\Gamma_2} = 0 \quad \forall \bar{v} \in V_0. \tag{12.12}$$

Obviously (12.12) expresses the principle of virtual work for a structure resulting from the initial one by eliminating the superpotential constraints on Γ_3 and by applying the forces $\mu = \{\mu_i\}$ on Γ_3. Because of the linearity of (12.12) the solution $\bar{u}$ of it can be written as the sum $\bar{u}_{(1)} \in V_0$ and $\bar{u}_{(2)} \in V_0$ where $\bar{u}_{(1)}$ and $\bar{u}_{(2)}$ are solutions of the two variational equalities

$$a(\bar{u}_{(1)}, \bar{v}) - l(\bar{v}) + a(u_0, \bar{v}) = 0, \quad \forall \bar{v} \in V_0 \tag{12.13}$$

and

$$a(\bar{u}_{(2)}, \bar{v}) - [\mu, \bar{v}]_{\Gamma_3} = 0 \quad \forall \bar{v} \in V_0 \tag{12.14}$$

respectively. Here $\bar{u}_{(1)}$ and $\bar{u}_{(2)}$ are equilibrium configurations of two bilateral structures resulting from the initial one by ignoring the superpotential boundary conditions on Γ_3, and assuming that on certain parts of the boundary the load is zero; thus in the case of (12.13) the structure is loaded by the forces $\bar{p}$ in Ω and $\bar{T}$ on Γ_2, whereas on Γ_3 the loading is zero. Moreover the structure is subjected to an initial displacement field u_0 and is fixed on Γ_1. In the case of (12.14) the structure is loaded by a force $\mu = \{\mu_i\}$ on Γ_3 only and is fixed along Γ_1; the loading in Ω and on Γ_2 is zero. The solutions $\bar{u}_{(1)}$ and $\bar{u}_{(2)}$ are uniquely determined, as it is well known from the classical (bilateral) elasticity theory. For the bilateral structures the solutions $\bar{u}_{(1)}$ and $\bar{u}_{(2)}$ can be written in terms of Green's operator G, which is the same for both structures due to the same type of boundary conditions holding in each structure. Accordingly, we can write that

$$\bar{u}_{(1)} = G(\bar{l}), \quad \bar{u}_{(2)} = G(\mu), \quad u = \bar{u}_{(1)} + \bar{u}_{(2)}, \quad l = \{\bar{p}, \bar{T}, u_0\}. \tag{12.15}$$

We have to determine the unknown force distribution $\mu = \{\mu_i\} \in L$ on Γ_3. With respect to the linear elasticity problem described by (12.14) we apply Betti's theorem:

Assume that $\lambda = \{\lambda_i\} \in L$ on Γ_3 is a force distribution corresponding to a displacement field $\bar{v}_{(2)} \in V_0$ if $\bar{p} = 0$, $\bar{T} = 0$ on Γ_2 and $u_0 = 0$. Then we have that

$$[\lambda, \bar{u}_{(2)}]_{\Gamma_3} = [\mu, \bar{v}_{(2)}]_{\Gamma_3}. \tag{12.16}$$

But if $\bar{u}_{(1)}$ is the displacement field of (12.13), then we may easily verify that

$$[\lambda, \bar{u}_{(1)}]_{\Gamma_3} = [\mu, \bar{u}_{(1)}]_{\Gamma_3} = 0. \tag{12.17}$$

Indeed from (12.14) we obtain that

$$a(\bar{u}_{(2)}, \bar{u}_{(1)}) = [\mu, \bar{u}_{(1)}]_{\Gamma_3} \tag{12.18}$$

$$a(\bar{v}_{(2)}, \bar{u}_{(1)}) = [\lambda, \bar{u}_{(1)}]_{\Gamma_3}. \tag{12.19}$$

But $a(\bar{u}_{(2)}, \bar{u}_{(1)}) = a(\bar{v}_{(2)}, \bar{u}_{(1)}) = 0$ from the principle of virtual work since $\bar{u}_1$ results for Ω fixed at Γ_1 and subjected to $\bar{p}, \bar{T}, u_0$ and to zero forces on Γ_3, whereas $\bar{u}_{(2)}$ (resp. $\bar{v}_{(2)}$) are displacement fields for the same structure fixed at Γ_1 and subjected only to forces μ (resp. λ) on Γ_3 and having $\bar{p} = 0$, $\bar{T} = 0$ and $u_0 = 0$.
From (12.16) and (12.17) we obtain that

$$[\lambda, \bar{u}]_{\Gamma_3} = [\mu, \bar{v}]_{\Gamma_3}, \tag{12.20}$$

where $\bar{u} = \bar{u}_{(1)} + \bar{u}_{(2)}$ and $\bar{v} = \bar{u}_{(1)} + \bar{v}_{(2)}$. Every $\bar{v} \in V_0$ can be put in this form (consider 12.13 and (12.14) with μ_i replaced by λ_i) and therefore (12.20) holds for every $\bar{v} \in V_0$. Obviously we may write that

$$\bar{v}_{(2)} = G(\lambda). \tag{12.21}$$

Now (12.20) implies with (12.21), (12.15) and (12.17) that

$$\begin{aligned}
[\lambda, \bar{u}]_{\Gamma_3} &= [\mu, \bar{v}]_{\Gamma_3} = [\mu, \bar{u}_{(1)}]_{\Gamma_3} + [\mu, \bar{v}_{(2)}]_{\Gamma_3} \\
&= [\lambda, \bar{u}_{(1)}]_{\Gamma_3} + [\mu, \bar{v}_{(2)}]_{\Gamma_3} = \left[\lambda, [G(\bar{l})]\right]_{\Gamma_3} + \left[\mu, [G(\lambda)]\right]_{\Gamma_3}.
\end{aligned} \tag{12.22}$$

Now we introduce the bilinear symmetric (by Betti's theorem) form

$$\beta(\lambda, \mu) = \big[\mu, [G(\lambda)]\big]_{\Gamma_3}, \tag{12.23}$$

and the linear form

$$\bar{\gamma}(\lambda) = -\big[\lambda, [G(\bar{l})]\big]_{\Gamma_3}. \tag{12.24}$$

Assuming now that the tractions μ on Γ_3 are related to the displacement field u through the relation (12.3) we may write (cf. the definition of the generalized gradient in Ch.1) that

$$\tilde{j}^0(-\mu, -\lambda^\star) \geq -\lambda_i^\star(\bar{u}_i + u_{0i}) \quad \forall \lambda^\star \in L, \tag{12.25}$$

where $\tilde{j}^0(\cdot, \cdot)$ denotes the directional differential of Clarke. From (12.22) and (12.25) we obtain for $\lambda^\star = \lambda$ that

$$\int_{\Gamma_3} \tilde{j}^0(-\mu, -\lambda)d\Gamma \geq [-\lambda, (\bar{u} + u_0)]_{\Gamma_3} = \bar{\gamma}(\lambda) - \beta(\lambda, \mu) - [\lambda, u_0]_{\Gamma_3} = \gamma(\lambda) - \beta(\lambda, \mu),$$

$$\tag{12.26}$$

where

$$\gamma(\lambda) = \bar{\gamma}(\lambda) - [\lambda, u_0]_{\Gamma_3}. \tag{12.27}$$

The relation (12.26) holds for all $\lambda \in L$ and thus we are led to the following hemivariational inequality: Find $\mu \in L$ such as to satisfy

$$\beta(\mu, \lambda) - \gamma(\lambda) + \int_{\Gamma_3} \tilde{j}^0(-\mu, -\lambda)d\Gamma \geq 0 \quad \forall \lambda \in L. \tag{12.28}$$

Let us consider now the "substationarity" problem [Rock79]

$$\mu \in L, \quad 0 \in \bar{\partial}\Pi(\mu), \quad \Pi(\mu) = \frac{1}{2}\beta(\mu, \mu) - \gamma(\mu) + \int_{\Gamma_3} \tilde{j}(-\mu)d\Gamma. \tag{12.29}$$

Then every solution of (12.29) satisfies (12.28) but not conversely. Obviously (12.29) is equivalent to the multivalued boundary integral equation

$$\gamma - \frac{1}{2}\mathrm{grad}\beta(\mu, \mu) \in \bar{\partial}\left(\int_{\Gamma_3} \tilde{j}(-\mu)d\Gamma\right) \tag{12.30}$$

which is explicity written as

$$\frac{\partial}{\partial\mu}\left\{\big[-\mu, [G(\bar{l})] + u_0\big]_{\Gamma_3} - \frac{1}{2}\big[\mu, [G(\mu)]\big]_{\Gamma_3}\right\} \in \bar{\partial}\left(\int_{\Gamma_3} \tilde{j}(-\mu)d\Gamma\right). \tag{12.31}$$

We recall here that every local minimum and every saddle point of the energy Π is a substationarity problem. Also a local maximum, say $\mu_0 \in L$, is a substationarity point if Π is Lipschitzian around μ_0.

12.2 A Multivalued Boundary Integral Formulation with Respect to the Displacements on Γ_3.

In this Section we assume that on Γ_3 the nonmonotone possibly multivalued boundary conditions are expressed in the form (12.3a). Note that in the case of monotonicity studied in Ch. 10 we do not need to distinguish between (12.3) and (12.3a) since then j is convex and $\tilde{j}$ is the conjugate functional of j. But if convexity does not exist no appropriate definition of the "conjugacy operation" is possible which would enable us to invert (12.3) in order to get (12.3a).

Let us assume first that the displacements u on Γ_3 are given. Then we denote by Σ the set of all symmetric stress-tensors and let

$$\Sigma_1 = \{\tau | \tau = \{\tau_{ij}\}, \tau_{ij} = \tau_{ji} \in L^2(\Omega), \tau_{ij,j} + \bar{p}_i = 0 \text{ a.e. in } \Omega, T_i = \bar{T}_i \text{ a.e. on } \Gamma_2\}$$
$$(12.32)$$

be the statically admissible set. In (12.32) $\{T_i\}$ denotes the traction on Γ corresponding to the stress field $\{\tau_{ij}\}$. Let also $c = \{c_{ijhk}\}$ be the inverse tensor to $C = \{C_{ijhk}\}$, i.e.

$$\varepsilon_{ij} = c_{ijhk}\sigma_{hk} \qquad (12.33)$$

and let

$$A(\sigma, \tau) = (c\sigma, \tau) = \int_\Omega c_{ijhk}\sigma_{ij}\tau_{hk}d\Omega. \qquad (12.34)$$

For given displacements v on Γ_3 we can write the "principle" of complementary virtual work[8] for the structure in the form: find $\sigma = \sigma(v) \in \Sigma_1$ such that

$$A(\sigma, \tau) = [\bar{U}, T]_{\Gamma_1} + [v, T]_{\Gamma_3} \quad \forall \tau \in \Sigma_1. \qquad (12.35)$$

Let us now introduce a strain-field $\sigma_0 \in \Sigma_1$, i.e. a stress field satisfying the equations of equilibrium and the static boundary conditions on Γ_2 and let us introduce the new variables

$$\bar{\sigma} = \sigma - \sigma_0 \text{ and } \bar{\tau} = \tau - \tau_0 \qquad (12.36)$$

where $\bar{\sigma}, \bar{\tau} \in \Sigma_0$ and

$$\Sigma_0 = \{\tau | \tau = \{\tau_{ij}\}, \tau_{ij} = \tau_{ji} \in L^2(\Omega), \tau_{ij,j} = 0 \text{ a.e. in } \Omega, T_i = 0 \text{ a.e. on } \Gamma_2\}. \quad (12.37)$$

Thus (12.35) becomes: find $\bar{\sigma} = \bar{\sigma}(v) \in \Sigma_0$ such as to satisfy

$$A(\bar{\sigma}, \bar{\tau}) = [\bar{U}, \bar{T}]_{\Gamma_1} + [v, \bar{T}]_{\Gamma_3} + A(\sigma_0, \bar{\tau}) = 0, \quad \forall \bar{\tau} \in \Sigma_0. \qquad (12.38)$$

Note also that by the Green-Gauss theorem (10.74) holds and with a similar reasoning as there (10.75) or (10.76). Here we choose a σ_0 satisfying (10.75) i.e.

$$A(\sigma_0, \bar{\tau}) = 0 \qquad \forall \bar{\tau} \in \Sigma_0. \qquad (12.39)$$

in order to simplify all the arising expressions.

---

[8] (12.12) and (12.35) result by applying the Green-Gauss theorem.

The stress $\bar{\sigma}$ in (12.38) can be written as the sum $\bar{\sigma}_{(1)} + \bar{\sigma}_{(2)}$ where $\bar{\sigma}_{(1)}$ and $\bar{\sigma}_{(2)}$ are solutions of the variational equalities

$$A(\bar{\sigma}_{(1)}, \bar{\tau}) - [v, \bar{T}]_{\Gamma_3} = 0 \quad \forall \bar{\tau} \in \Sigma_0, \tag{12.40}$$

$$A(\bar{\sigma}_{(2)}, \bar{\tau}) - [\bar{U}, \bar{T}]_{\Gamma_1} = 0 \quad \forall \bar{\tau} \in \Sigma_0, \tag{12.41}$$

respectively. Both (12.40) and (12.41) respectively express the "principle" of complementary virtual work for bilateral structures resulting from the initial one in the following way: for (12.40) (resp. (12.41)) we consider the structure Ω under the action of "given" displacements v (resp. zero) on Γ_3, zero forces in Ω and Γ_2 and zero (resp. $\bar{U}$) displacements on Γ_1. Since these structures are linear elastic, $\bar{\sigma}_{(1)}$ and $\bar{\sigma}_{(2)}$ are uniquely determined. Therefore (12.40) and (12.41) imply that

$$\bar{\sigma}_{(1)} = H(v), \quad \bar{\sigma}_{(2)} = H(\bar{U}), \quad \bar{\sigma} = \bar{\sigma}_{(1)} + \bar{\sigma}_{(2)}, \quad \bar{\sigma} \in \Sigma_0 \tag{12.42}$$

where H is the Green's stress-displacement operator for the two problems (12.40) and (12.41). Both fictive bilateral structures corresponding to (12.40) and (12.41) have the same H-operator because of the same type of boundary conditions. Moreover we denote by $\tilde{H}$, as in Sect. 10.3, the operator transforming the displacement at the boundary into the traction $S = \{S_i\}$. Thus we may write that

$$\bar{S}_{(1)} = \tilde{H}(v), \quad \bar{S}_{(2)} = \tilde{H}(\bar{U}). \tag{12.43}$$

We have to determine the unknown displacement distribution $v = \{v_i\} \in N$ on Γ_3. Note that in the functional framework introduced previously $N = $ restriction of $[H^{1/2}(\tilde{\Gamma})]^3$ to Γ_3. Let $w = \{w_i\} \in N$ be another displacement distribution on Γ_3 corresponding to the stress field $\bar{\tau}_{(1)} \in \Sigma_0$ through (12.40). Moreover $v = \{v_i\} \in N$ corresponds to $\bar{\sigma}_1 \in \Sigma_0$. Applying Betti's theorem we can write that

$$[\bar{T}_{(1)}, v]_{\Gamma_3} = [\bar{S}_{(1)}, w]_{\Gamma_3}. \tag{12.44}$$

Recall that $\bar{S}_{(2)}$ results from (12.41). Then we have that

$$[\bar{S}_{(2)}, w]_{\Gamma_3} = [\bar{S}_{(2)}, v]_{\Gamma_3} = 0. \tag{12.45}$$

Indeed from (12.40) we have that

$$A(\bar{\sigma}_{(1)}, \bar{\sigma}_{(2)}) = [v, \bar{S}_{(2)}]_{\Gamma_3}, \tag{12.46}$$

and

$$A(\bar{\tau}_{(1)}, \bar{\sigma}_{(2)}) = [w, \bar{S}_{(2)}]_{\Gamma_3}. \tag{12.47}$$

But

$$A(\bar{\sigma}_{(1)}, \bar{\sigma}_{(2)}) = A(\bar{\tau}_{(1)}, \bar{\sigma}_{(2)}) = 0, \tag{12.48}$$

because of the principle of complementary virtual work; indeed $\bar{\sigma}_{(2)}$ is the stress field for structure Ω having zero forces in Ω and on Γ_2, zero displacements on Γ_3 and $\bar{U}$; displacements on Γ_1 and $\bar{\sigma}_{(1)}$ (resp. $\bar{\tau}_1$) is a stress field for the same structure with zero displacements on Γ_1 and v_i displacements on Γ_3. From (12.44) and (12.46), (12.47) we may write that

$$[\bar{T}, v]_{\Gamma_3} = [\bar{S}, w]_{\Gamma_3} \tag{12.49}$$

where $\bar{S} = \bar{S}_{(1)} + \bar{S}_{(2)}$ and $\bar{T} = \bar{T}_{(1)} + \bar{S}_{(2)}$. Note that (12.49) holds for every $\bar{\tau} \in \Sigma_0$ ($\bar{T}$ corresponds to $\bar{\tau}$). Moreover we can write analogously to (12.43) that

$$\bar{T}_{(1)} = \tilde{H}(w). \tag{12.50}$$

Relation (12.49) implies with (12.50), (12.45) and (12.43) that

$$\begin{aligned}[w, \bar{S}]_{\Gamma_3} &= [v, \bar{T}_{(1)}]_{\Gamma_3} + [v, \bar{S}_{(2)}]_{\Gamma_3} = \left[v, [\tilde{H}(w)]\right]_{\Gamma_3} \\ &\quad + [w, \bar{S}_{(2)}]_{\Gamma_3} = \left[v, [\tilde{H}(w)]\right]_{\Gamma_3} + \left[w, [\tilde{H}(\bar{U})]\right]_{\Gamma_3}.\end{aligned} \tag{12.51}$$

Now the bilinear symmetric (due to Betti's theorem) form

$$\delta(v, w) = \left[[\tilde{H}(v)], w\right]_{\Gamma_3}. \tag{12.52}$$

is introduced and the linear form

$$\bar{\zeta}(w) = \left[[\tilde{H}(\bar{U})], w\right]_{\Gamma_3} \tag{12.53}$$

and thus (12.51) implies that

$$[w, \bar{S}]_{\Gamma_3} = \delta(v, w) - \bar{\zeta}(w). \tag{12.54}$$

But (12.3a) implies by definition that

$$j^0(v, w^\star) \geq -S_i w_i^\star = -(\bar{S}_i + S_{0i})w_i^\star \quad \forall w_i^\star \in N. \tag{12.55}$$

From (12.54) and (12.55) we obtain for $w^\star = w$ that

$$\int_{\Gamma_3} j^0(v, w)d\Gamma \geq [\bar{S}, w]_{\Gamma_3} - [S_0, w]_{\Gamma_3} = -\delta(v, w) + \zeta(w), \tag{12.56}$$

where

$$\zeta(w) = \bar{\zeta}(w) - [S_0, w]_{\Gamma_3}. \tag{12.57}$$

Relation (12.56) holds for all $w \in N$ and thus we are led to the following hemivariational inequality: Find $v \in N$ such as to satisfy

$$\delta(v, w) - \zeta(w) + \int_{\Gamma_3} j^0(v, w)d\Gamma \geq 0 \quad \forall w \in N. \tag{12.58}$$

We can prove (cf. Prop. 1.15) the following proposition:

Proposition 12.1 Suppose that the superpotential $\xi \to j(\xi)$ locally satisfies a Lipschitz condition, is $\bar{\partial}$-regular and fullfils a growth assumption of the type: "for every

$f \in \bar{\partial} j(u)$ the estimate $|f(x)| \leq c(1 + |u(x)|^q)$ holds for $q \geq 1$". Then every solution of the hemivariational inequality (12.58) is a solution of the "substationarity" problem

$$v \in N, \quad 0 \in \bar{\partial}\tilde{\Pi}(v), \quad \tilde{\Pi}(v) = \frac{1}{2}\delta(v,v) - \zeta(v) + \int_{\Gamma_3} j(v)d\Gamma \qquad (12.59)$$

and conversely.

Proof: The proof is the same as the proof of prop. 4.1 in [Pan91] and therefore it is omitted here.

By the definition of the generalized gradient (12.59) is equivalent to the multivalued boundary integral equation

$$\zeta - \frac{1}{2}\mathrm{grad}\delta(v,v) \in \bar{\partial}\left(\int_{\Gamma_3} j(v)d\Gamma\right) \qquad (12.60)$$

which may be put in the form

$$\frac{\partial}{\partial v}\left\{ [[\tilde{H}(\bar{U})] + S_0, v]_{\Gamma_3} - \frac{1}{2}[[\tilde{H}(v)], v]_{\Gamma_3} \right\} \in \bar{\partial}\left(\int_{\Gamma_3} j(v)d\Gamma\right). \qquad (12.61)$$

12.3 On the Numerical Treatment of Nonmonotone (Zigzag) Multivalued Contact Laws. A New Efficient Algorithm.

Zigzag stress-strain or reaction-displacement laws appear in several engineering problems and they constitute the most general case of the nonmonotone multivalued laws. We can mention here the reaction-displacement (or relative displacement) diagram of fig. 12.1a which results in the tangential sense to the interface, if two bodies are glued by an adhesive material. This material has either brittle fracture (dotted line) or semibrittle fracture. In the case of brittle fracture the law has complete vertical branches, i.e. it is a multivalued law. The same effect may appear at the interface of sandwich beams and plates, as well as in composite materials between fibers and matrix and is called delamination effect. Also the nonmonotone variants of the well-known friction law of Coulomb (fig. 12.1b) or the adhesionless friction law of fig. 12.1c can also be mentioned. Similar is the situation with sawtooth stress-strain laws in contact problems of an elastic body with a support of reinforced concrete in tension (Scanlon's diagram) and in composite material structures (fig. 12.1d). For this type of laws and for their physical meaning we refer to [Pan84;88, Gr, Gil, Scw, Wil, Shi, Bani85;86;87a,b] among others. As already mentioned in Ch.1 nonmonotone possibly multivalued laws in structures and solids give rise to a new type of variational expressions in inequality form which the second author has called hemivariational inequalities (see e.g. [Pan85]). In the previous two Sections multivalued B.I.Es have been derived for the same problems. Their numerical treatment will be achieved through the corresponding boundary hemivariational inequalities. The theory of hemivariational inequalities leads to the result that local minima of the potential or the complementary energy of the structure represent equilibrium positions of the problem. It is possible however, that certain solutions of the problem may not be local minima but another more general type of points which make the potential energy "substationary" i.e. they satisfy the differential inclusion $0 \in \bar{\partial}\Pi(u)$, where $\Pi(\cdot)$ is the potential energy and u is the displacement vector. Analogous results hold for the complementary energy when it is expressed as a function of the stresses.

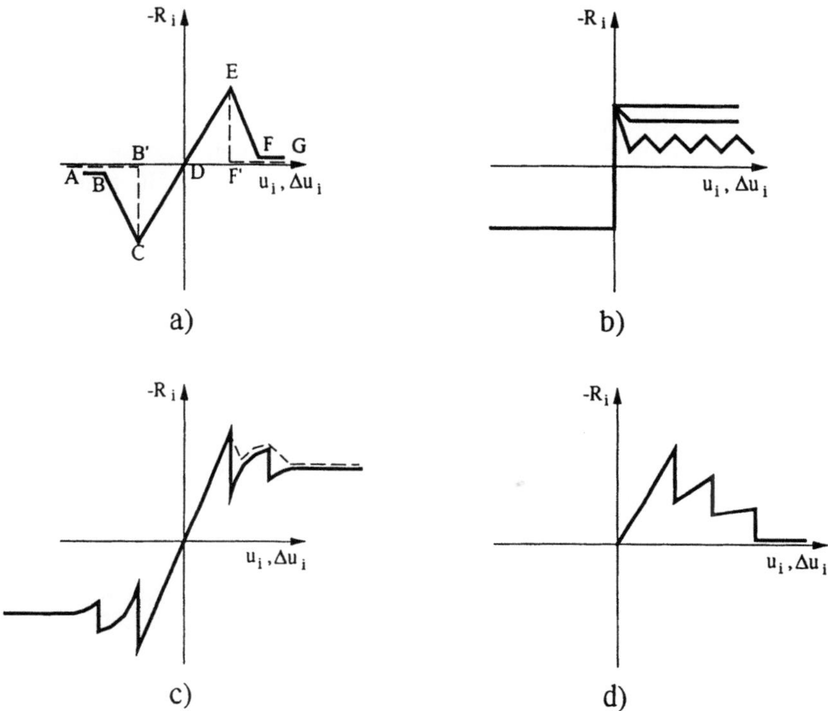

Fig. 12.1: Zigzag reaction (R_i)-displacement (u) or relative displacement (Δu_i) laws

The formulation of a nonmonotone problem as a hemivariational inequality has several advantages concerning the mathematical study of this problem [Pan91, Nan88; 89a,b], but this formulation does have analogous advantages concerning the numerical treatment, as until now the several attempts by the second author of the present book and his coworkers have shown. And we mean here the effective numerical treatment of realistic large scale engineering problems. Note also in this context that the numerical determination of all local minima of a nonconvex function is a still open problem in numerical optimization [Fle90]. Moreover, only a few results are known estimating the efficiency of the available nonconvex optimization algorithms for the determination of a local minimum, (cf. [Tza]). This direct approach may find some more applications in the case of multivalued B.I.Es or boundary hemivariational inequalities where we have an elimination of the internal DOFs. Several other approaches have been developed based on the variational formulation, as e.g. the quasidifferentiability approach [Stav91;92, Pan92] and the progressive damage model [Dug76;80, Pan87, Kol91, Fra]. At this point we should mention that numerically efficient equilibrium path tracing methods have been developed in [Cri82;86;88;91]. This last method is not based on a variational approach but on a direct enforcement of the nonmonotone stress-strain diagrams. We should mention here also the dissipation method (of e.g. [Fre88]), which has advantages from the standpoint of the physical interpretation of the nonmonotonicity.

However all the aforementioned methods do not treat efficiently the combinatorial character of the problem which results from the generalized gradient in the B.I.Es (12.30) and (12.60). Let us explain this fact here: Suppose that we have a structure containing certain elements obeying, the law of fig. 1a. For a given loading we want to determine a corresponding equilibrium state. In an equilibrium state some elements may have a stress and strain state on the branch AB, others on BC others on CDE. etc., i.e. it is not known a priori which part of the stress-strain law will be realized. For a very small number of structural elements obeying this law one could determine the solution by considering all possible branch combinations of the stress strain law. For a larger number of elements we could apply an incremental back-and forward trial and error procedure for the loading in order to arrive at a solution of the problem (cf. e.g. the methods of [Cri91]). Note that due to the lack of monotonicity no uniqueness is generally guaranteed.

A difficult problem is the determination of the elements having stress and strains on AB or on BC e.t.c. and this problem becomes more difficult if the law has complete vertical branches i.e. CB' or EF'; this is a problem of combinatorial nature. We recall that in the case of monotone laws (fig. 12.2), e.g. in the case of a structure having only cables (fig. 12.2a), which may become slack, the solution of the inequality constrained problem of minimum potential or complementary energy gives those cables which are slack (on AB) or under tension (on BC) [Pan76]. This solution is obtained through a convex programming (C.P.) algorithm. The minimum is uniquely determined, due to the monotonicity of the stress-strain law. This approach works well for the more complicated law of fig. 12.2b, as well as for its three-dimensional version, and determines simply by the use of a C.P. algorithm the solution of the problem. Note that for a continuous structure, i.e. if, for instance, a plate is glued to a support by means of a glue having the law of fig. 12.2b, we have to determine six regions on the plate according to the diagram of fig. 12.2b; thus a C.P. algorithm determines all the free boundaries between the six a priori unknown regions. In the case of the monotone laws of fig. 12.2a,c the convex minimization problem reduces to a quadratic programming (Q.P.) problem. Both C.P. and Q.P. problems may be treated by efficient algorithms yielding quickly a reliable solution of the problem [Fle].

In the present Section we present an algorithm which "replaces" a nonmonotone law problem by a sequence of problems involving monotone laws, and more specifically laws of the form of fig. 12.2c which lead to Q.P. problems. Thus the efficient treatment of the combinatorial character of the problem is successfully extended to nonmonotone laws and for large number of unknowns. This combinatorial task could be possibly fulfilled by the direct use of a nonconvex programming (N.C.P.) algorithm; but then no large scale problem could be treated . Moreover the N.C.P. algorithms do not have the convergence rate and the robustness of the C.P. and the Q.P. algorithms. Comparisons of the proposed algorithm with the path-following numerical methods of [Cri91] have shown that the present method has considerable advantages concerning first the treatment of the complete vertical branches and second the combinatorial character of the problem. However, the method of [Cri91] can be more easily embodied into general purpose F.E. computer programs and traces more accurately the equilibrium path. The case of unloading paths, reduces to an incremental use of the present algorithm, as well as, the extension of the present method (cf. e.g. [Pan88] p. 98) to 3D-nonmonotone laws.

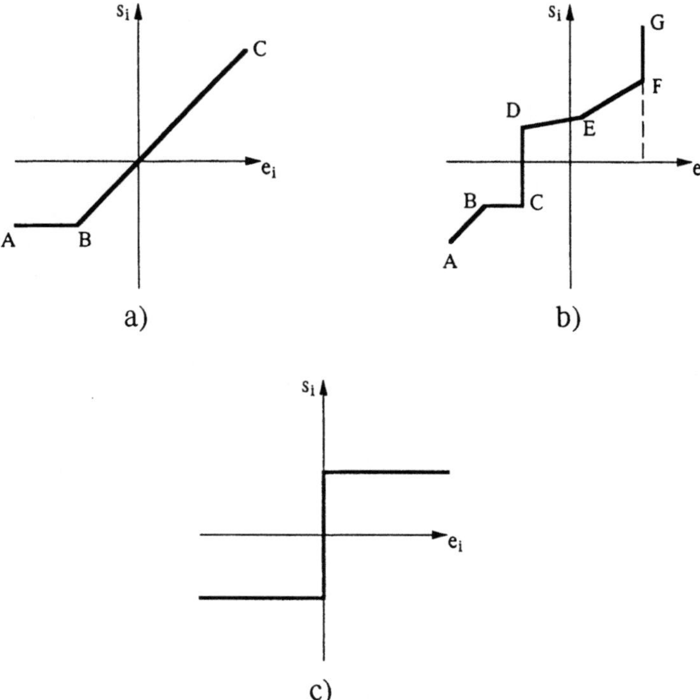

Fig. 12.2: Monotone multivalued stress-strain laws leading to global minima

Note that this algorithm, which is due to Mistakidis [Mis92a,b], is completely new in
the numerical analysis and has been applied until now to the numerical treatment of
hemivariational inequalities arising in realistic large scale engineering problems. Here
we have adapted this algorithm to boundary hemivariational inequalities or to nonmonotone multivalued B.I.Es.

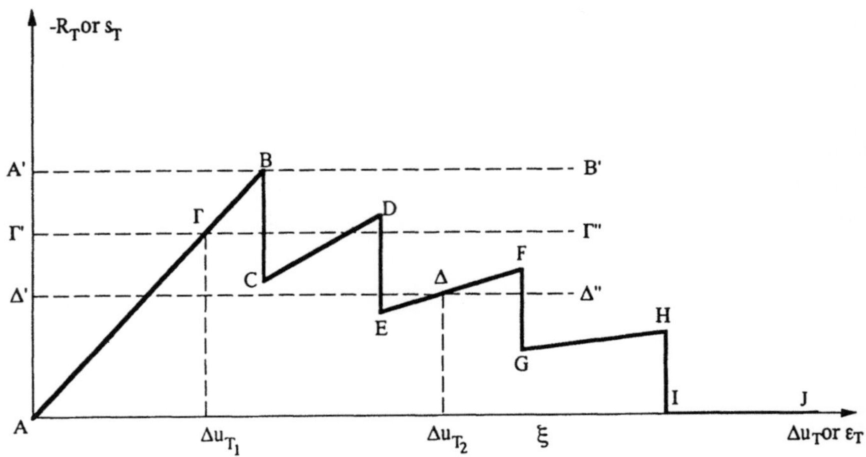

Fig. 12.3: Presentation of the algorithm

We shall show first the main idea of the present algorithm by means of fig. 12.3.
Let us assume that the diagram ABCD...J of fig. 12.3 represents the possible reaction-
displacement diagram of a nonmomotone friction law on the boundary of an elastic
body. First we assume that all the boundary points obey the fictitious friction law
AA'BB' . Then the problem has a unique solution, as it is known from the solution of
the variational inequality of the friction problem (cf. e.g. [Pan85]). Note that the arising
problem is a friction problem with a given normal force $S_N = C_N$, i.e. A' is prescribed
($AA' = q|C_N|$ where q is the friction coefficient). Suppose that the friction problem gives
as a result certain $\Delta u_T = \Delta u_{T_1}$ (tangential displacement or more generally a relative
tangential displacement). Now we solve a new friction problem, with given normal
force, obeying the law AΓΓ". Suppose that the solution of this problem offers as a
result a new Δu_{T_2} which gives rise to the new friction diagram AΔ'ΔΔ". This procedure
is continued until in all elements the differences $|R_{T,i-1} - R_{T,i}|$ and $|\Delta u_{T,i+1} - \Delta u_{T,i}|$
become small enough. This covers the case of multivalued laws, i.e. of diagrams with
complete vertical branches which are very common in contact problems with composite
material supports and in adhesive contact problems. Note also that within each step
the boundary points having the zigzag law may obey to different friction laws i.e. AA'
, AΓ' and AΔ' is different from point to point on the boundary. Concerning now the
solution of the friction problem we recall (see Ch.10) that we have either to solve the
multivalued B.I.E. (10.48) with respect to the boundary tractions $\{S_N, S_T\}$, or the
multivalued B.I.E. (10.88) with respect to the boundary displacements $\{u_N, u_T\}$. The

first is equivalent to a Q.P.P. with the subsidiary conditions $|S_T| \leq q|C_N|$, $S_N = C_N$, where q is the coefficient of friction, and the second to a convex programming problem of the type (10.84) where now

$$\Pi_2(v) = \frac{1}{2}\delta(v,v) - \zeta(v) + \int_{\Gamma_3} q|C_N|\,|v_T|d\Gamma - [C_N, v_N]_{\Gamma_3} \quad v = \{v_N, v_T\}. \qquad (12.62)$$

In the numerical applications we prefer to solve the B.I.E. (10.48) or equivalently the minimum problem with respect to the boundary tractions which is a Q.P.P. with only inequality subsidiary conditions; then the nondifferentiable term $\int_{\Gamma_3} q|C_N|\,|v_T|d\Gamma$ in (12.62) is avoided. We recall also that the minimum problem of Π_2 is equivalent to the following variational inequality: Find $\{u_N, u_T\} \in \tilde{\Lambda} = \{H^{1/2}(\tilde{\Gamma}), H_T(\tilde{\Gamma})\}\big|_{\Gamma_3}$ (cf. eq. 1.116)

$$\delta(u, v-u) + \int_{\Gamma_3} \mu|(C_N)|(|v_T| - |u_T|)d\Gamma \geq \zeta(v-u) + [C_N, v_N - u_N]_{\Gamma_3} \quad \forall v = \{v_N, v_T\} \in \tilde{\Lambda}$$

$$(12.63)$$

Let us go now back to the nonmonotone multivalued zigzag law of fig. 12.3. We consider the same linear elastic body Ω as in the previous friction problem under the same external actions; we assume also that on Γ_3 a normal force distribution C_N is prescribed and that in the tangential direction with respect to the boundary part Γ the reaction-displacement law of fig. 12.3 holds in the two directions of an intrinsic coordinate system on Γ (in the two dimensional case, $\Omega \subset \mathbf{R}^2$, this situation is simplified). Analogously to (12.58) an equilibrium position is characterized by the multivalued B.I.E. (12.60) or by the following hemivariational inequality problem: Find $u = \{u_N, u_T\} \in \tilde{\Lambda}$ such as to satisfy the inequality

$$\delta(u, v-u) + \int_{\Gamma_3} j^0(u_T, v_T - u_T)d\Gamma \geq \zeta(v-u) + [C_N, v_N - u_N]_{\Gamma_3} \quad \forall v = \{u_N, u_T\} \in \tilde{\Lambda}$$

$$(12.64)$$

Here $j(\cdot)$ is the nonconvex superpotential of the law of fig. 12.3 and $j^0(\cdot, \cdot)$ is the directional differential in the sense of Clarke which here is given by the formula

$$j^0(u_T, v_T) = \limsup_{\substack{h \to 0 \\ \lambda \to 0_+}} \frac{j(u_T + \lambda v_T + h) - j(u_T + h)}{\lambda}. \qquad (12.65)$$

In the case of the graph of the one-dimensional law of fig. 12.3, $\xi \to j(\xi)$ is the area between the horizontal axis and the graph until the point ξ of the horizontal axis. Note also that if Γ contains one interface or interfaces then in the term $\int_{\Gamma_3} j^0(u_T, v_T - u_T)d\Gamma$ in (12.64) u_T and v_T denote a relative tangential displacement at the interface.

The algorithm presented before can now be justified by comparing the variational inequality (12.63) with the hemivariational inequality (12.64). Let us write (12.64) in the form: Find $u \in \tilde{\Lambda}$ such that

$$\delta(u, v-u) \quad + \quad \int_{\Gamma_3} q|C_N|(|v_T| - |u_T|)d\Gamma \geq \zeta(v-u) + [C_N, v_N - u_N]_{\Gamma_3} \qquad (12.66)$$

$$+ \quad A(v_T, u_T, |v_T|, |u_T|) \quad \forall v \in \tilde{\Lambda}$$

where

$$A(v_T, u_T, |v_T|, |u_T|) = \int_{\Gamma_3} q|C_N|(|v_T| - |u_T|)d\Gamma - \int_{\Gamma_3} j^0(u_T, v_T - u_T)d\Gamma. \qquad (12.67)$$

Then the following iterative scheme is proposed: Find $u^{(\rho)} \in \tilde{\Lambda}$ such that

$$\delta(u^{(\rho)}, v - u^{(\rho)}) \; + \; \int_{\Gamma_3} q|C_N|(|v_T| - |u_T^{(\rho)}|)d\Gamma \geq \zeta(v - u^{(\rho)}) \qquad (12.68)$$

$$+ \; [C_N, v_N - u_N^{(\rho)}]_{\Gamma_3} + A(v_T, |v_T|, u_T^{(\rho-1)}, |u_T^{(\rho-1)}|) \quad \forall v \in \tilde{\Lambda}.$$

This iterative scheme is compatible with the fig. 12.3 as it is obvious from the geometrical meaning of the quantity A. Indeed A expresses the difference of the variation of the area variation under the two graphs, i.e. it is a measure of the difference of the reaction forces corresponding to the two laws the nonmonotone one and the frictional one.

The foregoing approximation scheme can be generalized by considering any type of monotone possibly multivalued law and not only the frictional law. Indeed let us assume that we have the hemivariational inequality (12.64). In order to approximate its solution by the solution of the variational inequality

$$u = \{u_N, u_T\} \in \tilde{\Lambda}, \quad \delta(u, v - u) + \Phi(v_T) - \Phi(u_T) \qquad (12.69)$$

$$\geq \zeta(v - u) + [C_N, v_N - u_N]_{\Gamma_3} \quad \forall v = \{v_N, v_T\} \in \tilde{\Lambda}$$

we consider the following iterative scheme: Find $u^{(\rho)} \in \tilde{\Lambda}$ such that

$$\delta(u^{(\rho)}, v - u^{(\rho)}) + \Phi(v_T) - \Phi(u_T^{(\rho)}) \geq \zeta(v - u^{(\rho)}) + [C_N, v_N - u_N^{(\rho)}]_{\Gamma_3} + B(v_T, u_T^{(\rho-1)}) \quad (12.70)$$

where

$$B(v_T, u_T^{(\rho-1)}) = \Phi(v_T) - \Phi(u_T^{(\rho-1)}) - \int_{\Gamma_3} j^0(u_T^{(\rho-1)}, v_T - u_T^{(\rho-1)})d\Gamma. \qquad (12.71)$$

The above scheme approximates the nonmonotone multivalued law of fig. 12.3 by the monotone law (10.3) (see also (10.10)) on Γ_3.

The numerical experiments we have performed have always shown a very good convergence. This is also reasonable from the physical point of view. The convergence proof can be performed along the lines of the existence proof of the solution of variational hemivariational inequalities given in [Pan91]. There are however some obscure points in the mathematical proof and therefore we avoid to present this attempt for proving the convergence of the proposed algorithm here. Moreover we close this section by the remark that the proposed iterative scheme may be applied to the definition and treatment of the three-dimensional extensions of the nonmonotone law of fig. 12.3 by approximating them by a sequence of friction problems with given normal force C_N, which are well defined for three-dimensional bodies.

12.4 A Numerical Application: The Nonmonotone Friction and the Adhesive Contact Problem with Debonding. A Fixed Point Type Algorithm.

We consider here the following problem. Suppose that we have a linear elastic body, $\Omega \subset \mathbf{R}^2$ with the regular boundary Γ consisting of Γ_1, Γ_2 and Γ_3. On Γ_3 a nonmonotone friction law of the type of fig. 12.4b holds at the points where the body remains in contact with a rigid or a deformable support (fig. 12.4d). Thus we can write the following law

$$\text{i) if } u_N < 0 \text{ then } S_N = 0 \text{ and } S_T = 0 \tag{12.72}$$

$$\text{ii) if } u_N = 0 \text{ then } S_N < 0 \text{ and } \{S_T, u_T\} \text{ are according to fig. 12.4b} \tag{12.73}$$

In the case of a deformable support with Winkler constant k, the last relation must be replaced by the relation

$$\text{iii) if } u_N \geq 0 \text{ then } S_N + k u_N = 0 \text{ and } \{S_T, u_T\} \text{ are according to fig. 12.4b} \tag{12.74}$$

We assume further that Ω contains certain interfaces denoted by Γ_4 in which an adhesive material guarantees a shear behavior according to the law of fig. 12.4c. Now Δu_T denotes the relative tangential displacement of the two sides of the interface. In the normal direction the unilateral contact law of fig. 12.4d holds. Thus we may write relations analogous to (12.72) and (12.73) with the difference that u_N and u_T are replaced by the relative displacements Δu_N and Δu_T. We denote this condition as (12.72a) and (12.73a). Of course Γ_3 or Γ_4 may be empty but not both of them. The boundary and interface condition defined by (12.72), (12.73) and (12.72a), (12.73a) do not lead directly to a hemivariational inequality.

If instead of the nonmonotone friction law a classical Coulomb's friction law would be given (see the dotted line BAA′B′ in fig. 12.4b) then we would apply the algorithm for the unilateral contact problem with Coulomb friction proposed in [Pan75] (cf. also Sect. 1.3.1 and Sect. 9.3). First the pure (i.e without friction) unilateral contact problem (normal algorithm) is solved with $S_T = S_T^{(i)}$ given on Γ_3 and Γ_4 and let $S_N^{(i)}$ be the obtained normal reaction. Then the pure friction problem (tangential algorithm) is solved assuming that on Γ_3 and Γ_4 $S_N = S_N^{(i)}$; let $S_T^{(i+1)}$ be the resulting tangential force on Γ_3 and Γ_4. Again the first subproblem is solved with $S_T = S_T^{(i+1)}$ and so on until the differences $|S_T^{(i)} - S_T^{(i+1)}|$ and $|S_N^{(i)} - S_N^{(i+1)}|$ and the differences of the corresponding displacements and/or relative displacements become small enough.

This fixed point type algorithm used for the classical unilateral contact problem with Coulomb friction is extended here for the unilateral contact problem with nonmonotone (zigzag) possibly multivalued friction boundary condition and/or adhesive contact interface condition. We formulate the following algorithm:

1st step: Calculate the structure if on Γ_3 and Γ_4 $S_T = S_T^{(0)}$ is given and the unilateral contact condition

$$S_N \leq 0, \quad u_N \leq 0, \quad S_N u_N = 0 \tag{12.75}$$

holds. We use a Q.P. algorithm in the normal direction (normal algorithm). Usually we assume that $S_T^{(0)} = 0$. Let us denote by $S_N^{(1)}$ the resulting normal force and let $u_N^{(1)}$

and $u_T^{(1)}$ be the corresponding normal and tangential displacement on Γ_3 (resp. relative displacement on Γ_4.

2nd step: This step is a slight variation of the algorithm prescribed previously. The difference consists in the fact that after each such step of the tangential algorithm, we go back to the normal contact algorithm to obtain a better estimation of the debonding region (i.e. after the calculation of Δu_{T_1} in fig. 12.3 and not after the determination of the tangential subproblem equilibrium solution e.g. of Δu_{T_5}. Indeed, because the contact and debonding regions can vary considerably from step to step it is time consuming to perform all the steps of the tangential algorithm for a given normal force. Also the numerical experience shows a high sensitivity of the whole algorithm in the case of the unstable contact regions. So we perform only one substep of the algorithm until the contact region is well estimated and apply all the substeps of the tangential algorithm afterwards. Suppose that the normal algorithm yields $u_T^{(i)}$. From $u_T^{(i)}$ and the $\{u_T, S_T\}$ law the corresponding $S_T^{(i)}$ is obtained; set it as $S_T^{(i,1)}$ where the second upper index denotes the number of substeps performed in the tangential algorithm. Then we solve the corresponding classical (or Coulomb) friction problem for $|S_T| \leq |S_T^{(i,1)}|$ for the given $S_N^{(i)}$ taken from the normal algorithm. If the debonding area is not well established, or if S_N varies considerably from step to step then we go back to the normal algorithm and so on. Otherwise we continue the substeps of the tangential algorithm: from the solution of the friction Q.P.P. a new $u_T^{(i,2)}$ is obtained. Then we go back to the $\{S_T, u_T\}$ law and we obtain $S_T^{(i,2)}$. Again the corresponding classical (or Coulomb) friction problem is solved for $|S_T| \leq |S_T^{(i,2)}|$ and a new $u_T^{(i,3)}$ is obtained. This procedure is continued until the algorithm converges, say after k-substeps to $u_T^{(i,k)}$. In all these substeps S_N is always taken equal to $S_N^{(i)}$.

3rd step: The 1st step is repeated for $S_T = S_T^{(i,k)}$ obtained from the solution of the k-substep in the 2nd-step. The whole procedure is repeated until the differences between the S_N's, S_T's, u_N's, and u_T's of two consecutive cycles become appropriately small. In order to avoid a landing on a branch like the IJ of fig. 12.3 a small modification is proposed in the 2nd step of the first cycle: $u_T^{(1)}$ is ignored and we take $S_T^{(1,1)}$ equal to the upper and/or lower bound of the nonmonotone law. Thus in the case of fig. 12.3 $S_T^{(1,1)}$ is defined by the point B, i.e. it is taken as equal to AA′ . With this $S_T^{(1,1)}$ the classical friction problem is solved with $S_N = S_N^{(1)}$ and $|S_T| \leq |S_T^{(1,1)}|$ and a new $u_T^{(1,2)}$ is obtained. Then from the $\{S_T, u_T\}$ law the corresponding $S_T^{(1,2)}$ results and so on as it was described before. In fig. 12.5 we give a diagram describing this algorithm which is also a fixed point type algorithm.

The mathematical convergence proof of this last algorithm with the debonding is still an open problem. However the numerical experiments we have performed have shown very good convergence properties.

According to the described theory, an appropriate FORTRAN code was developed on a HP-750 workstation at the Institute of the second author in Greece. The first step of the algorithm, i.e. the solution of the pure contact problem, was achieved using the direct stiffness method and a fast active set Q.P. technique (subroutine VE09 of the HARWELL Subroutine Library). The second step of the algorithm, i.e. the treatment of the friction problem was achieved by solving the friction B.I.E. with respect to the boundary traction. We have formulated the corresponding boundary Q.P.P. and we have applied the Hildreth and d'Esopo Q.P. algorithm, which, although slower in the first step, yields some results (inverse matrices, decompositions e.t.c.) which reduce the total execution time. Since the whole algorithm is time consuming we have applied a lot of classical numerical techniques optimizing the performance of the used finite element codes which have been used for the determination of the discrete versions of $\beta(\cdot,\cdot)$ and $\gamma(\cdot)$ in (10.48). Thus a profile reduction algorithm is used to reduce the bandwidth of the matrices, among others. Note that this part could be performed by means of a classical B.E.M. code (cf. Sect. 7.4). The numerical scheme proposed, was applied to the analysis of the structure depicted in fig. 12.6 which represents a joint with nonmonotone friction law and debonding holding on the interfaces. The discretization was performed using constant stress triangular elements. The structure has two interfaces with 31 couples of nods. It is loaded with forces in the normal and the tangential to the interfaces directions and is well determined kinematically. The nonmonotone friction law depicted in fig. 12.7a is assumed to hold on the interfaces. The six load cases of fig. 12.7b were considered, for values of friction coefficient q equal to 0.1 and 0.2. In order to obtain the two nonmonotone friction laws considered we have to multiply at each point of the boundary or the interface the vertical coordinates of the diagram of fig. 12.7a by $q|S_N|$ where S_N is the normal force at the point considered. The distribution of the contact forces is almost identical for all the cases and the results with respect to the 31 distinct positions on the one of the interfaces, are shown in figs 12.8a,b. The influence of the nonmonotone diagram on the results of the friction forces given in figs 12.9a,b is very considerable. When the tangential forces F_2 are small the behavior of the joint is linear and all the results lie on branch AB of fig. 12.7a. As the tangential forces increase, the equilibrium points on the diagram, move to the branches at the right (BCDE...) and finally the joint fails (e.g. fig. 12.9b, load case 5, points 22-31). It is easy to interpret the points of abrupt changes of the curves in figs 12.9a,b as the points where the strength of the joint passes from the one branch of the zigzag diagram to the other, thus exhibiting the progressive failure as the tangential displacement increases. It is important to note here the great influence of the friction coefficient on the results: this is clearly seen in the results of fig. 12.10, where the relative tangential displacements at each point is depicted for the various load cases.

The convergence of the algorithm is very fast. All the previous problems have been solved in about 7-14 cycles of the normal-tangential algorithm using a second order norm of the normal and tangential forces as a stopping criterion: the algorithm terminates when $\frac{\|S^{(i)}-S^{(i-1)}\|}{\|S^{(i)}\|}$ becomes smaller than 10^{-4}. In fig. 12.11 we give a diagram representing the convergence rate of the algorithm. The differences of the relative tangential displacements between two consecutive steps with respect to the step are depicted and the convergence rate seems clearly to be quadratic. This numerical application has been solved by Mistakidis [Mis92a,b].

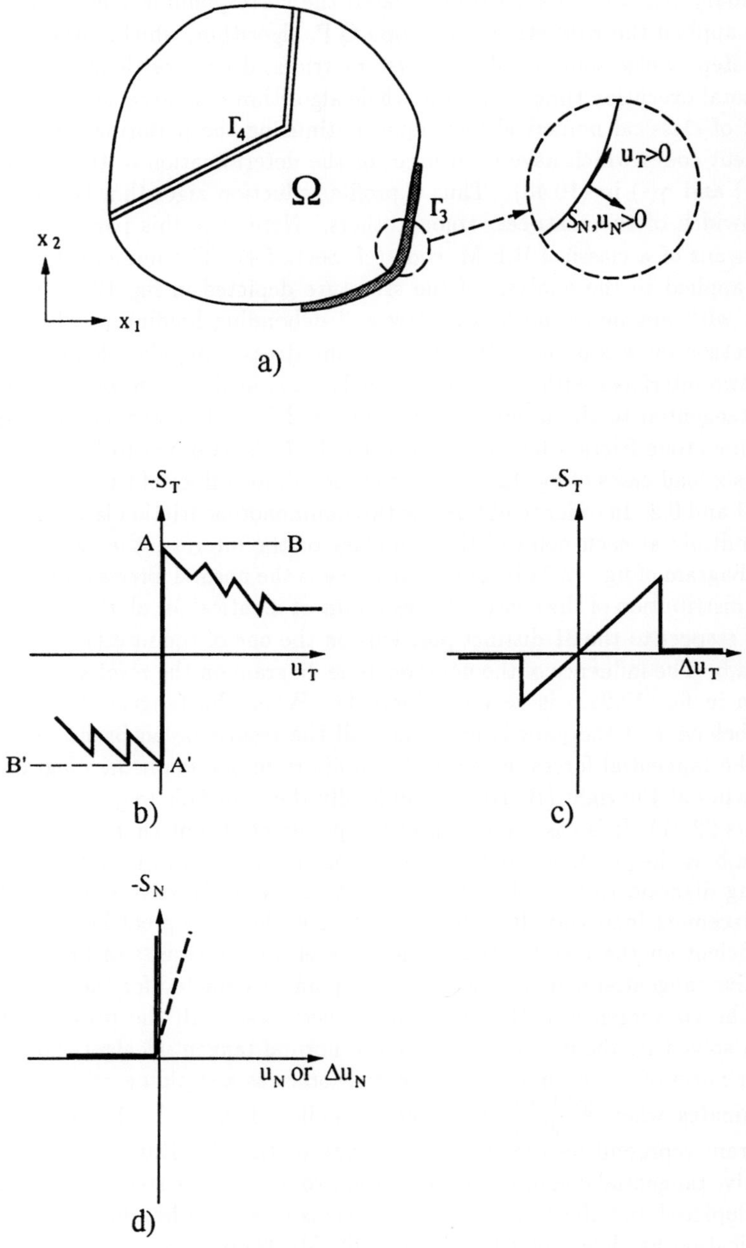

Fig. 12.4: Nonmonotonicity with debonding

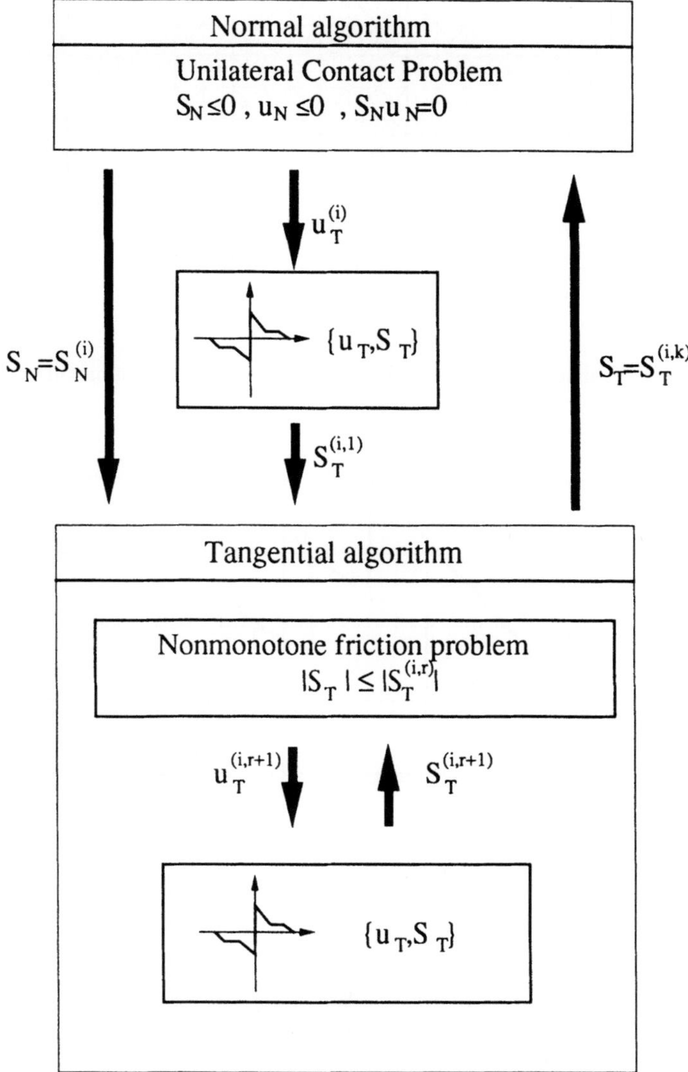

Fig. 12.5: The fixed point algorithm. The ith step of the algorithm

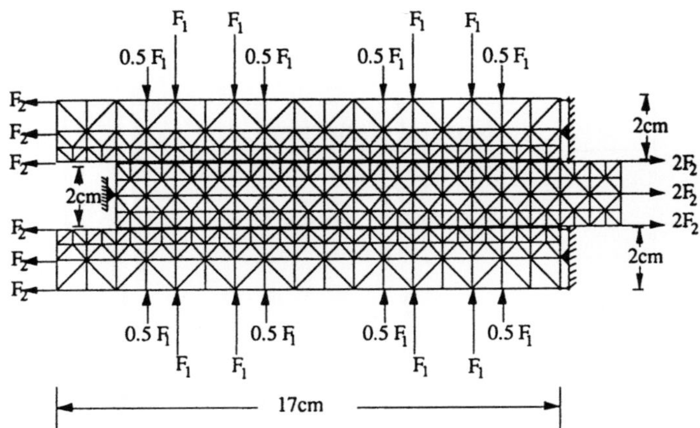

Fig. 12.6: F.E. discretization and loading of the analyzed structure

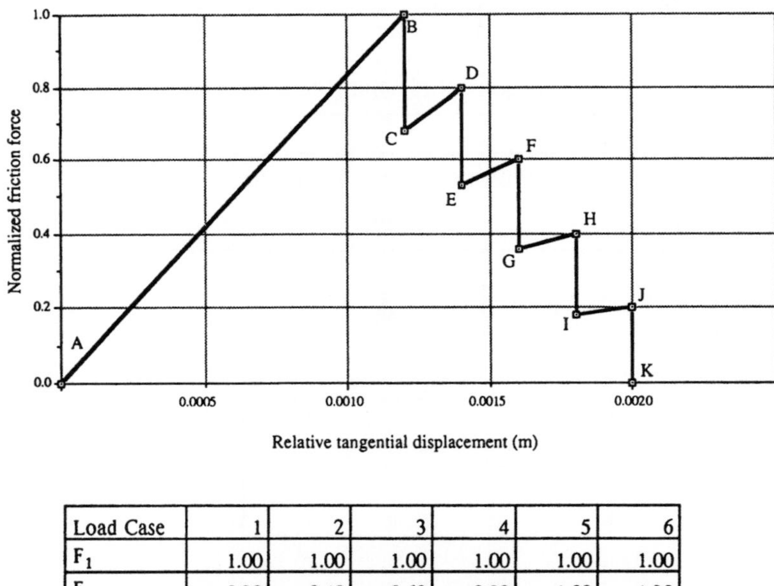

Relative tangential displacement (m)

Load Case	1	2	3	4	5	6
F_1	1.00	1.00	1.00	1.00	1.00	1.00
F_2	0.20	0.40	0.60	0.80	1.00	1.20

Fig. 12.7: a) The nonmonotone law. b) The load cases

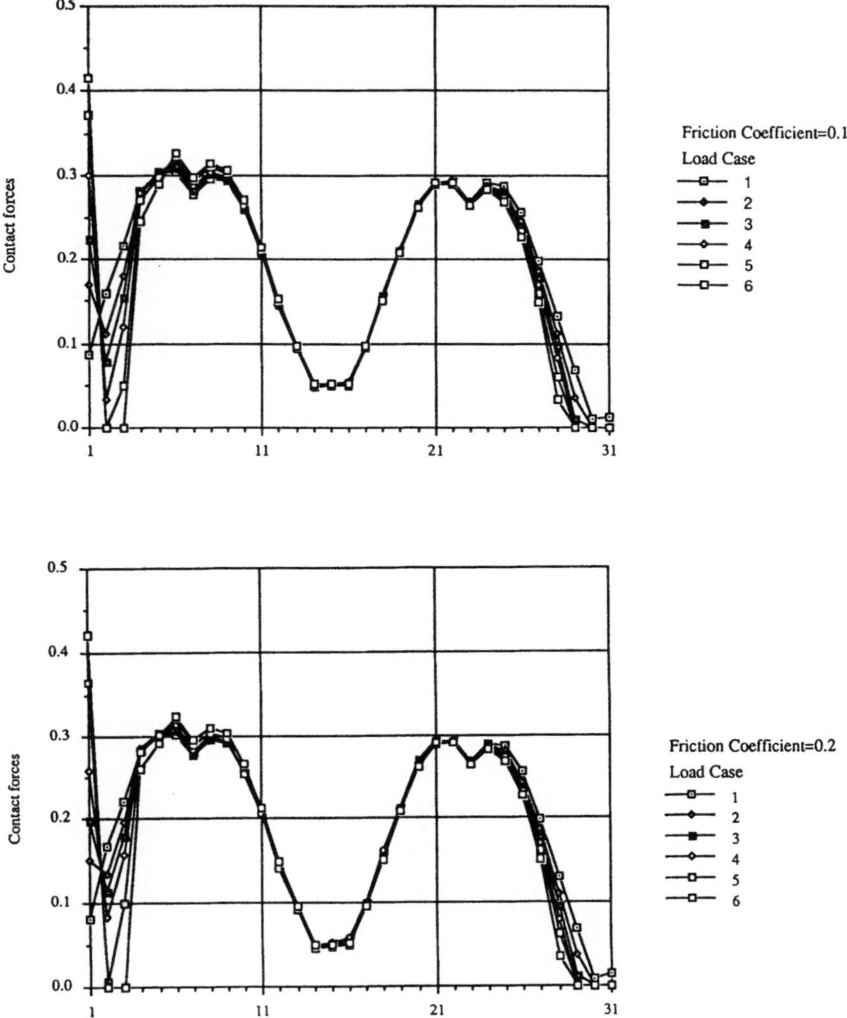

Fig. 12.8: a) Distribution of the normal forces on the interface. (Friction coefficient=0.1). b) Distribution of the normal forces on the interface. (Friction coefficient=0.2)

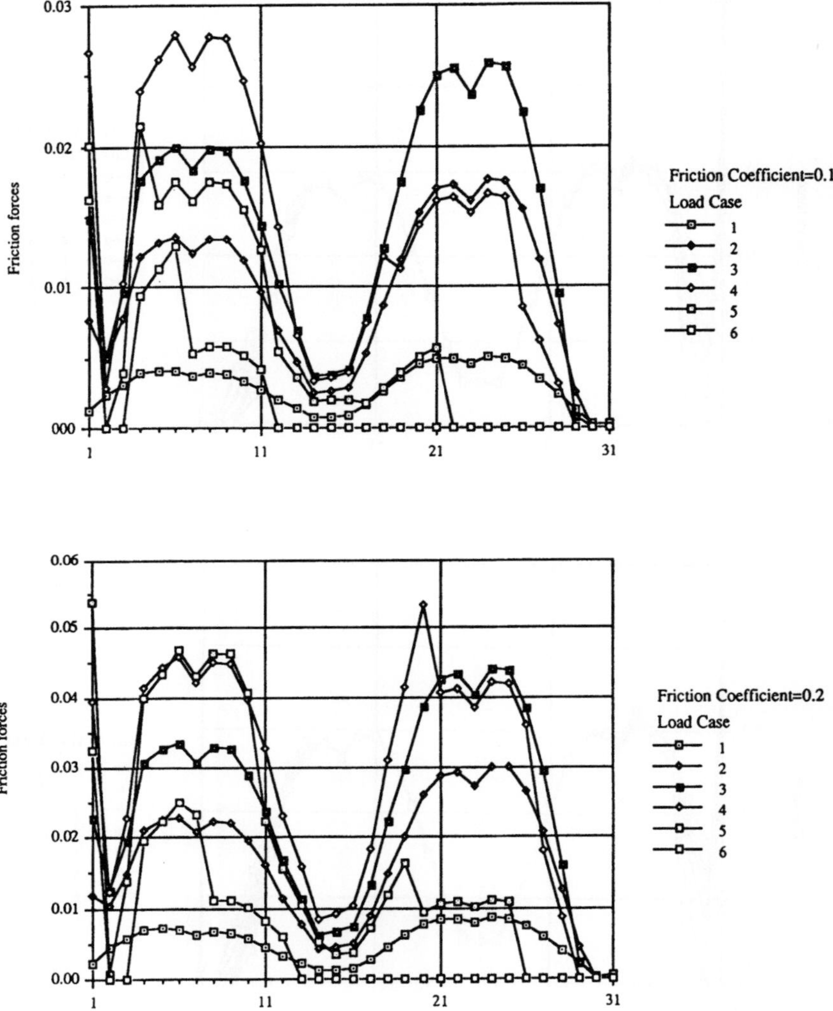

Fig. 12.9: a) Distribution of the frictional forces on the interface. (Friction coefficient=0.1). b) Distribution of the frictional forces on the interface. (Friction coefficient=0.2)

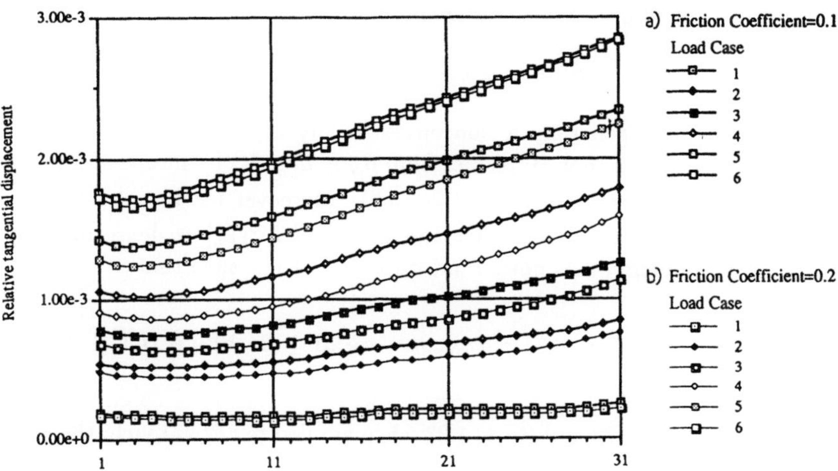

Fig. 12.10: Influence of the friction coefficient on the relative tangential displacements

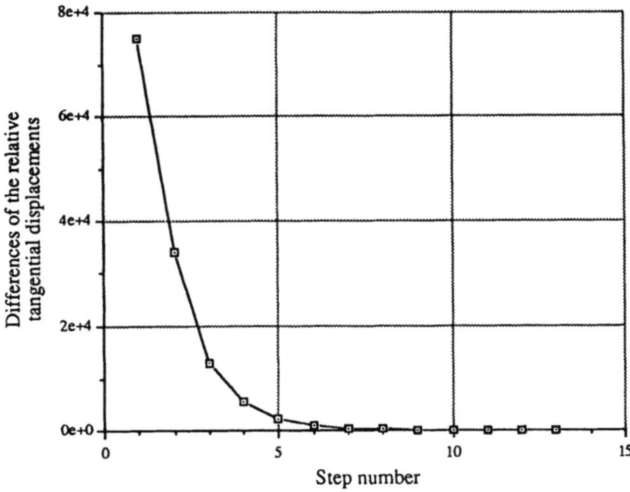

Fig. 12.11: Convergence rate of the algorithm

12.5 A Short Note on Certain Coercive and Semicoercive Nonconvex Unilateral Contact Problems

Let us put ourselves in the functional framework of (10.106) with the difference that now on Γ_3 instead of (10.105) the conditions

$$-S_N \in \bar{\partial} j_N(u_N) \quad \text{and} \quad S_T = C_T \tag{12.76}$$

hold, where now j_N is generally a nonconvex locally Lipschitz superpotential. We make also the assumption that j_N results from $\beta_N \in L^\infty_{\text{loc}}(\mathbb{R})$ by means of a procedure analogous to the one described in (1.62)÷(1.66). Moreover for $V = [H^{1/2}(\tilde{\Gamma})]\big|_{\Gamma_3}$ we may verify that $V \subset L^2(\Gamma)$ is compact and that $V \cap L^\infty(\Gamma_3)$ is dense in V for the $H^{1/2}$-norm. The problem reads now: Find $u_N \in H^{1/2}(\tilde{\Gamma})\big|_{\Gamma_3}$ such that

$$\delta(u_N, v_N - u_N) + \tilde{\zeta}(v_N - u_N) + \int_{\Gamma_3} j_N^0(u_N, v_N - u_N)d\Gamma \geq 0 \quad \forall v_N \in H^{1/2}(\tilde{\Gamma})\big|_{\Gamma_3} \tag{12.77}$$

Then Prop. 1.21, Prop. 1.22 and Prop. 1.23 when applied to (12.77) yield the following results [Pan93].

Proposition 12.2 Suppose that mes $\Gamma_1 \neq 0$ (coercive case) and that there is $\xi \in \mathbb{R}$ such that

$$\operatorname*{ess\,sup}_{(-\infty,-\xi)} \beta_N(\xi) \leq \operatorname*{ess\,inf}_{(\xi,\infty)} \beta_N(\xi). \tag{12.78}$$

Then (12.77) has at least one solution.

Proposition 12.3 Suppose that $\Gamma_1 = \emptyset$ (semicoercive case) and let

$$\beta_N(-\infty) \leq \beta_N(\xi) \leq \beta_N(+\infty) \quad \forall \xi \in \mathbb{R}. \tag{12.79}$$

Then for $r_N = r_i n_i$, $r = \{r_i\} \in \mathcal{R}_\Gamma, i = 1, 2, 3$ the inequality

$$\int_{\Gamma_3} [\beta_N(-\infty)r_{N+} - \beta_N(\infty)r_{N-}]d\Gamma \leq \tilde{\zeta}(r) \leq \int_{\Gamma_3} [\beta_N(\infty)r_{N+} - \beta_N(-\infty)r_{N-}]d\Gamma \; \forall r \in \mathcal{R}_\Gamma \tag{12.80}$$

is a necessary condition for the existence of a solution of (12.77). If (12.79) holds strictly the same happens also for (12.80).

Proposition 12.4 Supose that $\Gamma_1 = \emptyset$ (semicoercive case) and let

$$\beta_N(-\infty) < \beta_N(\infty). \tag{12.81}$$

Then if (12.80) holds as a strict inequality for every $r \in \mathcal{R}_\Gamma$, $r \neq 0$, (12.77) has at least one solution.

Chapter 13
Miscellanea

13.1 Unilateral Contact and Friction in Cracks. A General Indirect B.I.E.M. for Inequality Problems

Cracks in solids form interfaces where both the unilateral contact effect in the direction normal to the interface and the frictional effect in the tangential sense, appear in a coupled form. This nonclassical and highly nonlinear behaviour of the cracks has not yet been extensively studied. Concerning only the frictional contact in cracks, the works by Comninou and Dundurs [Com, Dun] are of importance; however they make an assumption concerning the contact zone, something not necessary in the framework of an inequality theory (cf. also [Dub89b] p.79). Note also in this context that, as shown by the second author [Pan80;85], the trial and error methods, e.g. assumptions on contact zones and/or adhesive zones, in large scale problems containing inequalities like the unilateral contact and the friction problem, may lead to erroneous results. Unilateral contact problems for cracks including friction have been studied by Dubourg et al [Dub88;89a]. In these papers the classical indirect B.I.E.M. (cf. Ch.2) based on dislocations has been combined with the algorithm of Kalker for the treatment of the unilateral contact and friction problem [Ka88a,b;90]. For this contact algorithm no convergence proof is until now given in the framework of the F.E.M. or the B.E.M.

Here we follow [The92] where the method of [Pan83b] has been adapted to crack problems. We apply the indirect B.I.E.M. for the modelling of the crack, but for the numerical treatment we combine this method with the inequality contact algorithm presented in [Pan75] (cf. Sect. 9.3). Bisbos [Bis90] has improved this algorithm for large scale 3D-unilateral contact and friction problems arising in industrial applications and Nečas et al [Neč80, Ja83;84] have shown the convergence of this algorithm for discretized and continuous problems.

Let us note here that there also exist other types of algorithms which may treat the inequality constraints, as e.g. the classical and augmented Lagrangian method [Bat, Wri, For], the perturbed Lagrangian method [Sim, Ju, Kik] and the penalty method (see e.g. Kikuchi-Oden [Kik]). Most of the above methods due to their variational or semivariational character satisfy the constraints in an "average" sense [Wri], instead of a "pointwise" sense. The method proposed here is more appropriate for a crack situation, i.e., responds to the increased accuracy requirements concerning the determination of the stress intensity factors. Therefore here the active set strategy of [Pan80] has been preferred. This strategy combined with the indirect B.I.E.M. permits the pointwise fullfilment of the constraints. In the case of cracks with unilateral contact and friction, the numerical implementation of the direct B.I.E.M. makes necessary the use of special singular boundary elements for the consideration of the arising crack singularities.

Let us consider a linear elastic body $\Omega \subset \mathbb{R}^3$ and let Γ be its boundary which is assumed to be smooth enough. The boundary is devided into three mutual disjoint parts Γ_1, Γ_2 and Γ_3. On Γ_1 the displacements $u = \{u_i\}$ are given whereas on Γ_2 the boundary tractions. On Γ_3 we assume that the unilateral contact and the friction boundary conditions hold. With the notation introduced in the previous Chapters and denoting by μ the friction coefficient, the boundary conditions on Γ_3 read:

i) If $u_N < 0$, then $S_N = 0$ and $S_T = \{S_{T_i}\} = 0$ $\hspace{2cm}$ (13.1)

ii) If $u_N = 0$, then $S_N \leq 0$ and then $\hspace{3cm}$ (13.2)

ii1) If $|S_T| < \mu|S_N|$, then $u_T = \{u_{T_i}\} = 0$ and $\hspace{1.5cm}$ (13.3)

ii2) If $|S_T| = \mu|S_N|$, then there exists $\lambda \geq 0$
$\hspace{1cm}$ such that $u_{T_i} = -\lambda S_{T_i}$ $i = 1, 2, 3$. $\hspace{2cm}$ (13.4)

Within the body Ω we have some formed cracks which we denote by AB. They may have any geometrical shape and their contour does not change during the loading. We assume unilateral contact and friction conditions holding at the interface. They have the following form:

i) If $[u_N] < 0$, then $S_N = 0$ and $S_T = \{S_{T_i}\} = 0$ $\hspace{2cm}$ (13.5)

ii) If $[u_N] = 0$, then $S_N \leq 0$ and then $\hspace{3cm}$ (13.6)

ii1) If $|S_T| < \mu|S_N|$, then $[u_T] = \{[u_{T_i}]\} = 0$ and $\hspace{1cm}$ (13.7)

ii2) If $|S_T| = \mu|S_N|$, then there exists $\lambda \geq 0$
$\hspace{1cm}$ such that $[u_{T_i}] = -\lambda S_{T_i}$ $i = 1, 2, 3$. $\hspace{2cm}$ (13.8)

Here $[u_N]$ and $[u_T] = \{[u_{T_i}]\}$ denote the relative displacements of the two crack sides. $[u_N]$ is negative if the crack tends to open. Note that (13.5) (resp. (13.6)) is satisfied at the points of the crack which do not remain (resp. are) in contact whereas (13.7) (resp. (13.8)) corresponds to the region of adhesive (resp. of sliding) friction. We want to determine for a given loading the stress, strain and displacement fields. Moreover the contact and noncontact regions and the adhesive and sliding friction regions should be determined on Γ_3 and on each crack. The body is assumed to be linear elastic and its strains and displacements are small enough.

The following method [Pan83] holds for any kind of boundary laws of the general subdifferential form (1.100), (1.101) both for convex and nonconvex superpotentials. Here let us formulate the indirect B.I.E.M. for the present problem. We imbed the body Ω in an infinite elastic medium and we determine layers of singular solutions for the equations of the elastostatics for the infinite elastic medium such that the boundary conditions on Γ_1, Γ_2, Γ_3 and the interface conditions on AB are satisfied. Suppose for instance that q is the vector of the unknown force and/or dislocation layer etc. on the boundary Γ and at the two sides of the interfaces AB in the infinite elastic medium. We can express $u, S, u_N, [u_N], S_N, u_T, [u_T], S_T$ in (13.1)$\div$(13.8) in terms of q through the appropriate operators (cf. Ch.2) and we have to "solve" the resulting system of equations and inequalities defined through the boundary, and the interface boundary condition. The method is also described in [Pan85;p.161].

The operators in the equalities and the inequalities may include the Cauchy principal value integral and kernels of various kinds. Their form depends on the unknown of the problem, i.e. whether the unknown q is a dislocation layer or a force layer or some other quantity. For the numerical treatment we have to apply a method appropriate for the treatment of the arising, integral operators, in this case the singular integral operators [Erd72, Kre75;81]. Note that other integral equation methods lead to regular integral operators [Oli68]. Here the resulting system of equalities and inequalities is treated as a purely unilateral problem – no trial and error varification of the inequalities.

When the contact and non-contact zones in the crack as well as the sliding and adhesive friction zones are determined by the method of [Pan75] the problem reduces to a singular B.I.E. formulation of a bilateral problem. Then the method of [Kre75] is used to the calculation of the stress intensity factors. As a general application we consider a halfplane with cracks which is referred to an orthogonal Cartesian system Ox_1x_2. Here Ox_2 is the boundary of the halfplane on which the loading acts; it is assumed to be horizontal. In order to formulate the corresponding integral operators it is sufficient to consider an unknown dislocation (or a load) distribution normally and tangentially to each crack since the fundamental solutions for the halfplane are known. Thus a dislocation with components $\{d_1(x_1'), d_2(x_2')\}$ at the point $(x_1', 0)$ generates the following stress and displacement field at a point (x_1, x_2)

$$\sigma_{ij} = \frac{2G}{\pi(k+1)} \sum_{\lambda=1}^{2} d_\lambda k_{ij}^\lambda(x_1, x_2, x_1'), \quad i, j = 1, 2, \tag{13.9}$$

$$u_i = \frac{1}{2G(k+1)} \sum_{\lambda=1}^{2} d_\lambda L_i^\lambda(x_1, x_2, x_1'), \tag{13.10}$$

where G is the shear modulus of the material, ν is Poisson's number, $k = 3 - 4/\nu$ (resp. $k = (3-\nu)/(1+\nu)$) for plane strain (resp. plane stress) problems and $k_{ij}^\lambda, L_i^\lambda, i, j = 1, 2$ are kernel functions [Hei78a]. The above expressions yield by forming the corresponding integrals (cf. Ch.2) and performing the necessary coordinate changes in the case of kinked or curved cracks, the stresses and displacements at any point, when we have dislocation distributions along a crack. Then the dislocation distributions $\{d_1, d_2\}$ are the unknowns of the problem, which results by substituting the corresponding integral operators into $(13.5) \div (13.8)$.

The appearance of singular integral operators compels us to apply for the numerical treatment the method developed in [Kre81] and [Erd] which expresses the unknown function $d(x), -1 < x < 1$, in the form

$$d(x) = w(x)F(x) \approx w(x) \sum_{k=0}^{n-1} c_k p_k(x), \tag{13.11}$$

where

$$w(x) = (1-x)^\alpha (1-x)^\beta \quad -1 < \alpha, \beta < 1,$$

$p_k, k = 0, \ldots, n-1$ are the Jacobi polynomials and c_k are constant coefficients, which have to be determined. In our problem, α, β take the values $\pm 1/2$ depending on the type of contact realized (e.g. sliding, adhesive etc.). Depending on the values of α, β

we shall have regular or singular behaviour of $d(x)$ at the ends $x = \pm 1$ of the crack. In order to apply numerical techniques suitable for the singularities we have to know at least approximately the different contact zones within each crack. Because, even in the case in which S_N, S_T etc. are expressed in terms of regular integral operators (cf. e.g. [Rie68]), a direct substitution of the arising integral expressions of S_N, S_T, $[u_N]$, $[u_{T_i}]$ into (13.5)÷(13.8) leads to a system of equations and inequalities for which one cannot obtain a numerical solution even in the simplest cases. Therefore we split the unilateral contact problem with friction into two subproblems: the unilateral contact problem with given tangential forces and the friction problem with given normal forces according to [Pan75]. Then for the determination of the interface free boundaries we apply to each inequality constrained subproblem the method of [Pan80] consisting in the "replacement" of the inequality unilateral contact or friction problems by a finite number of equality constrained subproblems. Note that the method of [Pan80] can be applied either with respect to the initial (S, u)-formulation of the problem or with respect to the d-formulation of the problem. Finally a classical bilateral problem is obtained whose solution gives the solution of the initial unilateral contact problem with friction; i.e. within each interface all the free boundaries have been obtained. Then the stress intensity factors k_I and k_{II} are obtained by estimating in (13.11), $F(1)$ [Kre81].

In order to improve the effectiveness of the algorithm, which determines the free boundaries within the crack concerning the number of unknowns (at most 3000 contact nodes in 3d-cracks for a HP-750 workstation), its automation level, its robustness and computing speed, we apply instead of the methods of [Pan75;80] the active set search algorithm by Bisbos [Bis90]. For numerical applications on crack problems the reader is referred to [The92]. Note here that singular integrals can be calculated by a wavelet approximation method [Dav] which seems to be in some sense "dual" to the fractal approximation method proposed in [Pdp91;92a].

13.2 Debonding and Delamination in Adhesively Bonded Cracks.

Within the framework of the previous section (with zero displacements on $\Gamma_1 \equiv \Gamma$, $\Gamma_2 = \Gamma_3 = \emptyset$) we assume that each crack has an adhesive behaviour both in the normal and in the tangential direction. This behaviour is described by laws of the form (12.4) and (12.4a) where u_N and u_T are replaced by the relative normal and tangential displacements of the two crack sides denoted by $[u_N]$ and $[u_T]$ respectively. Let us denote all cracks by γ. We shall follow a procedure fully analogous to the one of Ch. 12 with the difference that now instead of (12.14) the condition

$$a(\bar{u}_{(2)}, v) - \int_\gamma \left(\mu_N[v_N] + \mu_{T_i}[v_{T_i}] \right) ds = 0 \quad \forall \bar{v} \in V_0 \tag{13.12}$$

holds and instead of (12.40) the condition

$$A(\bar{\sigma}_{(1)}, \tau) - \int_\gamma \left([v_N]T_N + [v_{T_i}]T_{T_i} \right) ds = 0 \quad \forall \tau \in \Sigma_0, \tag{13.13}$$

holds. Then hemivariational inequalities analogous to (12.28) and (12.58) are obtained, which give rise to multivalued B.I.Es along the crack contours γ, with respect to the interface stresses μ_N and μ_T or with respect to the relative displacements of the two crack sides $[u_N]$ and $[u_T]$. The boundary integral equations along γ have the form

$$\frac{\partial}{\partial \mu} \left\{ - \int_\gamma \mu_i \big([G(\bar{l})]\big)_{,i} ds - \frac{1}{2} \int_\gamma \mu_i \big([G(\mu)]\big)_{,i} ds \right\} \qquad (13.14)$$

$$\in \bar{\partial} \left\{ \int_\gamma \big[\tilde{j}_N(-\mu_N) + \tilde{j}_T(-\mu_T)\big] ds \right\}$$

where $\mu = \{\mu_N, \mu_T\}$ and $\bar{l} = \{\bar{p}, \bar{T}\}$ (see (12.9) etc.), and $\big[G(\alpha)\big]_{,i}$ is a relative displacement of the two sides of γ in the i-direction due to the loading α, or the form

$$\frac{\partial}{\partial [v]} \left\{ \int_\gamma (\tilde{H}(\bar{U})_i + S_{0i}, [v]_i) ds - \frac{1}{2} \int_\gamma [v]_i \big(\tilde{H}([v])\big)_{,i} ds \right\} \qquad (13.15)$$

$$\in \bar{\partial} \left\{ \int_\gamma \big(j_N([v_N]) + j_T([v_T])\big) ds \right\}$$

where $[v] = \{[v_N], [v_T]\}$ and $\tilde{H}$ and S_0 are defined as in Ch.12 with the only difference that Γ_3 has to be replaced by γ.

The whole theory has been developed in [The91], where also numerical examples have been given. The numerical results have been obtained in [The91] by the bundle algorithm [Str]. Large numerical applications have been treated in [Pan93a] with the algorithm of Sect.12.4.

13.3 Fractal Interfaces and Boundaries. Unilateral Contact and Friction Problems

In the equality and inequality theories of B.I.Es we have made the assumption that the boundaries and the interfaces are appropriately smooth. In physical problems however we have boundaries and interfaces of fractal nature. Indeed the notion of fractals has been introduced in order to obtain an accurate description of the physical forms, as e.g. the geometry of cracked and crushed surfaces in deformable bodies, the geometry of the diffusion fronts, the percolation patterns, the geometry of bone cracks, the geometry of metals subjected to sandblasting or to meteoritic rain etc. (cf. e.g. [Man, Taka, Fed, Art, Bar]). Here we consider according to [Man, Wal] that a set $F \subset \mathbb{R}^n$ is a fractal set if F has fractional Hausdorff dimension $\dim F$, or if $\dim F$ is strictly larger than the topological dimension of F. Concerning the definition of the Hausdorff dimension see [Fal].

Here we shall investigate how the fractal nature of an interface or a boundary in a contact problem influences the application of the equality or the inequality B.I.E.M.

The answer is very simple: " All algorithms applied to problems having classical (i.e. nonfractal) geometries are coupled with a fixed point algorithm approximating the fractal geometry through classical geometries". In this respect and concerning generally the introduction of the fractal geometry hypothesis into mechanics and engineering theories we refer to the works of the second author and his coworkers [Pdp90a,b;91;92a,b,c, Pang92a,b].

Following [Bar] we consider that a fractal is the deterministic or the stochastic attractor of an "iterated function system" (I.F.S.). Then the fractal is the unique fixed "point" of an appropriately defined contraction mapping. Since this mapping W acts on the complete metric space $H(X)$ of the compact subsets of a metric space X, when $H(X)$ is endowed with the Hausdorff metric, this fixed point is a compact subset of the initial metric space X. Analogously the stochastic attractor is defined [Bar]. Moreover we consider that a fractal may be defined by means of a "fractal interpolation" i.e. interpolation of given data by a fractal "curve" or "surface". Again the fractal is defined as a fixed point of a given transformation T. Let us explain this construction more accurately. Suppose that in $\mathbb{R}^2$, for instance, we have a set of data (x_i, y_i) $i = 1, \ldots, N$. We want to find a fractal type interpolation function $y: [x_0, x_N] \to \mathbb{R}$, i.e. a fractal f such that $y(x_i) = y_i$ $i = 1, \ldots, N$. We consider the I.F.S. $\{\mathbb{R}^2; w_n, \ n = 1, \ldots, N\}$ defined by the "transformation"

$$(x, y) \to w_i(x, y) = \begin{bmatrix} a_i & 0 \\ c_i & d_i \end{bmatrix} \begin{bmatrix} x \\ y \end{bmatrix} + \begin{bmatrix} e_i \\ f_i \end{bmatrix} \quad i = 1, \ldots, N. \tag{13.16}$$

Let the factors d_n, called scaling factors, satisfy $0 \le d_n < 1$; they are free parameters of the problem. Moreover we have that

$$a_i = \frac{(x_i - x_{i-1})}{(x_N - x_0)}, \quad e_i = \frac{(x_N x_{i-1} - x_0 x_i)}{(x_N - x_0)} \tag{13.17}$$

$$c_i = \frac{(y_i - y_{i-1})}{(x_N - x_0)} - d_i \frac{(y_N - y_0)}{(x_N - x_0)} \tag{13.18}$$

$$f_i = \frac{(x_N y_{i-1} - x_0 y_i)}{(x_N - x_0)} - d_i \frac{(x_N y_0 - y_N x_0)}{(x_N - x_0)} \tag{13.19}$$

Then the following holds [Bar]: Let F be the attractor of the I.F.S. defined by (13.16)$\div$ (13.19). Then F is the graph of a continuous function $y: [x_0, x_N] \to \mathbb{R}$ interpolating the data $(x_i, y_i), i = 1, \ldots, N$. If C^0 is the set of all continuous functions $y: [x_0, x_N] \to \mathbb{R}$ then the sequence of functions $\tilde{y}_{m+1}(x) = (T\tilde{y}_m)(x)$, where the operator $T: C^0 \to C^0$ is defined by

$$T(\tilde{y}(a_i x + e_i)) = c_i x + d_i \tilde{y}(x) + f_i \quad i = 1, 2, \ldots, N \tag{3.20}$$

converges to the attractor F as $m \to \infty$. Furthermore if $x_0, \ldots, x_N$ are equally spaced then the $\dim F$ is given by the formula

$$\dim F = 1 + \frac{\ln \left(\sum_{i=1}^N |d_i| \right)}{\ln N}, \tag{13.21}$$

if the points $(x_i, y_i) i = 1, \ldots, N$ do not constitute a straight line (in this case $\dim F = 1$) and if $\sum_{i=1}^{N} |d_i| > 1$. Note that the proper choice of the parameters d_i may make $\dim F$ very close to 1 (line-like fractal) or very close to 2 (surface-like fractal).

Further we have studied a unilateral contact problem with Coulomb friction (Sect. 9.3), when the contact boundary or interface is of fractal nature. The fractal interface Φ is the fixed "point" of a transformation W or T resulting either from an I.F.S. or from the fractal interpolation of given data. Thus we may write that

$$\Phi = T\Phi \quad \text{and} \quad \Phi_{n+1} = T\Phi_n \quad \Phi_n \to \Phi \tag{13.22}$$

in the Hausdorff metric. Accordingly we shall apply for every Φ_n the algorithm [Pan75] for the splitting of the general problem into a pure unilateral contact problem with given $S_T = C_T$ and into a pure friction problem with given $S_N = C_N$. Then for each subproblem the multivalued B.I.E. with respect to Φ_n has been formulated and solved as indicated in Sect. 9.3. We repeat this procedure several times by increasing n and we claim that at the limit the solution of the fractal problem is obtained. The convergence proof is still open. Only several numerical experiments verify the correctness of the method followed.

As an example we consider here the structure of fig. 13.1 submitted to loading in its plane. The material is linear elastic with modulus of elasticity $E = 2,1 \times 10^5$ t/m^2 and Poisson's ratio $\nu = 0.33$. The thickness of the plate is 0.10 m. The interface is a fractal interpolating the data $(x_0, y_0) = (0,0)$, $(x_1, y_1) = (2.7, 4.0)$ and $(x_2, y_2) = (8.0, 0.0)$. Then the fractal interface is the graph of the attractor of the I.F.S. $\{\mathbb{R}^2; w_1, w_2\}$ where w_1 and w_2 are the transformations

$$w_1(x, y) = \begin{bmatrix} 0.3375 & 0.0 \\ 0.5 & 0.5 \end{bmatrix} \begin{bmatrix} x \\ y \end{bmatrix} + \begin{bmatrix} 0 \\ 0 \end{bmatrix} \tag{13.23}$$

$$w_2(x, y) = \begin{bmatrix} 0.6625 & 0.0 \\ -0.5 & 0.5 \end{bmatrix} \begin{bmatrix} x \\ y \end{bmatrix} + \begin{bmatrix} 2.7 \\ 4.0 \end{bmatrix}. \tag{13.24}$$

The free parameters d_i $i = 1, 2$ are $d_1 = d_2 = 0.5$. The coefficients of the multi-valued B.I.Es for each subproblem and for each Φ_n have been obtained by the unit loading (resp. displacement) method with respect to a classical bilateral direct boundary element scheme using linear interpolation elements. The required steps for the two-level (i.e. simple unilateral contact/simple friction) algorithm for $\Phi_1, \Phi_2, \ldots, \Phi_5$ are $7, 7, 9, 10, 15$ respectively. Each multivalued B.I.E. has been numerically solved by solving the equivalent Q.P.P. on the boundary by the Hildreth and d'Esopo Q.P. algorithm. The termination criterion was $|C_{T_i}^{(k+1)} - C_{T_i}^k| < 10^{-4}$ and $|C_N^{k+1} - C_N^{(k)}| < 10^{-4}$. Also the corresponding displacement fields were tested for the convergence. The procedure was repeated for each approximation of the fractal interface. It was sufficient to solve the problem for the five first approximations of the fractal interface; indeed the fourth and the fifth one give rise to almost the same stress and displacement fields.

The whole algorithm is extremely stable, although the complicated interfaces introduced by the approximations of the fractal interface do not offer an ideal basis for the solution of a unilateral contact problem with friction. At the first and second fractal approximations, the contact algorithm determines from the first step almost completely the true contact region. Thus a few steps were needed for the termination of the algorithm. The complexity of the third, fourth and fifth approximations of the fractal boundary caused an increase of the calculation time; the contact region determined in the first step has been improved only a little at the next steps where a few more constraints become active.

We notice that the differences become insignificant after the third approximation of the fractal boundary. For instance between the third (resp. fourth) and the fourth (resp. fifth) approximation on Φ the differences of the maximum stress is smaller that 5% (resp. 1%) in a curve parallel to Φ and lying 10 cm inside the boundary. It is also important to note that the approximation of the fractal Φ does not affect considerably the stress and displacement fields inside the body. Of course this is compatible with the St. Venant "principle" of classical elestostatics which holds actually for classical boundaries and for bilateral boundary conditions.

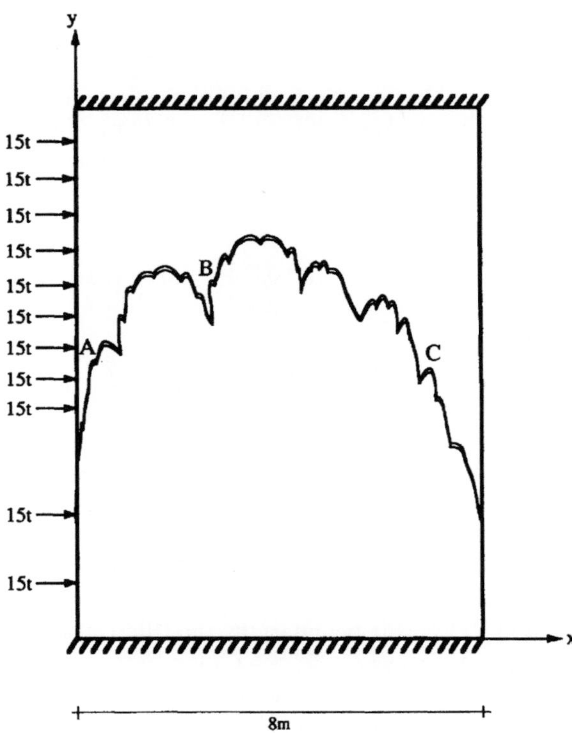

Fig. 13.1: Plane unilateral contact problem with Coulomb's friction for an interface of fractal type

The numerical results are quite reliable concerning the consideration of possible stress concentrations. Indeed the action of external and reentrant corners in each approximation has been appropriately taken into account by increasing the number of elements around the singular point. One should think also that in order to calculate the structure of fig. 13.1 it would be sufficient to use a random approximation of the boundary very close to the fractal interface Φ. From the numerical implementation of such a case we notice that the numerical results are "worse" than the results obtained by performing the exact approximation of the fractal boundary. In figs 13.2 and 13.3 we give the variation of the maximum stress and the variations of the displacement fields respectively. Moreover in fig. 13.4 (resp. fig. 13.5) the maximum difference of the normal (resp. the tangential) relative displacements between the two levels of the algorithm, for the fourth and fifth approximation of the interface are depicted with respect to k.

Major stress:

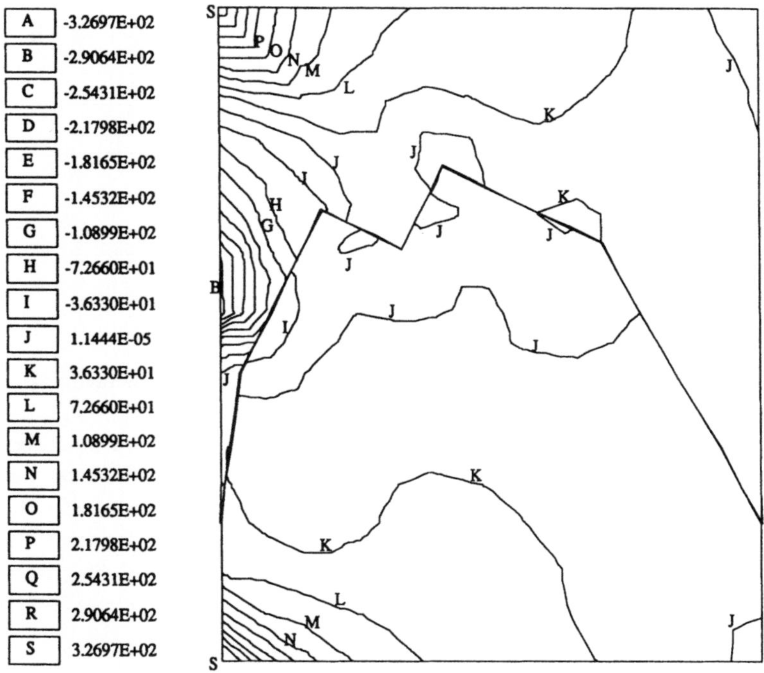

Fig. 13.2: Maximum stress of the fifth approximations of the fractal (stress in t/m²)

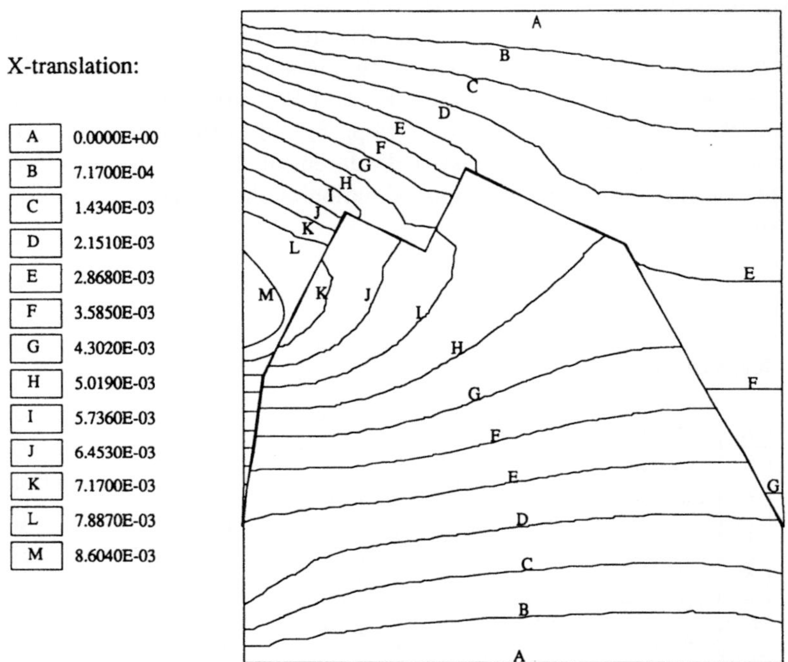

Fig. 13.3: X-translations for the fifth approximation of the fractal (in m)

The arising multivalued B.I.Es on fractal boundary or interface have again a unique solution for both the pure unilateral contact and the pure friction subproblem and also for the general problem studied in Ch.10. The same method is applied for the mathematical study of the existence of the solution with the difference that now we have to work on Besov spaces defined on the fractal boundary (or interface) according to the research results of Wallin and Jonson [Jon, Wal89]. Indeed the trace theorem for Γ fractal with dimension d, implies that the multivalued B.I.E. (10.88) (resp. (10.48)) must be defined on the Besov space $[B_\beta^{2,2}(\Gamma)]^3$, with $\beta = 1 - \frac{n-d}{2}$ where $\Omega \subset \mathbb{R}^n$, (resp. on its dual space).

Concerning finally the numerical treatment, there are two types of computer codes available: the first (resp. the second) calculates and plots stress and displacement fields etc. for bilateral contact (resp. unilateral contact with friction) problems of fractal geometry.[9]

---

[9]For information one can write to the second author of the present book.

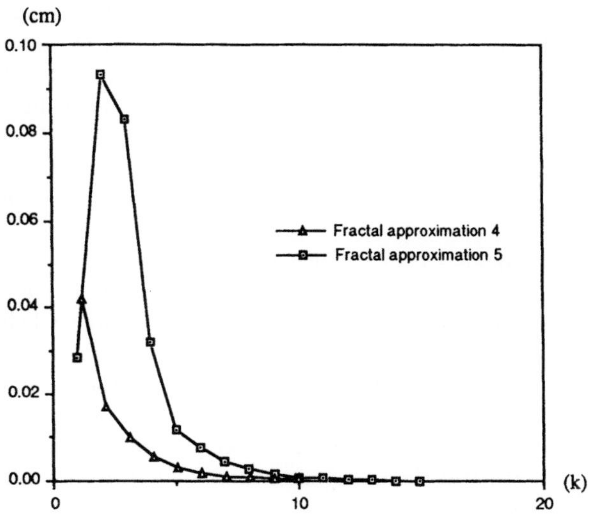

Fig. 13.4: Differences of the normal relative displacements between the pure contact and the pure friction subproblems with respect to k

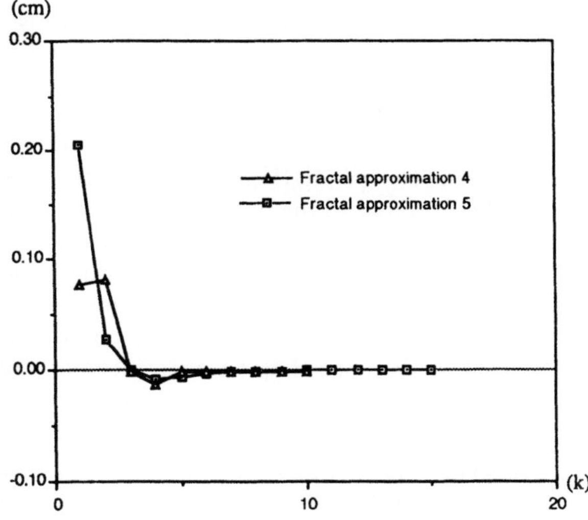

Fig. 13.5: Differences of the tangential relative displacements between the pure contact and the pure friction subproblems with respect to k

13.4 A Neurocomputing Approach to the Multivalued B.I.Es of the Inequality Contact Problems.[10]

The present Section refers to those inequality contact problems for which the multivalued B.I.Es (10.48) and (10.88) are such that their equivalent minimum problems lead in the discretized model to quadratic programming problems (Q.P.P). The main result of this Section is that in a neural network environment an inequality constrained Q.P.P needs the same computer memory for its calculation as a Q.P.P. without constraints, i.e. a linear equation system. This fact will give in the future a huge advantage to the inequality approach because this approach corresponds to more accurate physical models.

Neural network models are very efficient in computations where many assumptions must be satisfied in parallel. This is achieved, in contrast to the classical sequential computers, by using networks of analog neurons with nonlinear behaviour and with a high interconnectivity. The neurons are connected with links having variable weights. A neural network is defined by its node characteristics, the "learning" rules and the network topology. These rules control the improvement of the network performance through appropriate adaptive changes of the weights of the links. A computing machine based on neural networks has greater fault tolerance than a classical sequential computer.Indeed some neurons or links out of order do not affect considerably the whole performance of the network, as well as its learning properties.

The aforementioned artificial neural nets "immitate" the behaviour of biological nevrous systems. The study of real biological nets can be explained by the fact that the computational power of biological nervous systems is enormous and gives answers to complicated questions in a very short time. On the other hand most digital computers do not have these abilities even for very simple questions.

The aim of the present Section is to adapt to a neural network computing environment the equality and inequality B.I.Es for contact problems. The main idea is to use the neural network capability to solve optimization problems. Indeed this was one of the first applications, and constitutes one of the major advantages of the neural networks.

The research on artificial neural networks begins in the early forty's and continues with the works of Widrow and Hoff [Wid], Hopfield [Hop82], Hopfield and Tank [Hop85] among others (cf. e.g. [And]). From all the research which took place until now a basic result is that the parallel analog computation in a neural network is the most natural way to solve large scale optimization problems, as those arising in fields like speech and image recognition, etc. Moreover several nonlinearities which cost large computing time in a classical sequential computer may be treated very quickly in a computer based on the neural net concept. Indeed the simplest node (neuron) i of a neural network sums the weighted inputs, say $T_{ji}V_j$ (T_{ji} is the weight of the link or synapse between

---

[10]This Section and the next one contain some research results of the second author obtained in the framework of a large scale STRIDE program on "Massively Parallel, Neurocomputing Environment with Applications in Engineering and Medicine" supported by the E.E.C and the Greek Ministry of Industry and taking place in the Aristotle Univ. of Thessaloniki and the Democritus Univ. of Thrace, Greece

the i and the j neurons, V_j is the output of the j-neuron $j = 1,\ldots,n$), from all the n nodes with which it is connected and gives as an output $f_i(\sum_{\substack{j=1 \\ j \neq i}}^{n} T_{ji}V_j)$, where $f_i(.)$ is generally a nonlinear function. This output of the i-neuron is trasmitted to the other neurons and this procedure continues until a so-called "stable state" of the network is achieved, which corresponds to a local minimum of a characteristic network energy function. Here we shall "construct" a fictitious neural network apropriate for the treatment of the unconstrained and the inequality constrained Q.P.P. resulting in the B.I.E. approach to equality and to inequality contact problems. Then the proposed neural network and the process of transmission from neuron to neuron is simulated on a digital computer in order to obtain numerical results, because neural computers do not exist until now.

The fictitious analog network has parallel input and output channels and n-neurons with a large interconnectivity between them. These neurons are modelled as amplifiers having a general performance expressed through the functions f_i and the input resistor (resp. capacitor) ρ_i (resp. C_i) $i = 1,\ldots,n$. Then V_j is the output voltage of the amplifier j and u_j is the input voltage. A synapse of the neurons i,j is characterized by a conductance T_{ij} which connects the output of the neuron j with the input of the neuron i. Moreover each neuron receives an externally supplied input current I_i. The aforentioned circuit immitates in a quite optimal way the function of biological neural networks and its evolution with the time t is given by the equations [Hop85]

$$C_i\left(\frac{du_i}{dt}\right) = \sum_{j=1}^{n} T_{ij}V_j - \frac{u_i}{R_i} + I_i \tag{13.25}$$

$$V_j = F_j(u_j)\,. \tag{13.26}$$

Here $R_i^{-1} = \rho_i^{-1} + \sum_{j=1}^{N} R_{ij}^{-1}$ (parallel connection of the resistors and $R_{ij} = \frac{1}{T_{ij}}$. We assume for the sake of simplicity that R_i, C_i, have the same values R, C respectively for every neuron and that the response functions f_i of all the neurons are the same and equal to f. Then for given initial values of the neuron inputs u_i at $t = 0$, the integration of (13.25), (13.26) in a digital computer gives the numerical results for the fictitions network considered. It can be easily shown (cf. [Hop85]) that if $T_{ij} = T_{ji}$, the solution of (13.25), (13.26) converges after some time to solutions having constant the outputs V_i of all neurons. These solutions are called stable states and make stationary the quantity

$$E = -\frac{1}{2}\sum_{i,j=1}^{n} T_{ij}V_iV_j + \sum_{i=1}^{n}\left(\frac{1}{R_i}\right)\int_0^{V_i} f_i^{-1}(V_i)dV_i - \sum_{i=1}^{n} I_iV_i \tag{13.27}$$

which is called the Liapunov function of the system.

Thus the neural network approach to an unconstrained optimization problem consists in the determination of the quantities T_{ij}, I_i, R_i, and the functions f_i in (13.27) such that a local minimum of E, i.e. a stable state, will be identical to the solution of the minimum problem. From the solution of the differential equations (13.25), (13.26) the final stable states may be obtained in a digital computer. Roughly speaking if a neural computer would be available the calculation speed would be comparable with the electric field speed diminished by the polarization time of the electronic components of the computer.

Let us consider now the following two problems arising in contact problems (equality or inequality approach) after the discretization of the B.I.Es and of the corresponding minimum problems.

1. Find $x \in \mathbb{R}^n$ such that

$$\min \left\{ \frac{1}{2} x^T A x - b^T x \right\} \quad \text{or} \quad A x = b \tag{13.28}$$

2. Find $x \in \mathbb{R}^n$ such that

$$\min \left\{ \frac{1}{2} x^T A x - b^T x \,\middle|\, x \geq 0 \right\} \tag{13.29}$$

Here $A = \{a_{ij}\}$ is a given symmetric matrix and $b = \{b_i\}$ is a given vector. Some elementary variable changes lead to the only constraint $x \geq 0$ in (13.29).

In order to formulate the fictitious neural network we make the following substitutions $(i, j = 1, \ldots, n)$, where $A = \{a_{ij}\}$ and $b = \{b_i\}$:

$$T_{ij} = \begin{cases} -a_{ij} & \text{if } i \neq j \\ -a_{ij} + \frac{1}{R_i} & \text{if } i = j \end{cases} \quad V_i = x_i \tag{13.30}$$

$$I_i = b_i, \quad V_i = x_i, \quad V_i = f_i(u_i) = u_i \tag{13.31}$$

We may further assume for the sake of simplicity that $R_i = C_i = 1$. Then the circuit evolution is described by the differential equation

$$\frac{dV_i}{dt} = \sum_{j=1}^{n} T_{ij} V_j - V_i + I_i \tag{13.32}$$

At the stable state we have $\frac{dV_i}{dt} = 0$ and thus (13.32) becomes by means of (13.30) (13.31) the matrix equation $A x = b$.

For the minimum problem (13.29) the function f_i has the different form

$$V_i = f_i(u_i) = \begin{cases} u_i & \text{if } u_i > 0 \\ 0 & \text{if } u_i \leq 0 \end{cases}, \quad x_i = V_i \tag{13.33}$$

Thus the solution of the network fullfils automatically the constraints $x_i \geq 0$, $i = 1, \ldots, n$ of the problem (13.29). Accordingly a solution of the differential equation (13.32) with given initial conditions tends to a stable state of neural net, which is a solution of the inequality constrained problem (13.29). Since the numerical result in a real neurocomputer would be obtained from the corresponding circuit evolution problem the memory needed for the calculation of the stable state is independent of the performance function f_i of the neurons. In [The93] numerical applications are given concerning the closing and opening of cracks, i.e. the numerical solution of the multivalued B.I.E. (7.84), in a neural environment.

Closing this Section we would like to note that, besides the memory advantage, the neural network approach is an iterative approach converging to the solution and therefore round-off errors do not affect the solution. Moreover the method works without the storage of large matrices as it is the case in Q.P. algorithms on sequential computers.

13.5 A Supervised Learning Approach to the Parameter Identification in Contact Problems.

As shown in the previous Chapters the unilateral contact problem, or the contact problem with given friction, or with any type of monotone contact law, lead to a multivalued B.I.E. on the contact region, which is equivalent to a variational inequality and to a minimum problem. Therefore the corresponding parameter identification problem which is formulated as a minimum deviation problem leads to an optimal control problem governed by a variational inequality. For the mathematical theory of this type of nonclassical control problem we refer to [Yv, Pan77, Mig, Pan83c, Bar]. In the case of nonmonotone contact laws we are led to optimal control problems governed by hemivariational inequalities. For this type of problems certain existence results are given in [Pan89, Has89]. Further we examine the case of unilateral contact and of contact with given friction, where the B.I.E. is equivalent to a Q.P.P. In this respect we shall present a new formulation of the parameter identification problem as a supervised learning problem of a neural network. Everything in the sequel holds for the parameter identification problems for all systems which are governed by a variational inequality, or by an equivalent Q.P.P., or by an equivalent linear complementary problem (L.C.P) (cf. also [Pan85] Ch.10 and 11). In the previous Section we have not introduced any "learning rule" in the network, i.e. any rule for the updating of the weights of the neuronal links for a new calculation step according to the output of the previous step(s). This is not necessary if one simply wants to calculate a structure but it becomes necessary for the treatment of the parameter identification problem.

The parameter identification problem is the inverse problem, i.e. the problem in which a solution is prescribed and we ask for those elastic properties and/or loading and/or geometric quantities which will give a solution very close to a prescribed one. Both in the case of equality and inequality contact problems the inverse problem is not easy to solve numerically.

In a neural network environment the parameter identification problem may be treated very easily by formulating it as a "supervised learning" problem [Bea, Kha]. The learning is called "supervised" because it is guided by taking into account what we want to achieve. Any change in the elastic properties and/or the geometry of the structure gives rise to corresponding changes of the matrix A and b in problems (13.28) and (13.29). But matrix A is related to the synapses' weights T_{ij}. Accordingly we introduce a learning rule of the form

$$T_{ij}^{(k+1)} = T_{ij}^{(k)} + c_i r_{ij}^{(k)} \tag{13.34}$$

where $T_{ij}^{(k)}$ denotes the weight of the synapse ij at the k-step of the identification procedure ($T_{ij}^{(0)} = T_{ij}$, given by (13.30)) $r_{ij}^{(k)}$ is the reinforcement signal [Kha] and c

is the learning rate, which is a positive constant. Moreover the solution at the k-step is considered as an initial value for the calculation of the network at the $(k+1)$-step via the differential equation (13.32). This procedure of modifying the synapses' weights is continued until the wished result $\{x_i^\star\}$ in (13.28) or (13.29), or $\{V_i^\star\}$ in (13.32) is obtained. There are several types of learning rules [Bea]. We apply here the "perceptron learning algorithm", where for the adaption of the weights a learning rule analogous to the Widrow-Hoff learning rule [Bea, Kha] has been used.

The learning rule reads [Kha]

$$r_{ij}^{(k)} = \left(V_i^\star - V_j^{(k)}\right)u_i^{(k)} \tag{13.35}$$

and $0 \le c \le 1$. Here $V_j^{(k)}$ is the output of the j-neuron and $u_i^{(k)}$ the input of the i-neuron, whereas $V_j^\star$ is the wished response of the system. For more accurate calculation of c we refer to Beale and Jackson [Bea], as well as for the convergence proof of the perceptron algorithm. Note that this proof is very simple, compared with the existence proofs for the optimal control problem governed by a Q.P.P. [Pan83]. The idea of this Section can be extended to all other types of parameter identification and optimal control problems. Then a perceptor type proof replaces any other direct existence proof for the solution. For numerical results we refer to [The93].

References

[Ahm87a] Ahmad, S. and Manolis, G.D.: Dynamic analysis of 3-D Structures by a Transformed Boundary Element Method, Computational Mechanics 2 (1987) 185-196.

[An] Andersson, L.E.: A Global Existence Result for Quasistatic Contact Problem with Friction. LiTH-MAT-R- 89-00, Linköping Institute of Technology, Linköping, Sweden 1989.

[And80] Andersson, T., Fredriksson, B. and Allan-Persson, B.G.: The Boundary Element Method Applied to Two-Dimensional Contact Problems. In "New Developments in Boundary Element Methods", (ed. by Brebbia, C.A.) CML Publ, Southampton 1980.

[And81] Andersson, T.: The Boundary Element Method Applied to Two-Dimensional Contact Problems with Friction. In "Proc. Third Intern. Seminar on Boundary Element Methods", (ed.by Brebia, C.A.) Springer-Verlag, Berlin 1981.

[And83] Andersson, T. and Allan-Persson, B.G.: The Boundary Element Method Applied to Two-Dimensional Contact Problems. In "Progress in Boundary Element Methods 2", (ed. by Brebia, C.A.) Pentech Press, London 1983.

[An72] Antes, H.: Splinefunktionen bei der Lösung von Integralgleichungen, Numer. Math. 19 (1972) 116-126.

[An73] Antes, H.: Über die Integralgleichungen von Massonnet und Rieder, ZAMM 53 (1973) T64-T66.

[An84a] Antes, H.: On a Regular Boundary Integral Equation and a Modified Trefftz Method in Reissner's Plate Theory, Engng. Analysis 1 (1984) 149-153.

[An84b] Antes, H.: The Stress Functions of Point Loadings in Reissner's Plate Theory, Mech. Res. Comm. 11 (1984) 115-120.

[An85a] Antes, H.: Basic Geometrical Singularities in Reissner's Plate Theory, Mech. Res. Comm. 12 (1985) 295-302.

[An85b] Antes, H.: A Boundary Element Procedure for Transient Wave Propagations in Two-Dimensional Isotropic Elastic Media, Finite Elements in Analysis and Design 1 (1985) 313-322.

[An86a]   Antes, H.: Dual Complementary Variational Principles in Reissner's Plate Theory, Acta Mechanica 65 (1986) 13-25.

[An86b]   Antes, H.: An Indirectly Derived Integral Equation System for Simply Supported Reissner Plates, Mech. Res. Comm. 13 (1986) 63-69.

[An86c]   Antes, H. and von Estorff, O.: Dynamic Soil-Structure Interaction by BEM in the Time and Frequency Domain. In "Proc. 8th Europ. Conf. on Earthq. Engng. 2" 5.5/30-40 Lab. Nac. Eng. Civil, Lisbon, 1986.

[An87a]   Antes, H.: Influence Functions of Statical and Geometrical Singularities in Reissner Plates (in German), ZAMM 67 (1987) T174-T176.

[An87b]   Antes, H. and von Estorff, O.: On Causality in Dynamic Response Analysis by Time-dependent B.E.M., Earthqu. Eng. Struct. Dyn. 15 (1987) 865-870.

[An87c]   Antes, H. and von Estorff, O.: The Effect of Non-convex Boundaries on Time-Domain Boundary Element Solutions. In "Boundary Elements IX, Vol. 1. Mathematical and Computational Aspects" (C.A. Brebbia, W.L. Wendland, G. Kuhn, eds.) Springer Verlag, Berlin, New York 1987.

[An87d]   Antes, H. and von Estorff, O.: Transient Behaviour of Strip Foundations Resting on Different Soil Profiles by Time Domain B.E.M. In "Developments in Geotechnical Engineering" (Cakmak, A.S., ed.) Ground Motion and Engineering Seismology, Elsevier, Amsterdam 1987.

[An87e]   Antes, H.: Time Domain Boundary Element Solutions of Hyperbolic Equations for 2-D Transient Wave Propagation. "In Panel Methods in Fluid Mechanics with Emphasis on Aerodynamics" (J. Ballmann, R. Eppler, W. Hackbusch, eds.) Vieweg, Braunschweig 1987.

[An88a]   Antes, H. and von Estorff, O.: Seismic Response Amplification due to Topographic Influences, Proc. 9th World Conf. Earthq. Engng., Vol. III, Tokyo / Kyoto 1988.

[An88b]   Antes, H.: A Multiple Grid Boundary Elememt Procedure for the 2-D Helmholtz Equation, In "The Mathematics of Finite Elements and Application IV-Mafelap 1987" (J. R. Whiteman, ed.), Academic Press, London 1988.

[An88c]   Antes, H.: Applications of the Boundary Element Method in Elasto-dynamics and Fluid Dynamics (in German), Math. Methoden in der Technik 9, Teubner, Stuttgart 1988.

[An88d]   Antes, H. and Steinfeld, B.: Unilateral Contact in Dynamic Soil- Structure Interaction by a Time Domain Boundary Element Method. In "Boundary Elements X, Vol 4: Geomechanics, Wave Propagation and Vibrations", (ed. by Brebbia, C.A.), Springer Verlag, Berlin 1988.

[An89a]   Antes, H. and Meise, Th.: Scalar Wave Propagation - Calculation Capabilities of a 3-D Time Domain Boundary Element Method, In "Advances in Boundary Elements", (ed. by C.A. Brebbia and J.J. Connor), Vol. 3, Stress Analysis, Proc. 11th B.E.M. Conf., Springer Verlag, Berlin, New York, 1989.

[An89b]   Antes, H. and von Estorff, O.: Ausbreitung Transienter Akustischer Wellen. Untersuchungen mit einer Zeitschritt-Randelementmethode, Ing. Arch. 59 (1989) 17-31.

[An89c]   Antes, H. and von Estorff, O.: Dynamic Response Analysis of Rigid Foundations and of Elastic Structures by Boundary Element Procedures, Soil Dyn. Earthq. Engng. 8 (1989) 68-74.

[An89d]   Antes, H. and Steinfeld, B.: Seismic Wave Response of Elastic Structures in Unilateral Soil Contact. Proc. 2nd National Cong. on Mechanics, Athens, Greece, 1989.

[An90a]   Antes, H. and Meise, T.: 3-D Sound Generated by Moving Sources, Proc. IABEM-90 (T. Cruse, L. Morino, eds.) Rome, Springer Verlag 1991.

[An90b]   Antes, H. and Meise, Th.: Coping with Non-Convex Domains in the Boundary Element Analysis, Comp. Mech. 6 (1990) 47-53.

[An90c]   Antes, H.: Eine vollständige Randintegralformulierung für den dynamischen Kontakt schwerer elastischer Strukturen, ZAMM 70 (1990) T708-T709.

[An90d]   Antes, H. and Steinfeld, B.: The Influence of Friction on Unilateral Dynamic Soil-Structure Contact. In "Comp. Engng. with Boundary Elements 2: Solid and Computational Problems." (ed. by Brebbia, C.A., Cheng, A.) CML Publ., Southampton 1990.

[An90e]   Antes, H. and Steinfeld, B.: Dynamic Response of Elastic Structures in Unilateral Soil Contact with Friction. Proc. EURODYN. (ed. by Krätzig, W.) Rotterdam: Balkema 1990.

[An91a]   Antes, H.: Applications in Environmental Noise. In "Advances in B.E.M. in Acoustics" (R. Ciskowski, C.A. Brebbia, eds.), Elsevier (Appl. Science), London 1991.

[An91b]   Antes, H. and von Estorff, O.: On FEM-BEM Coupling for Fluid- Structure Interaction Analyses in the Time Domain, Int. J. Numer. Meth. Engng. 31 (1991) 1151-1168.

[And]   Anderson, J. and Rosenfeld, E.: Neurocomputing. Foundations of Research, The MIT Press, Cambridge Mass, 1988.

[Art]   Artemiadis, N.: The Geometry of Fractals, Proc. Nat. Academy of Athens 63 (1988) 479-500.

[Atl78]   Atluri, S.N. and Grannell, J.J.: Boundary Element Methods (BEM) and Combination of BEM-FEM, Rep. No. GIT-ESM-SA-78-16, Center for the Advancement of Computational Mechanics, Georgia Inst. of Technology, Atlanta 1978.

[Au79]   Aubin, J.P. and Clarke, F.H.: Shadow Prices and Duality for a Class of Optimal Control Problems, SIAM J. Control Optimization 17 (1979) 567-586.

[Au89]   Aubin, J.P. and Ekeland, I.: Applied Nonlinear Analysis, J. Wiley, N.York 1984.

[Au90]   Aubin, J.P. and Frankowska, H.: Set-valued Analysis, Birkhäuser Verlag, Basel 1990.

[Au91]   Aubin, J.P.: Viability Theory, Birkhäuser Verlag, Basel 1991.

[Ba84]   Baiocchi, C. and Capelo, A.: Variational and Quasivariational Inequalities: Applications to Free Boundary Problems, J.Wiley and Sons, Chichester, 1984.

[Ba86]   Baiocchi, C., Gastaldi, F. and Tomarelli, F.: Some Existence Results on Noncoercive Variational Inequalities, Ann. Scuola Norm. Sup. Pisa cl. Sci., IV. 13 (1986) 617-659.

[Ba88]   Baiocchi, C., Buttazzo, G., Gastaldi, F. and Tomarelli, F.: General Existence Results for Unilateral Problems in Continuum Mechanics, Arch. Rational Mech. Anal. 100 (1988) 149-189.

[Bal89]   Balas, J., Sladek, J. and Sladek, V.: Stress Analysis by Boundary Element Methods, Studies in Appl. Mech. 23, Elsevier, Amsterdam 1989

[Ball]   Ballmann, J. and Kim, K.: Numerische Simulation Mechanischer Wellen in Geschichteten Elastischen Körpern, ZAMM 70 (1990) T204-T206.

[Bah63a]   Banaugh, R.P. and Goldsmith, W.: Diffraction of Steady Acoustic Waves by Surfaces of Arbitrary Shape, J. Acoust. Soc. Amer. 35 (1963) 1590-1601.

[Bah63b]   Banaugh, R.P. and Goldsmith, W.: Diffraction of Steady Elastic Waves by Surfaces of Arbitrary Shape, J. Applied Mech. 30 (1963) 589-597.

[Ban76]   Banerjee, P.K.: Integral Equation Methods for Analysis of Piecewise Non-homogeneous Three-Dimensional Elastic Solids of Arbitrary Shape, Int. J. of Mechanical Sciences 18 (1976) 293-303.

[Ban77]   Banerjee, P.K. and Butterfield, R.: Boundary Element Methods in Geomechanics. In "Finite Elements in Geomechanics", (ed. by Gudehus, G.) J. Wiley and Sons, London 1977.

[Ban78]   Banerjee, P.K and Mustoe, G.C.W.: The Boundary Element Method for Two-Dimensional Problems of Elastoplasticity. In "Recent Advances in Boundary Element Methods", (ed. by Brebbia, C.A.) Pentech Press, London 1978.

[Ban79]   Banerjee, P.K. and Butterfield, R. (eds.): Developments in Boundary Element Methods, Vol. 1, Elsevier (Applied Science), London 1979.

[Ban81]   Banerjee, P.K. and Butterfield, R.: Boundary Element Methods in Engineering Science, McGraw-Hill, London 1981.

[Ban82]   Banerjee, P.K. and Shaw, R.P. (Eds.): Developments in Boundary Element Methods, Vol. 2, Elsevier (Applied Science), London 1982.

[Ban84]   Banerjee, P.K. and Mukherjee, S. (eds.): Developments in Boundary Element Methods, Vol. 3., Elsevier (Applied Science), London 1984.

[Ban86]   Banerjee, P.K. and Watson, J.O. (eds.): Developments in Boundary Element Methods, Vol. 4, Elsevier (Applied Science), London 1986.

[Ban87]   Banerjee, P.K., Ahmad, S. and Manolis, G.D.: Advanced Elasto-Dynamic Analysis. In "Boundary Element Methods in Mechanics", (ed. by Beskos, D.E.) North-Holland, Amsterdam 1987.

[Bani85]   Baniotopoulos, C.C.: Analysis of Structures for "Complete" Consititutive Laws, Doctoral Dissertation. Scientific Annual of the Faculty of Technology of the Aristotle University, Nr. 27 of the Θ' Issue, Thessaloniki 1985.

[Bani87]   Baniotopoulos, C.C. and Panagiotopoulos, P.D.: A Hemivariational Approach to the Analysis of Composite Material Structures. In "Engineering Applications of New Composites" (ed. by S.A. Paipetis and G.C. Papanicolaou) Omega Publ., London 1987.

[Bani89a]   Baniotopoulos, C.C. and Panagiotopoulos, P.D.: A Contribution to the Analysis of Composite Material Structures, ZAMM 69, 1984, T 489-491.

[Bani89b]   Baniotopoulos, C.C.: Free Boundary Value Problems Arising in Composite Masonry Structures. In "Mathematical Models for Phase Change Problems". (ed. by F. Rodrigues) International Series of Numerical Mathematics, vol. 88, Birkhäuser Verlag, Basel 1989.

[Bar83]   Barbu, V.: Optimal Control of Variational Inequalities, Pitman, London 1983.

[Bar88]   Barnsley, M.: Fractals Everywhere, Academic Press, Boston, N. York, 1988.

[Bar89a]   Bardzokas, D, Parton, V.Z. and Theocaris, P.S.: The Plane Problem of the Theory of Elasticity for an Orthotropic Field with a Defect, Dokl. Akad. Nauk USSR 309 (1989) 1072-1077.

[Bar89b]   Bardzokas, D, Parton, V.Z. and Theocaris, P.S.: The General Case of the Plane Problem of the Theory of Elasticity for Multiply-Connected Fields, PMM 53 (1989) 485-496.

[Bat]   Bathe, K.J. and Mijailovich, S.: Finite Element Analysis of Frictional Contact Problems. Journal de Mécanique Théor. et Appl. 7 (special issue) (1988) 31-46.

[Bea]   Beale, R. and Jackson, T.: Neural Computing. An Introduction, Adam Hilger, Bristol 1990.

[Ben72]   Benjumea, R. and Sikarskie, D.L.:On the Solutions of Plane, Orthotropic Elasticity Problems by an Intergral Equation Method, J. of Appl. Mech. 39 (1972) 801-808.

[Bes84a]   Beskos, D.E., Krauthammer, T. and Vardoulakis, I. (eds): Dynamic Soil-Structure Interaction Proc. of the Int. Symp. on Dynamic Soil-Structure Interaction. A.A. Balkema, Rotterdam 1984.

[Bes86]   Beskos, D.E., Dasgupta, B. and Vardoulakis, I.G.: Vibration Isolation Using Open or Filled Trenches. Part I: 2-D Homogeneous Soil, Comp. Mech. 1, (1986) 43-63.

[Bes87a]   Beskos, D.E. (ed.): Boundary Element Methods in Mechanics, North- Holland, Amsterdam 1987.

[Bes87b]   Beskos, D.E.: Boundary Element Methods in Dynamic Analysis, Appl. Mech. Rev. 40 (1987) 1-23.

[Bes88]   Beskos, D.E. (ed.): Boundary Element Methods in Structural Analysis, American Society of Civil Engineers, New York 1988.

[Bes91]   Beskos, D.E. (ed.): Boundary Element Analysis of Plates and Shells, Springer Verlag, Berlin 1991.

[Bes92]   Beskos, D.E.: Wave Propagation Through Ground, In: Boundary Element Techniques in Geomechanics, (G.D.Manolis and T.G.Davies, eds.) Computational Mechanics Publications, Southampton 1992.

[Bet]   Betti, E.: Teoria dell' Elasticita, Il Nuovo Ciemento 7 (1872) 10-18.

[Bez78]   Bezine, G.: Boundary Integral Formulation for Plate Flexure with Arbitrary Boundary Conditions, Mech. Res. Comm. 5 (1978) 197-206.

[Bez84]   Bezine, G. and Fortune, D.: Contact Between Plates by a New Direct Boundary Integral Equation Formulation, Int. J. Solids Structures 20 (1984) 739-746.

[Bis86]   Bisbos, C.: A Cholesky Condensation Method for Unilateral Contact Problems, S.M.Archives 11 (1986) 1-23.

[Bis90]   Bisbos, C.: A New Algorithm for 3-D Unilateral Frictional Contact Problems. Report for INPRO (Innovationsgesellschaft für die Produktion in der Automobilindustrie) Berlin, 1990.

[Boi]   Boieri, P., Gastaldi, F. and Kinderlehrer, D.: Existence, Uniqueness and Regularity Results for the Two Bodies Contact Problem, Appl. Math. Optim., 15 (1987) 251-277.

[Böh]   Böhm, J.M.: Eine Inkrementelle Formulierung für Festkörperkontakt mit Reibung, Doct. Dissertation, RWTH Aachen 1987.

[Bra]   Brandt, A.: Multilevel Adaptive Techniques. Rep. RC 6026, IBM, J.J. Watson Res. Center, Yorktown Heigth, 1976.

[Bre77a]   Brebbia, C.A.: Approximate Methods In Mathematical Modelling. in: "Proceedings of the 1st International Conference on Mathematical Modelling, Dept. of Engineering Mechanics, Univ. of Missouri-Rolla" (ed. by Avula, X.J.R.) 1977.

[Bre77b]   Brebbia, C.A. and Dominguez, J.: Boundary Element Methods for Potential Problems, Appl. Math. Model. 1 (1977) 372-378.

[Bre78a]   Brebbia, C.A. and Butterfield, R.: Formal Equivalence of Direct and Indirect Boundary Element Methods, Appl. Math. Model. 2 (1978) 132-134.

[Bre78b]   Brebbia, C.A.: The Boundary Element Method for Engineers, Pentech Press, London 1978.

[Bre80a]   Brebbia, C.A. and Walker, S.: Boundary Element Techniques in Engineering, Newnes-Butterworths, London 1980.

[Bre80b]   Brebbia, C.A. (ed.): New Developments in Boundary Element Methods, CML Publication, Southampton 1980.

[Bre81a]   Brebbia, C.A. (ed.): Boundary Element Methods, Springer-Verlag, Berlin 1981.

[Bre81b]   Brebbia, C.A. (ed.): Progress in Boundary Element Methods, Vol. 1, Pentech Press, London 1981.

[Bre82]   Brebbia, C.A. (ed.): Boundary element Methods in Engineering, Springer-Verlag, Berlin 1982.

[Bre83a]   Brebbia, C.A. (ed.): Progress in Boundary Element Methods, Vol. 2, Pentech Press, London 1983.

[Bre83b]   Brebbia, C.A., Futagami, T. and Tanaka, M. (eds.): Boundary Elements, Springer Verlag, Berlin 1983.

[Bre84a]   Brebbia, C.A., Telles, I.C.F und Wrobel, L.C.: Boundary Element Techniques - Theory an Applications in Engineering, Springer Verlag, Berlin, Heidelberg, New York 1984.

[Bre84b]   Brebbia, C.A. (ed.): Boundary Elements VI, Springer Verlag, Berlin 1984.

[Bre84c]   Brebbia, C.A. (ed.): Topics in Boundary Element Research, Vol. 1, Springer Verlag, Berlin 1984.

[Bre85]   Brebbia, C.A. (ed.): Topics in Boundary Element Research, Vol. 2, Springer Verlag, Berlin 1985.

[Bre87]   Brebbia, C.A. and Long, S.Y.: Boundary Element Analysis of Plates Using Reissner's theory. In: "Boundary Elements IX Vol. 2: Stress Analysis Applications" (eds.: Brebbia, C.A., Wendland, W.L., Kuhn, G.) Springer Verlag, Berlin 1987.

[Bre88]   Brebbia, C.A. (ed.): Boundary Elements X, Springer Verlag, Berlin 1988.

[Bre90]   Brebbia, C.A., Tanaka, M. and Honma, T. (eds.): Boundary Elements XII, Springer-Verlag, Berlin 1990.

[Bre91a]   Brebbia, C.A. and Gipson, G.S.(eds.): Boundary Elements XIII, Springer-Verlag, Berlin 1991.

[Bre91b]   Brebbia, C.A. (ed.): Boundary Element Technology VI, Elsevier (Appl. Science), London 1991.

[Brez73]   Brézis, H.: Problèmes Unilatéraux. J.Math. Pures et Appl. 51 (1972) 1-168.

[But70]   Butterfield, R. and Banerjee, P.K.: The Problem of Pile Reinforced Half-Space, Geotechnique 20 (1970) 100-103.

[But71]   Butterfield, R. and Banerjee, P.K.: The Elastic Analysis of Compressible Piles and Pile Groups, Geotechnique 21 (1971) 43-60.

[But]   Butler, G.F.: A Note on Improving the Attenuation Given by a Noise Barrier, J. Sound Vibration 32 (1974) 367-369

[Cam]   Campos, L.T., Oden, J.T. and Kikuchi, N.: A Numerical Analysis of a Class of Contact Problems With Friction in Elastostatics, Comp. Meth. appl. Mech. Eng. 34 (1982), 821-845.

[Cer]   Cerrolaza, M. and Alarcon, E.: A Bi-Cubic Transformation for the Numerical Evaluation of the Cauchy Principal Value Integrals in Boundary Methods. Int. J. Numer. Meth. in Engng. 28 (1989) 987-999.

[Ch]   Chang, K.C.: Variational Methods for Non-Differentiable Functionals and their Applications to Partial Differential Equations. J.Math.Anal. Appl. 80 (1981) 102-129.

[Cho68]   Chopra, A.K.: Earthquake Behavior of Reservoir-Dam Systems, Eng. Mech. Div. ASCE 94 (1968) 1475-1499

[Cho73]   Chopra, A.K. and Chakrabarti, P.: Earthquke Analysis of Gravity Dams Including Hydrodynamic Interaction, Earthqu. Eng. Struct. Dyn. 2 (1973) 143-160.

[Cho83]   Chopra, A.K. and Fenves, G.: Effects of Reservoir Bottom Absorption on Earthquake Response of Concrete Gravity Dams, Earthqu. Eng. Struct. Dyn. 11 (1983) 809-829.

[Chr75a]   Christiansen, S.: Integral Equations without a Unique Solution can be Made Useful for Solving some Plane Harmonic Problems, J. Inst. Maths Applic. 16 (1975) 143-159.

[Chr75b]   Christiansen, S. and Hansen, F.: A Direct Integral Equation Method for Computing the Hoop Stress at Holes in Plane Isotropic Sheets, Journal of Elasticity 5 (1975), 1-14.

[Cis91]   Ciskowski, R.D. and Brebbia, C.A. (Eds.): Boundary Element Methods in Acoustics, Elsevier (Appl. Science), London 1991.

[Clar73]   Clarke, F.H.: Necessary Conditions for Nonsmooth Problems in Optimal Contorl and the Calculus of Variations, Ph.D.Thesis, University of Washington, Seattle 1973.

[Clar75]   Clarke, F.H.: Generalized Gradients and Applications, Trans. A.M.S. 205 (1975) 247-262.

[Clar81]   Clarke, F.H.: Generalized Gradients of Lipschitz Functionals, Advances in Math. 40 (1981) 52-67.

[Clar83]   Clarke, F.H.: Optimization and Nonsmooth Analysis, Wiley, New York 1983.

[Co]   Cocu, M.: Existence of Solutions of Signorini Problems with Friction, Int. J. Engng. Sci. 22 (1984) 567-575

[Col78]   Cole, D.M., Kosloff, D.D. and Minster, J.B.: A Numerical Boundary Integral Equation Method for Elastodynamics, I. Bull. Seismol. Soc. Am. 68 (1978) 1331-1357.

[Col]   Collatz, L.: Funktionalanalysis und Numerische Mathematik, Springer Verlag, Berlin, Heidelberg 1968

[Com]   Comninou, J.: Interface Crack with Friction in the Contact Zone, J. Appl. Mech. 44 (1977) 780-781.

[Comi]   Comi, C. and Maier, G.: Extremum, Convergence and Stability Properties of the Finite-Increment Problem in Elastic-Plastic Boundary Elament Analysis, Int. J. Solids and Struct. 29 (1992) 249-270.

[Cri82]   Crisfield, M.A.: Accelarated Solution Techniques and Concrete Cracking. Comp. Meth. in Appl. Mech. and Eng. 33 585-607 (1982)

[Cri86]   Crisfield, M.A.: Snap-Through and Snap-Back Response in Concrete Structures and the Dangers of Under-Integration, Intern. J. for Num. Meth. in Eng. 22, 751-767 (1986).

[Cri88]   Crisfield, M.A. and Wills, J.: Solution Strategies and Softening Materials, Comp. Meth. in Appl. Mech. and Eng. 66, 267-289(1988).

[Cri91]   Crisfield, M.A.: Nonlinear Finite Element Analysis of Solids and Structures, Vol.1, J.Wiley and Sons, N.York, 1991.

[Cro73]   Crouch, S.L.: Two-Dimensional Analysis of Near-Surface Single Seam Extraction, Intern. Journal of Rock Mech. and Mining Sc., Geomechanics Abstracts 10 (1973) 85-96.

[Cro76]   Crouch, S.L.: Solution of Plane Elasticity Problems by the Displacement Discontinuity Method, I and II, Internat, J. for Num. Meth. in Eng. 10 (1976) 301-343.

[Cro83]   Crouch, S.L. and Starfield, A.M.: Boundary Element Methods in Solid Mechanics, George Allen and Unwin, London 1983.

[Cru68a]   Cruse, T.A. and Rizzo, F.J.: A Direct Formulation and Numerical Solution of the General Transient Elastodynamic Problem, I., J. Math. Anal. Appl. 22 (1968) 244-259.

[Cru68b]   Cruse, T.A.: A Direct Formulation and Numerical Solution of the General Transient Elastodynamic Problem, II, Math. Anal. Appl. 22 (1968) 341-355.

[Cru69]   Cruse, T.A.: Numerical Solutions in Three-Dimensional Elastostatics, Int. J. Solids and Struct. 5 (1969) 1259-1274.

[Cru71]   Cruse, T.A. and Van Buren, W.: Three-Dimensional Elastic Stress Analysis of a Fractured Speciment with an Edge Crack, Intern. J. Fract. Mech. 7 (1971) 1-15.

[Cru73]   Cruse, T.A.: Application of the Boundary Integral Equation Method to Three-Dimensional Stress Analysis, Comp. and Struct. 3 (1973) 509-527.

[Cru75]   Cruse, T.A. and Rizzo, F.J. (eds.): Boundary Integral Equation Method: Computational Applications in Applied Mechanics, AMD-11, American Society of Mechanical Engineers, New York 1975.

[Cur86]   Curnier, A.: A Theory of Friction, Intern. J. Solids Struct., 20 (1986) 637-647.

[Cur88]   Curnier, A. and Alart, P.: A Generalized Newton Method for Contact Problems with Friction, J. Méc. Théor. Appl. 7 (Special issue) (1988) 67-82.

[Dav]   David, Guy: Wavelets and Singular Integrals on Curves and Surfaces, Springer Verlag (Lect. Notes in Math. Nr.1465), Berlin 1991.

[DeD]   De Donato, O., and Maier, G.: Mathematical Programming Methods for the Inelastic Analysis of Reinforced Concrete Frames Allowing for Limited Rotation Capacity, Intern. J. for Num. Meth. in Eng. 4 (1972) 307-329.

[Del74]   Delves, L.M. and Walsh, I.: Numerical Solution of Integral Equations, Oxford University Press, London, England, 1974.

[De]   Demkowicz, L. and Oden, J.T.: On some Existence and Uniqueness Results in Contact Problems with Nonlocal Friction. Nonl. Anal. 6 (1982) 1075-1093.

[Del85]   Delves, L.M. and Mohamed, J.L.: Computational Methods for Integral Equations, Cambridge Univ. Press, Cambridge 1985.

[Dem76]   De May, G.: Calculation of Eigenvalues of Helmholtz Equation by an Integral Equation, Int. J. Num. Meth. Engng. 10 (1976) 59-66.

[Dob]   Doblare, M.: Computational Aspects of the Boundary Element Method, Chapt. 4 in Topics in Boundary Element Research, Vol. 3: Computational Aspects (ed.: C.A. Brebbia) Springer Verlag, Heidelberg 1987.

[Dom78a]   Dominguez, J.: Dynamic Stiffness of Rectangular Foundations. Dept. of Civil Engineering, Report No. R78-20. Massachusetts Institute of Technology, Cambridge 1978.

[Dom78b]   Dominguez, J.: Response of Embedded Foundations to Travelling Waves. Dept. of Civil Engineering, Report No. R78-24. Massachusetts Institute of Technology, Cambridge 1978.

[Dom81]   Dominguez, J.,Alarcon, E.: Elastodynamics. In "Progress in Boundary Element Methods, Vol. 1" (ed by Brebbia, C.A.) Halsted Press, New York 1981.

[Dom85]   Dominguez, J.: Applications of Boundary Element Methods in Elastodynamics. In "BETECH 85", (ed. by Brebbia, C.A. and Noye, B.J.) Springer
Verlag, Berlin 1985.

[Dom87]   Dominguez, J. and Abascal, R: Dynamics of Foundations. In "Topics in
Boundary Element Research, Vol. 4" (ed. by Brebbia, C.A.), Springer
Verlag, Berlin 1987.

[Dom]   Dominguez, J. and Abascal, R.: Non-linear Effects due to the Contact Conditions on the Dynamic Response of Embedded Foundations. In "Comp.
Engng. with Boundary Elements 2: Solid and Computational Problems".
(ed. by Brebbia, C.A.; Cheng, A.) Southampton: CML, Publ. 1990.

[Dra88]   Dravinski, M.: Elastodynamics. In "Boundary Element Methods in Structural Analysis", (ed. by Beskos, D.E.) American Society of Civil Engineers,
New York 1988.

[Du]   Du, Q., Yao, Z. Song, G.: Solutions of Some Plate Bending Problems Using
Boundary Element Method, Appl. Math. Modelling 8 (1984) 15-21.

[Dub88]   Dubourg, M.C., Mouwakeh, M. and Villechaise, B.: Interaction Fissure-
Contact. Etude Theorique et Expérimentale. J. de Mécanique Théor. et
Appl. 7 (1988) 623-643.

[Dub89a]   Dubourg, M.C. and Villechaise, B.: Unilateral Contact Analysis of a Crack
with Friction. Eur. J. Mech. A/Solids 8 (1989) 309-319.

[Dub89b]   Dubourg, M.C.: Le Conatact Unilateral avec Frottement le Long de Fis-
sures de Fatigue dans les Liaisons Mecaniques. Doct. Dissertation, Institut
Nat.Sc.Appl. Lyon 1989, Nr.891 SAL 0088.

[Dug76]   Dougill, J.W.: On Stable Progressively Fracturing Solids, J. of Appl. Math.
and Phys. 27 (1976) 423-437.

[Dug80]   Dougill, J.W. and Rida, M.A.M.: Further Consideration of Progressively
Fracturing Solids, ASCE Eng. Mechanics, 106 (1980) 1021-1038.

[Dun]   Dundurs, J. and Comninou, M.: Some Consequences of the Inequality
Conditions in Contact and Crack Problems, J. of Elast. 9 (1979) 131-137.

[Dup]   Dupuis, G. and Probst, A.: Etude d'une Structure élastique Soumise des
Conditions Unilatérales, J.de Mécanique 6 (1967) 1-41.

[Duv71]   Duvaut, G. and Lions, J.L.: Un Probléme d' élasticité avec Frottement, J.
de Mécanique 10 (1971) 409-420

[Duv72]   Duvaut, G. and Lions, J.L.: Les Inéquations en Mécanique et en Physique.
Dunod, Paris 1972.

[Duv80]   Duvaut. G.: Equilibre d' un Solide élastique avec Contact Unilateral et Frottement de Coulomb, C.R. Acad. Sc. Paris 290 (1980) 263-265.

[Eke]   Ekeland, I. and Temam, R.: Convex Analysis and Variational Problems. North Holland, Amsterdam and American Elsevier, New York 1976.

[Erd72]   Erdogan, F. and Gupta, G.D.: On the Numerical Solution of Singular Integral Equation, Quart. Appl. Math. 30 (1972) 525-534.

[Erd73]   Erdogan, F., Gupta, G.D. and Cook, T.S.: Numerical Solution of Integral Equations, In "Mechanics of Fracture, Vol. 1, Methods of Analysis and Solutions of Crack Problems", (ed. by Sih, G.C.) Noordhoff Publishing Co., Leyden 1973.

[Eri]   Eringen, A.C. and Suhubi, E.S.: Elastodynamics, Vol. II, Academic Press, New York 1975

[Fah91]   Al-Fahed, A.M., Stavroulakis, G.E. and Panagiotopoulos, P.D.: Hard and Soft Fingered Robot Grippers. The Linear Complementarity Approach, ZAMM, 71(1991) 257-265.

[Fah92]   Al-Fahed, A.M. and Panagiotopoulos, P.D.: Multifingered Frictional Robot Grippers: A New Type of Numerical Implementation, Comp. and Struct. 42 (1992) 555-562.

[Fal]   Falconer, K.J.: The Geometry of Fractal Sets, Cambridge Univ. Press, Cambridge, 1985.

[Fed]   Feder, J.: Fractals, Plenum Press, N. York, 1988.

[Fei]   Feijóo, R.A., Barbosa, H.J.C. and Zouain, N.: Numerical Formulations for Contact Problems with Friction. J. de Méc. Théor. Appl. 7 (Special issue) (1988) 129-144.

[Fel]   Felippa, C.A. and Park K.C.: Direct Time Integration Methods in Nonlinear Structural Dynamics. Comp. Meth. Appl. Mech. Eng., 17/18, 277-313, 1979.

[Fich63]   Fichera, G.: The Signorini Elastostatics Problem with Ambiguous Boundary Conditions, Proc. Int. Conf. Application of the Theory of Functions in Continuum Mechanics, Vol. I, Tbilisi 1963.

[Fich64]   Fichera, G.: Problemi Elastostatici con Vincoli Unilaterali: il Problema di Signoini con Ambigue Condizioni al Contorno, Mem. Accad. Naz. Lincei, VIII 7 (1964) 91-140.

[Fich72]   Fichera, G.: Boundary Value Problems in Elasticity with Unilateral Constraints. In "Encyclopedia of Physics" (ed. by S.Flügge) Vol. VI a/2. Springer-Verlag, Berlin 1972.

[Fil]   Filippi, P.J.T.: Integral Equations in Acoustics, in Theoretical Acoustics and Numerical Techniques (ed. by P.J.T. Filippi) CISM Courses and Lectures No. 277, 1-49, Springer, Wien-New York, 1983

[Fle90]   Fletcher, R.: Practical Methods of Optimization, 2nd edition, J. Wiley & Sons, Chichester, N.York, 1990.

[Flo]   Floegl, H. and Mang, H.A.: Tension Stiffening Concept Based on Bond Slip., ASCE (ST 12), 108 (1982) 2681-2701.

[For]   Fortin, M. and Glowinski, R.: Méthodes de Lagrangien Augmenté. Dunod, Paris 1982.

[For69]   Forbes, D.J. and Robinson, A.R.: Numerical Analysis of Elastic Plates and Shallow Shells by an Integral Equation Method, Structural Research Series, Report No. 345, Univ. of Illinois, Urbana, Illinois 1969.

[Fr]   Frémond, M.: Contact with Adhesion. In "Nonsmooth Mechanics and Applications" (ed. by J.J.Moreau and P.D.Panagiotopoulos), CISM Courses and Lect. Nr 302, Springer Verlag, Wien, N.York 1988.

[Fra]   Frantzisconis, G.: Progressive Damage and Constitutive Behaviour of Geomaterials Including Analysis and Implementation, Ph.D. Thesis, University of Arizona 1986.

[Fre03]   Fredholm, I.: Sur une classe d'Equations Fonctionelles, Acta Mathematica, 27 (1903) 365-390.

[Fre06]   Fredholm, J.: Solution d'un Problème Fondamental de la Theorie de l'élasticité. Arkiv for Matematik., Astronomi och Fysik, 28 (1906) 3-8.

[Fre71]   Frémond, M.: Etude de Structures Viscoélastiques Stratifiées soumises des Charges Harmoniques et de Solides élastiques Reposant sur ces Structures. Thése de doctoral d'Etat, Univ. Paris VI, 1971.

[Fri62]   Friedman, M.B. and Shaw, R.: Diffraction of Pulses by Cylindrical Obstacles of Arbitrary Cross Section, J. Appl. Mech. 29 (1962) 40-46.

[Gak66]   Gakhov, F.D.: Boundary Value Problems, Pergamon Press, Oxford 1966.

[Gas88a]   Gastaldi F.: Ramarks on a Noncoercive Contact Problem with Friction in Elastostatics, Publ. No. 649, Istituto Anal. Num. C.N.R, Pavia 1988.

[Gas88b]   Gastaldi, F. and Martins J.A.C.: A Noncoercive Steady-Sliding Problem with Friction, Publ. No. 650, Istituto Anal. Num. C.N.R.. Pavia 1988.

[Gee83]   Geers, T.L.: Boundary Element Methods for Transient Response Analysis. In "Computational Methods for Transient Analysis", (ed. by Belytschko, T. Hughes, T.J.R.) North-Holland, Amsterdam 1983.

[Gil]   Gilbert, R.I. and Warner, R.F.: Tension Stiffening in Reinforced Concrete Slabs, ASCE J. Struct. Div. 104 (1978) 1885-1900.

[Gir]   Girkmann, K.: Flächentragwerke. Springer-Verlag, Wien 1963.

[Gl]   Glahn, H.: Die Anwendung der Riederschen Integralgleichungsmethode auf ebene Flächentragwerke unter Besonderer Berüchsichtigung Mehrfach Zusammenhängender Gebiete Sowie Gemischter Randbedingungen. Habilitationsschrift, TH Darmstadt 1979.

[Gla]   Glahn, H.: Eine Integralgleichung zur Berechnung Gelenkig Gelagerter Platten, Ing. Arch. 44 (1975) 189-198.

[Glo]   Glowinski, R., Lions, J.L. and Trémolidres, R.: Analyse Numérique des Inéquations Variationnelles, Dunod, Paris 1976.

[Go]   Golub, G. and van Loan, C.: Matrix Computations, North Oxford Acad. Publ., Oxford 1983.

[Gol]   Gol'dshtein, R.V. and Spector, A.A.: Variational Method of Investigation of Three-Dimensional Mixed Problems of a Plane Cut in an Elastic Medium in the Presence of Slip and Adhesion of its Surfaces. PMM U.S.S.R. 47, (1984) 232-239.

[Gol]   Gol'dshtein, R.V. and Spector, A.A.: Variational Methods of Solution and Analysis of Spatial Contact and Mixed Problems with Friction. In "Mechanics of Deformable Solids", (ed. by A.Yu.Ishlinski) Allerton Press, N.York 1988.

[Gr]   Green, A.K. and Bowyer, W.H.: The Testing Analysis of Novel Top-Hat Stiffener Fabrication Methods for use in GRP Ships. In Proceedings of the 1st International Conference on Composite Structures (ed. by I.H. Marshall), Applied Science, Barking, Essex 1981.

[Graf]   Graffi, D.: Sul Teorema a di Reciprocita Nella Dinamica dei Corpi Elastici, Mem. Accad. Sci. Bologna 18 (1947) 103-109

[Grü]   Grüters, H.: Berechnung des Spannungszustandes in Homogenen Scheiben aus Anisotrop-Elastischem Material, ZAMM 52 (1972) T71-T128.

[Ha]   Hamel, G.: Theoretische Mechanik, Springer Verlag, Berlin 1949.

[Hac]   Hackbusch, W.: Integralgleichungen - Theorie und Numerik, LAMM 68, Teubner Verlag, Stuttgart 1989.

[Hac85]   Hackbusch, W.: Multigrid Methods and Applications, Springer Verlag, N.York 1985.

[Had]   Hadamard, J.: Lectures on Cauchy's Problem in Linear Partial Differential Equations, Yale Univ. Press, New Haven 1923.

[Han]   Hansen, E.: Numerical Solution of Integro-Differential and Singular Integral Equations for Plate Bending Problems, J. of Elasticity 6 (1976) 39-56.

[Har69]   Harrington, R.F., Kontoppidan, K., Abrahamsen, P. and Albertsen, N.C.: Computation of Laplacian Potentials by an Equivalent- Source Method, Proceedings of the Institution of Electrical Engineers 116 (1969) 1715-1720.

[Ha81a]   Hartmann, F.: Elastostatics. In "Progress in Boundary Element Methods, Vol. 1" (ed. by C.A. Brebbia) Halsted Press, New York 1981.

[Ha81b]   Hartmann, F.: The Somigliana Identity on Piecewise Smooth Surfaces, J. of Elasticity 10 (1981) 403-423.

[Ha82]   Hartmann, F.: Elastic Potentials on Piecewise Smooth Sufaces, J. of Elasticity 12 (1982) 31-50.

[Has82]   Haslinger, J. and Hlavaček, I.: Approximation of the Signorini Problem with Friction by a Mixed Finite Element Method, J.Math.Anal. 86 (1982) 99-122.

[Has84]   Haslinger, J. and Panagiotopoulos, P.D.: The Reciprocal Variational Approach to the Signorini Problem with Friction. Approximation Results. Proc. Royal Soc. of Edinburgh 98A (1984) 365-383.

[Has]   Haslinger, J.: Approximation of Contact Problems. Shape Optimization in Contact Problems. In "Nonsmooth Mechanics and Applications" (ed by J.J.Moreau and P.D.Panagiotopoulos),CISM Vol. 302, Springer Verlag, Wien, N.York 1988.

[Has89]   Haslinger, J. and Panagiotopoulos, P.D.: Optimal Control of Hemivariational Inequalities. In "Control of Boundaries and Stabilization" (ed. by J. Simon), Lect. Notes in Control and Information Sciences, Springer Verlag, N.York 1989.

[Hei69]   Heise, U.: Eine Integralgleichungsmethode zun Lösung des Scheibenproblems mit Gemischten Randbedingungen, Doct. Dissertation, RWTH Aachen 1969.

[Hei75]   Heise, U.: Formulierung und Ordnung Einiger Integralverfahren für Probleme der Ebenen und Raumlichen Elastostatik unter Besonderer Berücksichtigung Mechanischer Gesichtspunkte, Habil. Thesis, Aachen 1975.

[Hei76]   Heise, U.: Non-Integral Terms in Integral Equations in the Plane and Three-Dimensional Theory of Elasticity, Mech. Res. Comm. 3 (1976) 119-124.

[Hei78a]   Heise, U.: Application of the Singularity Method for the Formulation of Plane Elastostatical Boundary Value Problems as Integral Equations, Acta Mechanica 31 (1978) 33-69

[Hei78b]   Heise, U.: Numerical Properties of Integral Equations in which the Given Boundary Values and the Sought Solutions are Defined on Different Curves, Comp. Struct. 8 (1978) 199-205.

[Hei85]   Heise, U.: The Spectra of some Integral Operators for Plane Elasto-Statical Boundary Value Problems, J. of Elasticity 8 (1985) 47-79.

[Hes62]   Hess, J.L.: Calculation of Potential Flow About Bodies of Revolution Having Axes Perpendicular to the Free Stream Direction, J. Aerospace Sc. 29 (1962) 726-742.

[Hes64]   Hess, J.L. and Smith, A.M.O.: Calculation of Nonlifting Potential Flow About Arbitrary Three-Dimensional Bodies, J. of Ship Res. 8 (1964) 22-44.

[Hl]   Hlavaček, J., Haslinger, J., Nečas, J. and Lovisek, J.: Solution of Variational Inequalities in Mechanics, Springer Verlag, N.York, Berlin 1988.

[Hop82]   Hopfield, J.J.: Neural Networks and Physical Systems with Emergent Collective Computational Abilities, Proc. of the Nat. Acad. of Sciences 79(1982) 2554-2558.

[Hop85]   Hopfield, J.J. and Tank, D.W.: "Neural" Computations of Decisions in Optimization Problems, Biol. Cybern. 52(1985) 141-152.

[Höf]   Höfling, O.: Physik II, Teil I: Mechanik, Wärme, Dümmler, Bonn 1976

[Hro84]   Hromadka II, T.V.: The Complex Variable Boundary Element Method, Springer-Verlag, Berlin 1984.

[Hut78]   Hutchinson, J.R.: Determination of Membrane Vibrational Characteristics by the Boundary Integral Equation Method. In "Recent Advances in Boundary Element Methods" (ed. by Brebbia, C.A.) Pentech Press, London 1978.

[Hün]   Hünlich, H. and Naumann, J.: On General Boundary Value Problems and Duality in Linear Elasticity, I, II. Apl. Matematiky 23 (1978) 208-229 and 25 (1980) 11-32.

[Iga]   Igarashi, S. and Takiziwa, E.I.: On the Aquation of Deflection of a Thick Plate, Ing. Arch. 54 (1984) 465-475.

[Ing84]   Ingham, D.B. and Kelmanson, M.A.: Boundary Integral Equation Analysis of Singular Potential and Biharmonic Problems, Springer-Verlag, Berlin 1984.

[Ioa76]   Ioakimidis, N.I. and Theocaris, P.S.: On the Numerical Evaluation of Cauchy Principal Value Integrals, Proc. Intern. Congrecess of Applied Mathematics, Thessaloniki, Greece. pp. 273-288, August 1976. Published also in: Revue Roumaine des Sciences Techniques - Serie de Mecanique appliqué, 22 (1977), 803-818.

[Ioa77a] Ioakimidis, N.I. and Theocaris, P.S.: On the Numerical Solution of a Class of Singular Integral Equations, J. of Math. Physical Sciences (India) (1977).219-235.

[Ioa78a] Ioakimidis, N.I. and Theocaris, P.S.: A Note on Stress Intensity Factors for Single Edge V-Notched Plates in Tension, Eng. Fract. Mech. 10 (1978) 685-686.

[Ioa78b] Ioakimidis, N.I. and Theocaris, P.S.: Numerical Solution of Cauchy Type Singular Integral Equations by Use of the Lobatto-Jacobi Numerical Integration Rule, Apl. Matematiky 23 (1978) 439-478.

[Ioa79a] Ioakimidis, N.I. and Theocaris, P.S.: Numerical Determination of a Class of Generalized Stress Intensity Factors, Intern. J. Num. Meth. Eng. 14, (1979) 949-959.

[Ioa79b] Ioakimidis, N.I. and Theocaris, P.S.: On the Numerical Solution of Singular Integrodifferential Equations, Quart. Appl. Math. .37 (1979) 325-331.

[Ioa79c] Ioakimidis, N.I. and Theocaris, P.S.: A Remark on the Numerical Evaluation of Stress Intensity Factors by the Method of Singular Integral Equations, Intern. J. Num. Meth. Eng. 14 (1979) 1710-1714.

[Ioa80a] Ioakimidis, N.I. and Theocaris, P.S.: On the Selection of Collocation Points for the Numerical Solution of Singular Integral Equations with Generalized Kernels Appearing in Elasticity Problems, Comp. and Struct. 11 (1980) 289-295.

[Ioa80b] Ioakimidis, N.I. and Theocaris, P.S.: A Comparison Between the Direct and the Classical Numerical Methods for the Solution of Cauchy Type Singular Integral Equations, SIAM J. Num. Anal. 17 (1980) 115-118.

[Ioa80c] Ioakimidis, N.I. and Theocaris, P.S.: On the Convergence of Two Direct Methods of Solution of Cauchy Type Singular Integral Equations of the First Kind, Nordish Tidskrift for Informations Behanding (BIT) 20 (1980) 83-87.

[Ioa82] Ioakimidis, N.I.: Application of Finite Part Integrals to the Singular Integral Equations of Crack Problems in Plane and 3-D Elasticity, Acta Mechanica 45 (1982) 31-47.

[Iva76] Ivanov, V.V.: The Theory of Approximate Methods and their Application to the Numerical Solution of Singular Integral Equations, Noordhoff Publishing Co., Leyden 1976.

[Ja83]   Jarusek, J.: Contact Problems with Bounded Friction. Coercive Case. Czech. Math. J. 33 (1983) 254-278.

[Ja84]   Jarusek, J.: Contact Problems with Bounded Friction. Semicoercive Case. Czech. Math. J. 34 (1984) 619-629.

[Jas63a]   Jaswon, M.A.: Integral Equation Methods in Potential Theory: I, Proc. of the Royal Soc. of London 275 (A) (1963) 23-32.

[Jas63b]   Jaswon, M.A. and Ponter, A.R.S.: An Integral Equation Solution of the Torsion Problem, Proc. of the Royal Soc. of London 273 (A) (1963) 237-246.

[Jas67]   Jaswon, M.A., Maiti, M. and Symm, G.T.: Numerical Biharmonic Analysis and some Applications, Intern. J. of Solids and Struct. 3 (1967) 309-332.

[Jas68]   Jaswon, M.A. and Maiti, A.: An Integral Formulation of Plate Bending Problems, J. Eng. Math. 2 (1968) 83-93.

[Jas77]   Jaswon, M.A. and Symm, G.T.: Integral Equation Methods in Potential Theory and Elastostatics, Academic Press, London 1977.

[Jas84]   Jaswon, M.A.: A Review of the Theory, In "Topics in Boundary Element Research" (ed. C.A. Brebbia), Springer-Verlag, Berlin 1984.

[Jea85]   Jean, M. and Pratt, E.: A System of Rigid Bodies with Dry Friction, Int. J. Eng. Sci., 23 (1985), 497-513.

[Jea87]   Jean, M. and Moreau, J.J.: Dynamics in the Presence of Unilateral Contacts and Dry Friction; a Numerical Approach. In "Unilateral Problems in Structural Analysis 2" (ed. G. Del Piero and F.Maceri), CISM Courses and Lectures No 304, Springer-Verlag, Wien 1987, 151-196.

[Jin]   Jin, H., Runesson, K. and Samuelsson, A.: Application of the Boundary Element Method to Contact Problems in Elasticity with a Nonclassical Friction Law. In "Boundary Elements IX, Vol 2: Stress Analysis Applications". (ed. by Brebbia, C.A.; Wendland, W.L.; Kuhn, G.) Springer Verlag, Berlin, 1986.

[Jon]   Jonsson, A. and Wallin, H.: Function Spaces on Subset of $\mathbb{R}^3$, Math. Rep. Vol.2, Harwood Acad. Publ., Chur, London 1984.

[Ju]   Ju, J.W. and Taylor, R.L., A Perturbed Lagrangian Formulation for the Finite Element Solution of Nonlinear Frictional Contact Problems, J. Méc. Théor. et Appl. 7 (Special issue) (1988) 1-14.

[Ka88a]   Kalker, J.J.: The Quasistatic Contact Problem with Friction for Three Dimensional Elastic Bodies, J.Méc.Theor.et Appl. 7 (Special issue) (1988) 55-66.

[Ka88b]   Kalker, J.J.: Contact Mechanical Algorithms, Comm. in Applied Num. Methods 4 (1988) 25-32.

[Ka90]   Kalker, J.J.: Three Dimensional Elastic Bodies in Rolling Contact, Kluwer Acad. Publ., Dordrecht 1990.

[Kal57]   Kalandiya, A.I.: Approximate Solution of a Class of Singular Integral Equationa, Dokl. Akad. Nauk SSSR 125 (1959) 715-718.

[Kar84]   Karabalis, D.L. and Beskos, D.E.: Dynamic Response of 3-D Rigid Surface Foundations by Time Domain Boundary Element Method, Earthq. Engng. Struct. Dyn. 12 (1984) 73-93.

[Kar86]   Karabalis, D.L. and Beskos, D.E.: Dynamic Response of 3-D Rigid Sembedded Foundations by the Boundary Element Method, Comp. Meth. Appl. Mech. Eng. 56 (1986) 91-120.

[Kar87a]   Karabalis, D.L. and Beskos, D.E.: Dynamic Soil-Structure Interaction. In "Boundary Element Methods in Mechanics", (ed. by Beskos, D.E.) North-Holland, Amsterdam 1987.

[Kar87b]   Karabalis, D.L. and Beskos, D.E.: Three-Dimensional Soil-Structure Interaction by Boundary Element Methods. In "Topics in Boundary Element Research, Vol. 4", (ed. by Brebbia, C.A.), Springer-Verlag, Berlin 1987.

[Kar]   Karam, V.J. and Telles, J.C.F.: On Boundary Elements for Reissner's Plate Theory, Eng. Anal. 5 (1988) 21-27.

[Kat]   Katsikadelis, J.T., Massalas, C.V. and Tzivanidis, G.J.: An Integral Equation Solution of the Plane Problem of the Theory of Elasticity, Mech. Res. Comm 4 (1977) 199-208.

[Kay]   Kaya, A.C. und Erdogan, F.: On the Solution of Integral Equations with Strongly Singular Kernels, Quart. of Applied Math. 45 (1987) 105-122.

[Kaz]   Kazantzakis, J. and Theocaris, P.S.: The Evaluation of Certain Two-Dimensional Singular Integrals Used in Three Dimensional Elasticity, Intern. J. Sol. and Struct. 15 (1979) 203-207.

[Kel29]   Kellog, O.D.: Foundations of Potential Theory, Springer Verlag, Berlin 1929; also published by Dover, New York 1953.

[Ker70]   Kermanidis, Th.: Eine Integralgleichungsmethode zur Lösung der Rotationssymmetrischen Randwertprobleme der Elastizitätstheorie, Insbesondere des Torsionsproblems. Doct, Dissertation, RWTH Aachen 1970.

[Ker73]   Kermanidis, Th.: Eine Integralgleichungsmethode zur Lösung des Torsions-Problems des Umdrehungskörpers, Acta Mechanica 16 (1973) 175-181.

[Ker75]   Kermanidis, T.: A Numerical Solution for Axially Symmetrical Elasticity Problems, Int. J. Solids Structures 11 (1975) 193-500.

[Ker76]   Kermanidis, Th.: Kupradze's Functional Equation for the Torsion Problem of Prismatic Bars. Part 1/2, Comp. Meth. in Appl. Mech. and Engng. 7 (1976) 39-46 and 249-259.

[Kha]   Khanna, Tarun: Foundations of Neural Networks. An Introduction. Addison-Wesley Pub.Co., New York 1990.

[Kik]   Kikuchi, N. and Oden, J.T.: Contact Problems in Elasticity A Study of Variational Inequalities and Finite Element Methods, SIAM Publ., Philadelphia 1988.

[Kin]   Kinderlehrer, D.: Remarks About Signorini's Problem in Linear Elasticity. Ann. Scuola Norm. Sup. Pisa cl. Sci., IV, 8 (1981), 605-645.

[Kit]   Kitahara, M.: Boundary Integral Equation Methods in Eigenvalue Problems of Elastodynamics and Thin Plates, Studies in Applied Mechanics 10, Elsevier, Amsterdam.

[Kla84]   Klarbring, A.: Contact Problems with Friction Using a Finite - Dimensional Description and the Theory of Linear Complementarity. Linköping Studies in Science and Technology, Thesis No. 20, Linköping Institute of Technology, Linköping, Sweden 1984.

[Kla87]   Klarbring, A.: Contact Problems with Friction by Linear Complementarity. In "Unilateral Problems in Structural Analysis, 2", CISM Courses and Lectures, No. 304 Springer Verlag, Wien, N.York 1987. (ed. by Del Piero, G., Maceri, F.)

[Kla88]   Klarbring, A. and Björkman, G.: A Mathematical Programming Approach to Contact Problem with Friction and Varying Contact Surface. Comp. and Struct. 30 (1988) 1185-1198.

[Kla90a]   Klarbring, A.: Examples of Non-Uniqueness and Non-Existence of Solutions to Quasistatic Contact Problems with Friction, Ing. Arch. 60 (1990) 529-541.

[Kla90b]   Klarbring, A., Mikelić, A. and Shillor, M.: Duality Applied to Contact Problems with Friction, Appl. Math. Optim. 22 (1990) 211-226.

[Kla90c]   Klarbring, A.: Derivation and Analysis of Rates Boundary-Value Problems of Frictional Contact, European J. of Mech. A/Solids 1 (1990) 53-85.

[Kob82a]   Kobayashi, S. and Nishimura, N.: Dynamic Analysis of Underground Structures by the Integral Equation Method. In "Numerical Methods in Geomechanics", (ed. by Eisenstein, Z.) A.A. Balkema, Publ., Rotterdam 1982.

[Kob82b]   Kobayashi, S. Nishimura, N.: On the Indeterminancy of BIE Solutions for the Exterior Problems of Time-Harmonic Elastodyna- mics and Incompressible Elastostactics. In "Boundary Element Methods in Engineering", (ed. by Brebbia, C.A.) Springer Verlag, Berlin 1982.

[Kob85]   Kobayashi, S.: Fundamentals of Boundary Integral Equation Methods in Elastodynamics. In "Topics in Boundary Element Research, Vol. 2", (ed. by Brebbia, C.A.) Springer Verlag, Berlin 1985.

[Kob87]   Kobayashi, S.: Elastodynamics. In "Boundary Element Methods in Mechanics", (ed. by Beskos, D.E.) North-Holland, Amsterdam 1987.

[Kr]   Kress, R.: Linear Integral Equations, Springer Verlag, Berlin 1989.

[Kre75]   Krenk, S.: On Quadrature Formulas for Singular Integral Equations of the First and the Second Kind, Quart. Appl. Math. 33 (1975) 225-232.

[Kre81]   Krenk, S.: Polynomial Solutions to Singular Integral Equation. Applications to Elasticity Theory. Doct. Thesis, RIS Nat. Lab. Denmark 1981.

[Kuh]   Kuhn, G. and Hildenbrand, J.: Ein Vergleich der Innerhalb der "Direkten" REM Wichtigsten Numerischen Lösungsverfahren, ZAMM 70 (1990) T722-T725.

[Kol91]   Koltsakis, E.K.: Theoretical and Numerical Study of Structures with Non-monotone Boundary Conditions. Application to Adhesion Joint, Doct. Dissertation, Dept. of Civil Eng., Aristotle University, Thessaloniki 1991.

[Kum77]   Kumar, V. and Mukherjee, S.: A Boundary Integral Equation Formulation for Time-Dependent Inelastic Deformation in Metals, Intern. J. Mech. Sc. 19 (1977) 713-724.

[Kup63]   Kupradze, V.D.: Potential Methods in the Theory of Elasticity. Israel Program for Scientific Translations, Jerusalem 1965 (Medoty Potentsiala v Teorii Uprugosti, Gosudarstvennoe Izdatel' stvo Fiziko Matematicheskoi Literatury, Moskva 1963).

[Kup68]   Kupradze, V.D. and Gegelia, T.G., Baschelejschwili, M.O. and Burtschuladze, T.V.: Three-Dimensional Problems of the Mathematical Theory of Elasticity and Thermoelasticity (in Russian), Publishing House of the Univ. of Tbilisi 1968, North Holland Publ. Co. Amsterdam 1979.

[Kwa88]   Kwak, B.M. and Lee, S.S.: A Complementarity Problem Formulation for Two-Dimensional Frictional Contact Problem. Comp. Struct. 28 (1988) 469-480.

[Kwa91]   Kwak, B.M.: Complementarity Problem Formulation of Three-Dimensional Frictional Contact. ASME J. Applied Mech. 58 (1991) 134-140.

[Lan]   Lanczos, C.: The Variational Principles of Mechanics, University of Toronto Press, Toronto 1966.

[Lau07]   Lauricella, G.: Sull' Integrazione delle Equazioni dei Corpi Elastici Isotropi, Rendiconto Accademia dei Lincei, Vol. XV, ser. 5, fasc. 8, 426-432, Rome. Il Nuovo Cimento 55, 13 (1907) pp. 104-119, 155-174, 237, 262, 501-518.

[Laz]   Lazaridis, P.P. and Panagiotopoulos, P.D.: Boundary Variational Principles for Inequality Structural Analysis Problems and Numerical Applications. Comp. Struct. 25 (1987) 35-49.

[Lee71]   Lee, J.J.: Wave Induced Oscillation in Harbors of Arbitrary Geometry. Journ. of Fluid Mech. 45 (1971) 375-394.

[Lem]   Lemke, C.A.: Some Pivot Schemes for the Linear Complementarity Problem, Math. Progr. Stud. 7 (1978) 15-35.

[Li86]   Liolios, A.A.: A Linear Complementarity Approach for the Signorini Problem with Friction, ZAMM 66 (1986) 349-352.

[Li87]   Liolios, A.A.: Upper and Lower Solution Estimates In Unilateral Viscoelastodynamics, Acta Mech. 66 (1987) 275-278.

[Li88]   Liolios, A.A.: Seismic Interaction Between Adjacent Structures: A Linear Complementarity Approach for the Unilateral Elastoplastic Softening Contact with Friction, In "Structural Dynamics and Earthquake Engineering", (ed. by Counadis A.N. and Krätzig W.B.) Athens 1988.

[Li89]   Liolios, A.A.: A Linear Complementarity Approach for the Non-Convex Dynamic Problem of Unilateral Contact with Friction Between Adjacent Structures, ZAMM 69, (1989) 420-422.

[Li91]   Liolios, A.A.: A Numerical Estimation for the Influence of Modifications to Seismic Interaction Between Adjacent Structures, In "Earthquake Resistant Construction and Design", (ed. by Savidis S.A.), A.A. Balkema Publ., Rotterdam 1991.

[Lig77]   Liggett, J.A.: Locations of Free Surface in Porous Media. Journ. of the Hydraulics Division, ASCE 103, (HY 4), (1977) 353-365.

[Lig83]   Liggett, J.A. and Liu, P.L.-F.: The Boundary Integral Equation Method for Porous Media Flow, George Allen and Unwin, London 1983.

[Lio67]   Lions, J.L. and Stampacchia, G.S.: Variational Inequalities. Comm. Pure and Appl. Math. XX (1967) 493-519.

[Lio71]   Lions, J.L.: Quelques méthodes de Résolution des Problèmes aux Limites non Linéaires. Dunod/Gauthier-Villars, Paris 1969.

[Ma82]   Mansur, W.J., Brebbia, C.A.: Formulation of the Boundary Element Method for Two-Dimensional Transient Scalar Wave Propagation Problems, Appl. Math. Modelling 6 (1982) 307-311.

[Ma83a]   Mansur, W.J.: A Time-Stepping Technique to Solve Wave Propagation Problems Using the Boundary Element Method, PhD. Thesis, Univ. Southampton 1983

[Ma83b]   Mansur, W.J. and Brebbia, C.A.: Transient Elastodynamics Using a Time-Stepping Technique. In "Boundary Elements" (C.A. Brebbia et al., eds.) Springer Verlag, Berlin 1983.

[Maek]   Maekawa, Z.: Noise Reduction by Screens, Mem. Faculty of Engng., Kobe Univ. 11 (1985) 29-53

[Mai87]   Maier, G. and Polizzotto C.: A Galerkin Approach to Boundary Element Elastoplastic Analysis, Comput. Meth. Appl. Mech. Engrg. 60 (1987) 175-194.

[Mai91]   Maier, G., Diligenti, M. and Carini, A.: A Variational Approach to Boundary Element Elastodynamic Analysis and Extension to Multidomain Problems. Comp. Meth. in Appl. Mech. and Eng. 92 (1991) 193-213.

[Man]   Mandelbrot, B.: The Fractal Geometry of Nature W.H. Freeman and Co., New York, 1972.

[Man80]   Manolis, G.D. and Beskos, D.E.: Dynamic Stress Concentration Studies by the Boundary Integral Equation Method. In "Innovative Numerical Analysis for the Engineering Sciences", (eds. Shaw, R.P. et al.) Univ. of Virginia Press, Charlottesville 1980.

[Man81]   Manolis, G.D. and Beskos, D.E.: Dynamic Stress Concentration Studies by Boundary Integrals and Laplace Transform, Int. J. Num. Meth. Engng. 17 (1981) 573-599.

[Man83a]   Manolis, G.D. and Beskos, D.E.: Dynamic Response of Lined Tunnels by an Isoparametric Boundary Element Method, Comp. Meth. Appl. Mech. Engng. 36 (1983) 291-307.

[Man83b]   Manolis, G.D.: A Comparative Study on Three Boundary Element Method Approaches to Problems in Elastodynamics, Int. J. Num. Meth. Engng. 19 (1983) 73-91.

[Man88]   Manolis, G.D. and Beskos, D.E.: Boundary Element Methods in Elastodynamics, Unwin Hyman Publishing Co., London, 1988.

[Mar86]   Martins, J.A.C.: Dynamic Frictional Contact Problems Involving Metallic Bodies. Ph.D.Dissertation, University of Texas at Austin 1986.

[Mar]   Martins, J.A.C. and Oden, J.T.: Existence and Uniqueness Results for Dynamic Contact Problems with Nonlinear Normal and Friction Interface Laws, Nonlinear Anal. 11 (1987) 407-428.

[Mas49]   Massonnet, C.: Resolution Graphomecanique des Problèmes Génèraux de l'élasticité Plane, Bull. Centre Et. Rech. Essais Sc. Genie Civil 4 (1949) 169-180.

[Mas65]   Massonnet, C.E.: Numerical Use of Integral Procedures. In "Stress Analysis", (ed. by Zienkiewicz, O.C. and Holister, G.S.) John Wiley and Sons, London 1965.

[Maz]   Maz'ja, V.: Sobolev Spaces, Springer Verlag, Berlin 1985.

[McCo]   McCormick, S.(ed): Multigrid Methods, SIAM Publ., Philadelphia, 1987.

[McD72]   McDonald, B.H. and Wexler, A.: Finite-Element Solution of Unbounded Field Problems, IEEE Transactions on Microwave Theory and Techniques MTT-20 (1972) 841-847.

[Meis]   Meise, Th.: Calculation of Scalar Wave Propagation in 3-D Frequency and Time Domain (in German), Doctoral Thesis, Techn. Rep. 90-6, Inst. KIB, Ruhr-Universitat Bochum, Germany 1990

[Men73]   Mendelson, A.: Boundary-Integral Methods in Elasticity and Plasticity, NASA Technical Note TN D-7418, Washington D.C. 1973.

[Mic26]   Miche, R.: Le calcul Pratique de Problèmes élastiques à deux Dimensions par la Méthode des Equations Intégrales. Proc. Second Int. Congr. Techn. Mech. Zürich, 126-130, 1926.

[Mig]   Mignot, F.: Contôrle Dans les Inequations Variationnelles Elliptiques, J.Funct. Anal., 22 (1976) 130-185.

[Mik57]   Mikhlin, S.G.: Integral Equations, Pergamon Press, Oxford 1957.

[Mik65a]   Mikhlin, S.G.: Approximate Solutions of Differential and Integral Equations, Pergamon Press, Oxford 1965.

[Mik65b]   Mikhlin, S.G.: Multidimensional Singular Integrals and Integral Equations, Pergamon Press, Oxford 1965.

[Mis92a]   Mistakidis, E. and Panagiotopoulos, P.D.: On the Numerical Treatment of Nonmonotone (zigzag) Friction and Adhesive Contact Problems with Deboding. Approximation by Monotone Subproblems, 1992, (Submitted for publication).

[Mis92b]   Mistakidis, E.: Theoretical and Numerical Study of Structures with Nonmonotone Boundary and Constitutive laws. Algorithms and Applications, Doct. Dissertation, Dept. of Civil Eng. Aristotle Univ. Thessaloniki (1992).

[Mit83]   Mitsopoulou, E.: Unilateral Contact, Dynamic Analysis of Beams by a Time-Stepping Quadratic Programming Procedure, Meccanica 18 (1983) 254-265.

[Mit]   Mitsopoulou, E.N. and Doudoumis, I.N.: A Contribution to the Analysis of Unilateral Contact Problems with Friction. Solid Mech. Arch. 12 (1987) 165-186.

[Mit91]   Mitsopoulou, E., Panagiotopoulos, P.D. and Zervas P.A.: Dynamic Boundary Integral "Equation" Method for Unilateral Contact Problems. Eng. Anal. with Bound. Elements 8 (1991) 192-199.

[Mit93]   Mitsopoulou, E., Panagiotopoulos, P.D. and Zervas P.A.: A Boundary Integral Equation Approach to Dynamic Inequality Problems and Applications. Eng. Anal. with Bound. Elements (to appear).

[Mon]   Monteiro Marques, M.D.P.: Inclusões Differenciais e Choques inelasticos, Doct. Dissertation, Faculty of Sciences, Univ. of Lisbon, 1988.

[Mor67]   Moreau, J.J.: Fonctionnelles Convexes. Séminaire sur les équations aux Dérivées Partielles. Collége de France, Paris 1967.

[Mor68]   Moreau, J.J.: La Notion du Surpotentiel et les Liaisons Unilitérales on Elastostatique, C.R.Acad.Sci. Paris 167A (1968) 954-957.

[Mor86]   Moreau, J.J.: Une Formulation du Contact Frottement Sec; Application au Calcul Numérique, C.R. Acad. Sci. Paris, Sér. II, 302 (1986), 799-801.

[Mor88a]   Moreau, J.J. and Panagiotopoulos, P.D.(eds): Nonsmooth Mechanics and Applications, CISM Vol. 302, Springer Verlag, Wien 1988.

[Mor88b]   Moreau, J.J., Panagiotopoulos, P.D. and Strang, G.(eds): Topics in Nonsmooth Mechanics, Birkhäuser Verlag, Basel, Boston 1988.

[Mor88c]   Moreau, J.J.: Unilateral Contact and Dry Friction in Finite Freedom Dynamics. In Nonsmooth Mechanics and Applications (ed. by J.J.Moreau and P.D.Panagiotopoulos),CISM Vol. 302, Springer Verlag, Wien, N.York 1988.

[Muk78]   Mukherjee, S. and Kumar, V.: Numerical Analysis of Time-Dependent Inelastic Deformation in Metallic Media Using the Boundary Integral Equation Method, Journ. of Appl. Mech. 45 (1978) 785-790.

[Muk82]   Mukherjee, S.: Boundary Element Methods in Creep and Fracture, Applied Science, London 1982.

[Mur]   Murty, K.G.: Linear Complementarity. Linear and Nonlinear Programming. Heldermann Verlag, Berlin 1988.

[Mus53a]   Muskhelishvili, N.I.: Singular Integral Equations, Noordhoff Publishing Co., Groningen 1953.

[Mus53b]   Muskhelishvili, N.I.: Some Basic Problems of the Mathematical Theory of Elasticity, Noordhoff Publishing Co., Groningen 1953.

[Nan89a]   Naniewicz, Z.: On Some Nonconvex Variational Problems Related to Hemivariational Inequalities, Nonlin. Analysis 13 (1989) 87-100.

[Nan89b]   Naniewicz, Z. and Wozniak, C.Z.: On the Quasi-Stationary Models of Debonding Processes in Layered Composites, Ing. Archiv 60 (1989) 31-40.

[Nan88]   Naniewicz, Z.: On Some Nonmonotone Subdifferential Boundary Conditions in Elastostatics. Ing. Archiv 58 (1988) 403-412.

[Neč]   Nečas, J.: Les Méthodes Directes en Théorie des équations Elliptiques, Masson, Paris and Academia Prague 1967.

[Neč80]   Nečas, J., Jarusek, J. and Haslinger, J.: On the Solution of the Variational Inequality to the Signorini Problem with Small Friction, Bulletino U.M.I. 17B (1980) 796-811.

[New68]   Newton, D.A. and Tottenham, H.: Boundary Value Problems in Thin Shallow Shells of Arbitrary Plan Form, Journ. of Enging. Math. 2 (1968) 211-224.

[Ngu]   Nguyen Dang Hung and Géry de Saxcé: Frictionless Contact of Elastic Bodies by Finite Element Method and Mathematical Programming Technique, Comp. and Struct. 11, 55, 1980.

[Nit]   Nitsiotas, G.: Elastostatics. Linear Theory Vol. I, II Thessaloniki 1978 (in Greek).

[Niw76]   Niwa, Y, Kobayashi, S. and Fukui, T.: An Application of the Integral Equation Method to Seepage Problems, Theor. and Appl. Mech. 24 (1976) 479-486.

[Niw80]   Niwa, Y, Fukui, T, Kato, S. and Fujiki, K.: An Application of the Integral Equation Method to Two-Dimensional Elastodynamics, Theor. Appl. Mech. 28 (1980) 281-290.

[Niw81a]   Niwa, Y., Kobayashi, S. and Kitahara, M.: Eigenfrequency Analysis of a Plate by the Integral Equation Method, Theor. Appl. Mech. 29 (1981) 287-307.

[Niw81b]   Niwa, Y., Kobayashi, S. and Kitahara, M.: Analysis of the Eigenvalue Problems of Elasticity by the Boundary Integral Equation Method, Theor. Appl. Mech. 30 (1981) 335-356.

[Niw82a]   Niwa, Y., Kobayashi, S. and Kitahara, M.: Applications of the Boundary Integral Equation Method to Eigenvalue Problems of Elastodynamics. In "Boundary Element Methods in Engineering", (ed. by Brebbia, C.A.) Springer Verlag, Berlin 1982.

[Niw82b]   Niwa, Y., Kobayashi, S. and Kitahara, M.: Eigenfrequency Analysis of a Dam by the Boundary Integral Equation Method. In "Numerical Methods in Geomechanics", (ed. by Eisenstein, Z.) A.A. Balkema, Publ., Rotterdam 1982.

[Niw82c]   Niwa, Y., Kobayashi, S. and Kitahara, M.: Determination of Eigenvalues by Boundary Element Methods. In "Developments in Boundary Element Methods-2", (ed. by Banerjee, P.K. Shaw,R.P.) Applied Science Publishers, London 1982.

[Niw85]   Niwa, Y. and Hirose, S.: Three-Dimensional Analysis of Ground Motion by Integral Equation Method in Wave Number Domain. In "Numerical Methods in Geomechanics", (ed. by Kawamoto, T. Ichikawa, Y.) 1985, A.A. Balkema, Publ., Rotterdam 1985.

[Niw86]   Niwa, Y. and Hirose, S.: Application of the BEM to Elastodynamics in a Three-Dimensional Half-Space. In "Recent Applications in Computational Mechanics", (ed. by Karabalis, D.L.) American Society of Civil Engineers, New York 1986.

[Od]   Oden, J.T. and Pires, E.: Contact Problems in Elastostatics with Non-Local Friction Laws. TICOM Report 81-12, University of Texas at Austin, 1981.

[Ode]   Oden, J.T. und Reddy, J.N.: On Dual-Complementary Variational Principles in Mathematical Physics, Int. J. Eng. Sci. 12 (1974) 1-29.

[Ode83]   Oden, J.T. and Pires, E.B.: Nonlocal and Nonlinear Friction Laws and Variational Principles for Contact Problems in Elasticity, ASME J. Appl. Mech. 50 (1983) 67-76.

[Ode85]   Oden, J.T. and Martins, J.A.C.: Models and Computational Methods for Dynamic Friction Phenomena, Comp. Meth. Appl. Mech. Eng. 52 (1985) 527-634.

[Oli68]   Oliveira, E.R.A.: Plane Stress Analysis by a General Integral Method. Proceedings of the ASCE Eng. Mech. Div. 94 (1968) 79-101.

[Pa85]   Panagiotopoulos, P.D.: Boundary Integral "Equation" Methods for the Signorini-Fichera Problem. In "Boundary Elements VII", (ed. by Brebbia C.A., Maier, G.) CML Publ., Southampton 1985.

[Pa87]   Panagiotopoulos, P.D.: Boundary Integral Equation Methods for the Friction Problem, Eng. Analysis 4 (1987) 100-105.

[Pah]   Pahnke, U.: Zur Berechnung von Scheiben und Platten mit Ausspringeden Echen, ZAMM 52 (1972) T142-T143.

[Pan75]   Panagiotopoulos, P.D.: A Nonlinear Programming Approach to the Unilateral Contact – and Friction – Boundary Value Problem in the Theory of Elasticity, Ing. Archiv. 44 (1975) 421-432.

[Pan76]   Panagiotopoulos, P.D.: A Variational Inequality Approach to the Inelastic Stress-Unilateral Analysis of Cable Structures. Comp. and Struct. 6 (1976) 133-139.

[Pan77]   Panagiotopoulos, P.D.: Optimal Control in the Unilateral Thin Plate Theory, Archives of Mechanics 29 (1977) 25-39.

[Pan80]   Panagiotopoulos, P.D. and Talaslidis, D.: A Linear Analysis Approach to the Solution of Certain Classes of Variational Inequality Problems in Structural Analysis. Int. J. Solids and Struct. 16 (1980) 991-1006.

[Pan81]   Panagiotopoulos, P.D.: Non-Convex Superpotentials in the Sence of F.H. Clarke and Applications, Mech. Res. Comm. 8 (1981) 335-340.

[Pan82]   Panagiotopoulos, P.D.: Non-Convex Energy Functionals. Application to Non-convex Elastoplasicity, Mech. Res. Comm. 9 (1982) 23-29.

[Pan83a]   Panagiotopoulos, P.D.: Nonconvex Energy Functions. Hemivariational Inequalities and Substationarity Principles, Acta Mechanica 42 (1983) 160-183.

[Pan83b]   Panagiotopoulos, P.D.: A Boundary Integral Inclusion Approach to Unilateral B.V.Ps in Elastostatics, Mech. Res. Comm. 10 (1983) 91-96.

[Pan83c]   Panagiotopoulos, P.D.: Optimal Control and Parameter Identification of Structures with Convex and Non-Convex Strain Energy Density. Applications to Elastoplasticity and Contact Problems. Solid Mech. Arch. 8 (1983) 363-411.

[Pan84]   Panagiotopoulos, P.D. and Baniotopoulos, C.C.: A Hemivariational Inequality and Substationarity Approach to the Interface Problem. Theory and Prospects of Applications, Engineering Analysis 1 (1984) 20-31.

[Pan85]   Panagiotopoulos, P.D.: Inequality Problems in Mechanics and Applications. Convex and Nonconvex Energy Functions. Birkhäuser Verlag, Basel, Boston 1985. Russian Translation MIR Publ. Moscow 1989.

[Pan87]   Panagiotopoulos, P.D. and Lazaridis, P.: Boundary Minimum Principles for the Unilateral Contact Problem, Int. J. Solids and Struct. 23 (1987) 1465-1484.

[Pan87a]   Panagiotopoulos, P.D. and Koltsakis, E.K.: Interlayer Slip and Delamination Effect: A Hemivariational Inequality Approach, Canadian Society for Mech. Engineering 11 (1987) 43-52.

[Pan87b]   Panagiotopoulos, P.D.: Ioffe's Fans and Unilateral Problems: A New Conjecture. In "Unilateral Problems in Structural Analysis 2", (ed. by G. del Piero, F.Maceri), CISM Courses and Lectures 304. Springer Verlag, Wien, N.York 1987.

[Pan88]   Panagiotopoulos, P.D.: Nonconvex Superpotentials and Hemivariational Inequalities. Quasidifferentiability in Mechanics. In "Nonsmooth Mechanics and Applications" (ed. by J.J. Moreau, P.D. Panagiotopoulos), CISM Courses and Lectures Nr. 302, Springer Verlag, Wien, N.York 1988.

[Pan89]   Panagiotopoulos, P.D. and Haslinger, J.: Optimal Control of Systems Governed by Hemivariational Inequalities. In "Mathematical Models for Phase Change Problems", (ed. by J.F.Rodriques) Birkhäuser Verlag, Basel, Boston 1989.

[Pan91]   Panagiotopoulos, P.D.: Coercive and Semicoercive Hemivariational Inequalities, Nonlin. Anal. 16 (1991) 209-231.

[Pan91]   Panagiotopoulos, P.D.: The B.I.E.M. for Inequality Problems, Math. Comput. Modelling 15 (1991) 257-267.

[Pan92]   Panagiotopoulos, P.D. and Stavroulakis, G.: New Types of Variational Principles Based on the Notion of Quasidifferentiability, Acta Mechanica (to appear).

[Pan93]   Panagiotopoulos, P.D. and Haslinger, J.: On the Dual Reciprocal Variational Approach to the Signorini-Fichera Problem. Convex and Nonconvex Generalizations, ZAMM 1992 (to appear).

[Pan93a]   Panagiotopoulos, P.D., Mistakidis, E. and Koltsakis, E.: Debonding and Sliding in Adhesively Bonded Cracks - A Numerical Algorithm (to appear).

[Pang92a]   Panagouli, O.K., Panagiotopoulos, P.D. and Mistakidis, E.S.: On the Numerical Solution of Structures with Fractal geometry: The F.E. Approach. Meccanica (to appear).

[Pang92b]   Panagouli, O.K.: Fractal Geometry, in Structural Analysis. Doct. Dissertation, Dept. of Civil Eng., Aristotle University, Thessaloniki 1992.

[Par]   Paris, F. and Garrido, J.A.: On the Use of Discontinuous Elements in Two-Dimensional Contact Problems. In "Boundary Elements VII", (ed. by Brebbia, C.A.; Maier, G.) CML Publ., Southampton 1985.

[Par82]   Parton, V.Z. and Perlin, P.I.: Integral Equations in Elasticity, MIr Publ., Moscow 1982.

[Pdp90a]   Panagiotopoulos, P.D.: The Mechanics of Fractals, Proc. Acad. of Athens 65 (1990) 185-212.

[Pdp90b]   Panagiotopoulos, P.D.: On the Fractal Nature of Mechanical Theories, ZAMM 70 (1990) 258-260.

[Pdp91]   Panagiotopoulos, P.D.: Fractal Approximation in the Theory of Elasticity, ZAMM 71 (1991) T658-T659.

[Pdp92a]   Panagiotopoulos, P.D.: Fractals and Fractal Approximation in Structural Mechanics, Meccanica 27 (1992) 25-33.

[Pdp92b]   Panagiotopoulos, P.D.: Fractal Geometry in Solids and Structures, Int. J. Solids and Structures, 29 (1992) 2159-2175.

[Pdp92c]   Panagiotopoulos, P.D., Mistakidis, E.S. and Panagouli, O.K.: Fractal Interfaces with Unilateral Contact and Friction Conditions, Comp. Meth. in Appl. Mech. Eng. 1992 (to appear).

[Pir]   Pires, E.: Analysis of Nonclassical Friction Laws for Contact Problems in Elastostatics. Ph.D.Dissertation, University of Texas at Austin 1982.

[Prö]   Prössdorf, S. and Silbermann, B.: Numerical Analysis for Integral and Related Operator Equations, Birkhäuser Verlag, Basel 1991.

[Pog66]   Pogorzelski, W.: Integral Equations and Their Applications, Pergamon Press Inc., New York, and PWN-Polish Scientific Publishers, Warszawa, Poland, 1966.

[Pol88]   Polizzotto, C.: An Energy Approach to the Boundary Element Method. Part I: Elastic Solids, Part II: Elastic-Plastic Solids. Comput. Meth. Appl. Mech. Engrg. 69 (1988) 167-184, and 263-276.

[Pot]   Potier-Ferry, M.: Problèmes Semi-coercifs. Applications aux Plaques de von Kármán, J.Math. Pures et Appl. 53 (1974) 331-346.

[Pp87]   Panagiotopoulos, P.D.: Multivalued Boundary Integral Equations for Inequality Problems. The Convex Case. Acta Mechanica 70 (1987) 145-167.

[Pp87]   Panagiotopoulos, P.D.: Boundary Integral Equations for Inequality Problems. The Nonconvex Case. Acta Mechanica 72 (1989) 152-168.

[Pra28]   Prager, W.: Die Druckverteilung an Körpern in Ebener Potentialstromung, Physikalische Zeitschrift 29 (1928) 865-869.

[Ra]   Rabinovich, V.L. and Spektor, A.A.: Solution of Some Classes of Three-Dimensional Contact Problems with an Unknown Boundary. Izv. AN SSSR. Mekhanika Tverdogo Tela 20 (1985) 93-100.

[Rab]   Rabier, P., Martins, J.A.C., Oden, J.T. and Campos, L.: Existence and Local Uniqueness of Solutions to Contact Problems in Elasticity with Nonlinear Friction Laws, Int. J. Eng. Sci. 24 (1986) 1755-1768.

[Rao]   Raous, M., Chabrand, P. and Lebon, F.: Numerical Methods for Frictional Contact Problems and Applications, Journal de Méc. Théor. et Appl. 7 (1988) 111-128.

[Re45]   Reissner, E.: The Effect of Transverse Shear Deformation on the Bending of Elastic Plates, J. Appl. Mech. 12 (1945) A 69-77

[Ric73]   Riccardella, P.: An Implementation of the Boundary Integral Technique for Plane Problems of Elasticity and Elastoplasticity, Ph.D. Thesis, Carnegie Mellon University, Pittsburgh, Pennsylvania 1973.

[Rie62]   Rieder, G.: Iterationsverfahren und Operatorgleichungen in der Elastizitätstheorie, Abh. Braunschweig Wiss. Ges. 14 (1962) 109-343.

[Rie68]   Rieder, G.: Mechanische Deutung und Klassifizierung Einiger Integralverfahren der Ebenen Elastizitatstheorie, Bull. Acad. Pol. Sci. Ser. Sci. Techn. 16 (1968) 101-114.

[Rim67]   Rim, K. and Henry, A.S.: An Integral Equation Method in Plane Elasticity, NASA Contractor Report CR-779, 1967.

[Riz67]   Rizzo, F.J.: An Integral Equation Approach to Boundary Value Problems of Classical Elastostatics, Quarterly of Applied Mathematics 25 (1967) 83-95.

[Riz70]   Rizzo, F.J. and Shippy, D.J.: A Method of Solution for Certain Problems of Transient Heat Conduction, American Institute of Aeronautics and Astronautics Journal 8 (1970) 2004-2009.

[Riz71]   Rizzo, F.J. and Shippy, D.J.: An Application of the Correspondence Principle of Linear Viscoelasticity Theory, SIAM Journal of Appl. Math. 21 (1971) 321-330.

[Riz77]   Rizzo, F.J. and Shippy, D.J.: An Advanced Boundary Integral Equation Method for Three-Dimensional Thermoelasticity, Int. J. for Num. Meth. in Eng. 11 (1977) 1753-1768.

[Riz]   Rizzo, F.J., Krishnasamy, G., Schmerr, L.W. and Rudolphi, T.J.: Hypersingular Boundary Integral Equations: Some Applications in Acoustic and Elastic Wave Scattering, J. Applied Mechanics (to appear).

[Rock]   Rockafellar, R.T.: Convex Analysis, Princeton Univ. Press, Princeton 1970.

[Rock79]   Rockafellar, R.T.: La théorie des Sous-Gradients et ses Applications à l'optimization. Fonctions Convexes et Non-convexes, Les Presses de l' Université de Montréal, Montréal 1979.

[Rock80]   Rockafellar, R.T.: Generalized Directional Derivatives and Subgradients of Non-convex Functions, Can.J.Math. XXXII (1980) 257-280.

[Ruo88]   Ruotsalainen, K. and Wendland, W.: On the Boundary Element Method for Some Nonlinear Boundary Value Problems, Numerische Mathematik 53 (1988) 299-314.

[Ruo89]   Ruotsalainen, K. and Saranen, J.: On the Collocation Method for a Nonlinear Boundary Integral Equation, Journal of Computational and Applied Mathematics 28 (1989) 339-348.

[Sa]   Sanchez-Palencia, E. and Suquet, P.: Friction and Homogeneization of a Boundary. In "Free Boundary Problems: Theory and Applications Vol. II", (ed. by Fasano A. and Primicerio M.) Pitman, London 1983.

[Scha]   Schatzman, M.: Problémes aux Limites Non-lineaires Noncoercives. Ann. Sc. Norm. Sup. Pisa XXVII, III (1973) 641-686.

[Sche]   Schenck, H.A.: Improved Integral Formulation for Acoustic Radiation Problems, J. Acoustic Soc. Amer. 44 (1968) 41-53

[Scw]   Schwartz, M.M.: Composite Materials Handbook, McGraw-Hill, New-York 1984.

[Sel]   Selvadurai, A.P.S. and Ap, M.C.: Response of Inclusions with Interface Separation, Friction and Slip. In "Boundary Elements VII", (ed. by Brebbia, C.A.; Maier, G.) Springer-Verlag, Berlin 1985.

[Sha62a]   Shaw, R.P. and Friedman, M.B.: Diffraction of Pulses by Deformable Cylindrical Obstacles of Arbitrary Cross Section. In "Proceeding of the 4th US National Congress of Applied Mechanics", (ed. by Rosenberg, R.M.) American Soc. of Mech. Eng., New York 1962.

[Sha62b]   Shaw, R.P.: Diffraction of Acoustic Pulses by Obstacles of Arbitrary Shape with a Robin Boundary Condition - Part A, J. of the Acoustical Soc. of America 41 (1962) 855-859.

[Sha70]   Shaw, R.P.: Methods of Solution for Water Wave Scattering Problems, In "Topics in Ocean Engineering II", (ed. by Bretschneider, C.), Gulf Publishing Co., Houston 1970.

[Shi]   Shidharan, N.S.: Elastic and Strength Properties of Continous/Chopped Glass Fiber Hybrid Sheet Modeling Compounds. In "Short Fiber Reinforced Composite Materials", (ed. by B.A. Sanders) ASTM Special Technical Publication 772, ASTM, Philadelphia 1982.

[Sil71]   Silvester, P. and Hsieh, M.S.: Finite Element Solution of 2-Dimensional Exterior Field Problems, Proc. of the Institution of Electrical Engineers 118 (1971) 1743-1747.

[Sim]   Simo, J.C., Wriggers, P. and Taylor, R.L.: A Perturbed Lagrangian Formulation for the Finite Element Solution of Contact Problems, Comp. Meth. Appl. Mech. Engng. 50 (1985) 163-180.

[Smi64]   Smirnov, V.I.: Integral Equations and Partial Differential Equations. A Course of Higher Mathematics, Vol. 4, Pergamon Press, Oxford 1964.

[Smi58]   Smith, A.M.O. and Pierce, J.: Exact Solution of the Neumann Problem, Calculation of Non-Circulatory and Axially Symmetric Flows About or Within Arbitrary Boundaries. In "Proceedings of the 3rd US National Congress of Applied Mechanics", (ed. by Haythornthweite, R.M.) The American Society of Mechanical Engineers, New York 1958.

[Sny75]   Snyder, M.D. and Cruse, T.A.: Boundary Integral Equation Analysis of Cracked Anisotropic Plates, Int. J. of Fracture 11 (1975) 315-328.

[Sok]   Sokolnikoff, I.S.: Mathematical Theory of Elasticity, McGraw-Hill, 1986.

[Spe]   Spector, A.A.: Variational Methods of Analysis for Certain Classes of Spatial Problems of Contact Between Elastic Bodies with Friction, Docl. Acad. Nauk. SSR 265, 1982 111-117.

[Spe82]   Spektor, A.A.: Variational Methods of Investigation of Certain Classes of Three-Dimensional Problems of Contact Between Elastic Bodies in the Presence of Friction, Dokl. Akad. Nauk SSSR 265 (1982) 592-596.

[Spe85]   Spektor, A.A.: Variational Method of Solving Three-Dimensional Contact Problems of the Nonstationary Interaction Between Elastic Solids with Friction, Dokl. Akad. Nauk. SSSR 285 (1985) 865-870.

[Spe87]   Spektor, A.A.: Variational Methods in Three-Dimensional Problems of Non-stationary Interaction of Elastic Bodies with Friction, PMM U.S.S.R., 51 (1987) 56-62.

[Spec]   Spector, A.A.: Variational Methods in Three-Dimensional Problems of Non- Stationary Interaction of Elastic Bodies with Friction, PMM U.S.S.R. 51 (1987) 56-62.

[Spy88]   Spyrakos, C.C.: Dynamic Behaviour of Foundations in Bilateral and Unilateral Contact, The Shock and Vibration Digest 20 (1988) 3-12.

[Stav91]   Stavroulakis, G.E.: Analysis of Structures with Interfaces. Formulation and Study of Variational- Hemivariational Inequality Problems, Doct. Dissertation, Dept. of Civil Eng., Aristotle University, Thessaloniki (1991).

[Stav92]   Stavroulakis, G.E.: Convex Decomposition for Nonconvex Energy Problems in Elastostatics and Applications, European J. of Mech. (to appear).

[Ste79]   Stern, M.: A General Boundary Integral Formulation for the Numerical Solution of Plate Bending Problems, Int. J. Solids Struct. 15 (1979) 769-782.

[Ste83]   Stern, M.: Boundary Integral Equation for Bending of Thin Plates. In "Progress in Boundary Element Methods, Vol. 2", (ed. by Brebbia C.A.) Pentech Press, London 1983.

[Str]   Strang, G.: Introduction to Applied Mathematics, Wellesley-Cambridge Press, Wellesley, Mass. 1986.

[Str]   Strodiot, J.J. and Nguyen, V.H.: On the Numerical Treatment of the Inclusion $0 \in \partial f(x)$. In "Topics in Nonsmooth Mechanics", (ed. by Moreau,J.J. Panagiotopoulos,P.D. Strand G.), Birkhäuser Verlag, Basel, Boston, 1988.

[Swe71]   Swedlow, J.L. and Cruse, T.A.: Formulation of Boundary Integral Equations for Three-Dimensional Elastoplastic Flow, Int. J. of Solids and Struct. 7 (1971) 1673-1681.

[Sym63]   Symm, G.T.: Integral Equation Methods in Potential Theory: II, Proc. of the Royal Soc. of London 275 (A) (1963) 33-46.

[Ta]   Tato, Y. Signorini's Problem With Friction in Linear Elasticity, Japan J. of Appl. Math. 4 (1987), 237-268.

[Tai74]   Tai. G.R.C. and Shaw, R.P.: Helmholtz Equation Eigenvalues for Arbitrary Domains, J. Acoust. Soc. Am. 56 (1974) 796-804.

[Taka]   Takayasu, H.: Fractals in the Physical Sciences, Manchester Univ. Press, Manchester 1990.

[Tak]   Takahashi, S.: Elastic Contact Analysis by Boundary Elements, Lect. Notes in Eng. Vol. 67 Springer Verlag, Berlin 1991.

[Tal]   Talaslidis, D. and Panagiotopoulos P.D.: Linear Finite Element Approach to Unilateral Contact Problem in Structural Dynamics. Int. J. Num. Meth. in Eng. 18 (1982) 1505-1520.

[Tas88]   Tassoulas, J.L.: Dynamic Soil-Structure Interaction. In "Boundary Element Methods in Structural Analysis", (ed. by Beskos, D.E.) American Society of Civil Engineers, New York 1988.

[Tel83]   Telles, J.C.F. : The Boundary Element Method Applied to Inelastic Problems, Springer Verlag, Berlin 1983.

[Tel]   Telega J.J.: Topics on Unilateral Contact Problems of Elasticity and Inelasticity. In "Nonsmooth Mechanics and Applications" (ed. by Moreau, J.J. Panagiotopoulos, P.D.), CISM Courses and Lectures No. 302, Springer Verlag, Wien 1988.

[The76]   Theocaris, P.S.: On the Numerical Solution of Cauchy-Type Singular Integral Equations. "Serdica" Bulgaricae Mathematicae Publicationes, 2 (1976) 252-275.

[The77a]   Theocaris, P.S. and Ioakimidis, N.I.: Numerical Integration Methods for the Solution of Singular Integral Equations, Quart. of Appl. Math., 35 (1977) 173-183.

[The77b]   Theocaris, P.S. and Ioakimidis, N.I.: Numerical Solution of Cauchy Type Singular Integral Equations, Proc. Nat. Acad. of Athens, 40 (1977) 1-39.

[The77c]   Theocaris, P.S. and Ioakimidis, N.I.: On the Numerical Solution of Cauchy Type Singular Integral Equations and the Determination of Stress Intesity Factors in Case of Complex Singularities, ZAMP 28 (1977) 1085-1098.

[The78a]   Theocaris, P.S. and Ioakimidis, N.I.: Application of the Gauss, Radau and Lobatto Numerical Integration Rules to the Solution of Singular Integral Equations, ZAMM 58 (1978) 520-522.

[The78b]   Theocaris, P.S. and Ioakimidis, N.I.: On the Gauss-Jacobi Numerical Integration Method applied to the solution of Singular Integral Equations. Bull. of the Calcutta Math. Soc., 71 (1978) pp.29-43.

[The79a]   Theocaris, P.S. and Ioakimidis, N.I.: A Method of Numerical Solution of Cauchy-Type Singular Integral Equations with Generalized Kernels and Arbitrary Complex Singularities. Int. Journal of Comp. Physics, 30 (1979) 309-323.

[The79b]   Theocaris, P.S. and Tsamasphyros, G.J.: Numerical Solution of Systems of Singular Integral Equations with Variable Coefficients, Applicable Analysis, 9 (1979) 37-52.

[The79c]   Theocaris, P.S., Chrysakis, A.C. and Ioakimidis, N.I.: Cauchy Type Integrals and Integral Equations with Logarithmic Singularities, Journal of Eng. Math. 13 (1979) 63-74.

[The79d]   Theocaris, P.S. and Ioakimidis, N.I.: A Remark on the Numerical Solution of Singular Integral Equations and the Determination of Stress-Intensity Factors. J. of Eng. Math. 13 (1979) 213-222.

[The80a]   Theocaris, P.S., Ioakimidis, N.I. and Kazantzakis, J.G.: On the Numerical Evaluation of Two-Dimensional Principal Value Integrals, Int. J. for Num. Meth. in Eng. .15 (1980) 629-634.

[The80b]   Theocaris, P.S., Ioakimidis, N.I. and Chrystakis, A.: On the Application of Numerical Integration Rules to the Solution of some Singular Integral Equations, Comp. Meth. in Appl. Mech. and Eng. 24 (1980) 1-11.

[The81a]   Theocaris, P.S. and Kazantzakis, J.G.: On the Numerical Evaluation of Two-and Tree-Dimensional Cauchy Principal Value Integrals, Acta Mechanica 39 (1981) 105-115.

[The81b]   Theocaris, P.S. and Kazantzakis, J.G.: Modified Quadrature Rules for the Numerical Evaluation of Certain Cauchy Principal Value Integrals. Revue Roumaine des Sciences Techniques, 26 (1981) 725-730.

[The81c]   Theocaris, P.S.: Numerical Solution of Singular Integral Equations I: Methods. ASCE Eng. Mech. 107 (1981) 733-752.

[The81d]   Theocaris, P.S.: Numerical Solution of Singular Integral Equations: Applications. ASCE Eng. Mech. 107 (1981) 753-771.

[The82a]   Theocaris, P.S., Tsamasphyros, G.J. and Mikroulis, G.: A Combined Integral-Equation and Photoelastic Method for Solving Contact Problems, Acta Mechanica 45 (1982) 215-231.

[The82b]   Theocaris, P.S. and Tsamasphyros, G.J.: A Numerical Method for Solving Singular Integrodifferential Equations with Variable coefficients, Proc. Nat. Acad. of Athens. 57 (1982) 581-595.

[The82c]   Theocaris, P.S., Tsamasphyros, G.J. and Theotokoglou, E.E.: A Combined Integral-Equation and Finite-Element Method for the Evaluation of S.I.Fs. Comp. Meth. in Appl. Mech. and Eng. 31 (1982) 117-128.

[The82d]   Theocaris, P.S. and Tsamasphyros, G.J.: A Photoelastic Point Matching Method for the Solution of Integral Equations in Contact Problems. Mech. Res. Comm., 9 (1982) 31-38.

[The82e]   Theocaris, P.S. and Tsamasphyros, G.J.: On the Solution of Systems of Singular Integral Equations with Variable Coefficients and Complex Weight Functions, Proc. Nat. Acad. of Athens. 57 (1982) 478-502.

[The83a]   Theocaris, P.S. and Bardzokas, D.: The Frictionless Contact of Cracked Elastic Bodies, ZAMM 63 (1983) 89-102.

[The83b]   Theocaris, P.S., Tsamasphyros, G.J. and Stassinakis, C.: A Numerical Solution of Singular Integral Equations without using Special Collocation Points, Int. J.for Num. Meth. in Eng., 19 (1983) 421-430.

[The83c]   Theocaris, P.S. and Kazantzakis, J.: Modified Quadrature Rules for the Numerical Evaluation of Certain Cauchy Principal Value Integrals, some comments by N.Ioakimidis and authors reply. Revue Roum. Sci. Techn., 28 (1983) 91-97.

[The83d]   Theocaris, P.S.: Modified Gauss-Legendre, Lobatto and Radau Curbature Formulas for the Numerical Evaluation of 2-D Singular Integrals, Int. J. Maths. and Math. Sciences, 6 (1983) 567-587.

[The84a]   Theocaris, P.S., Tsamasphyros, G.J. and Theotokoglou, E.E.: A Combination of the Finite-Element and Singular Integral-Equation Methods the Solution of the Generally Cracked Body, Int. J. for Num. Meth. in Eng., 11 (1984) 2065-2075.

[The84b]   Theocaris, P.S. and Tsamasphyros, G.J.: A Numerical Solution of Singular Integrodifferential Equations with Variable Coefficients, Appl. Math. and Computation 15 (1984) 47-59.

[The91]   Theocaris, P.S. and Panagiotopoulos, P.D.: On Debonding Effects in Adhesively Bonded Cracks - A Boundary Integral Approach, Arch. of Appl. Mech. (Former Ing. Archiv) 61 (1991) 578-587.

[The92]   Theocaris, P.S. and Panagiotopoulos, P.D.: On the Consideration of Unilateral Contact and Friction in Cracks. The Boundary Integral Method, Int. J. Num. Meth. Eng. 35 (1992) (in press).

[The93]   Theocaris, P.S. and Panagiptopoulos, P.D.: Neural Networks for Computing in Fracture Mechanics. Methods and Prospects of Applications. Comp. Meth. on Appl. Mech. and Eng. (to appear).

[Ton]   Tonti, E.: On the Mathematical Structure of a Large Class of Physical Theories, Rend. Accad. Nat. Lincei. Class. Sci. fis. mat. nat. 52 (1972), 48-56.

[Tont]   Tonti, E.: A Systematic Approach to the Search for Variational Principles. In "Variational Methods in Engineering" (ed. by C.A.Brebbia and H.Tottenham) Vol. I, Southampton Univ. Press, Southampton 1973.

[Tör]   Törnig, W., Gipser, A. and Kaspar, D.: Numerische Lösung von partiellen Differentialgleichungen der Technik. Math. Methoden in der Technik, Vol. 1, Teubner Verlag, Stuttgart 1985.

[Tra89]   Tralli, A. and Alessandri, C.: On B.E.M. Solutions of Elastic Frictionless Contact Problems. In: "Boundary Elements X, Vol 3: Stress Analysis", (ed. by Brebbia, C.A.) Springer Verlag, Berlin 1989.

[Tra90]   Alliney, S., Tralli, A. and Alessandri, C.: Boundary Variational Formulations and Numerical Solution Techniques for Unilateral Contact Problems. Comput. Mechanics 6 (1990) 247-257.

[Tre17]   Trefftz, E.: Über die Kontraktion kreisförmiger Flussigkeitsstrahlen, ZAAM 64 (1917) 34-61.

[Tri28]   Tricomi, F.: Equazioni Integrali contenenti il valor Principale di un Integrale doppio, Math. Zeit. 27 (1928) 87-133.

[Tri57]   Tricomi, F.: Integral Equations, Interscience Publ., London, N.York 1957 Reprint. Dover Publ, N.York 1985.

[Tsa77]   Tsamasphyros, G.J. and Theocaris, P.S.: On the Convergence of a Gauss Quadrature Rule for the Evaluation of Cauchy Type Singular Integrals, Nordisk Tidskrift for Informations Behandling (BIT) 17 (1977) 458-464.

[Tsa79]   Tsamasphyros, G.J. and Theocaris, P.S.: Curbature Formulas for the Evaluation of Surface Singular Integrals, Nordisk Tidskrift for Informations Behandling (BIT) 19 (1979) 368-377.

[Tsa80]   Tsamasphyros, G.J. and Theocaris, P.S.: Méthode générale de quadrature des integrales du type Cauchy, Balkan Congress of Applied Mathematics, Salonica, Greece, Vol. II, pp.570-581, 1976, Revue Roumaine des Sciences Techniques-Serie de Mécanique Appliqué 25 (1980) 839-856.

[Tsa81a]   Tsamasphyros, G.J. and Theocaris, P.S.: Equivalence and Convergence of Direct and Indirect Method for the Numerical Solution of Singular Integral Equations, Computing 27 (1981) 71-80.

[Tsa81b]   Tsamasphyros, G.J. and Theocaris, P.S.: Are Special Collocation Points Necessary for the Numerical Solution of Singular Integral Equations?, Intern. J. Fract., Rep. Current Research 17 (1981) R21-R24.

[Tsa82]   Tsamasphyros, G.J. and Theocaris, P.S.: A Recurrence Formula for the Direct Solution of Singular Integral Equations. Comp. Meth. in Appl. Mech. and Eng. 31 (1982) 79-89.

[Tsa83a]   Tsamasphyros, G.J. and Theocaris, P.S.: Integral-Equation Solutions for Half-Planes Bonded Together or in Contact and Containing Internal Cracks or Holes, Ing. Arch. 53 (1983) 225-241.

[Tsa83b]   Tsamasphyros, G.J. and Theocaris, P.S.: On the Convergence of Some Quadrature Rules for Cauchy-Principal-Value and Finite-Part Integrals, Computing 31 (1983) 105-114.

[Tza]   Tzaferopoulos, M.A.: Numerical Analysis of Structures with Monotone and Nonmonotone, Nonsmooth Material Laws and Boundary Conditions: Algorithms and Applications, Doct. Dissertation, Aristotle University, Dept. of Civil Eng., Thessaloniki 1991.

[Ven83]   Venturini, W.S.: Boundary Element Method in Geomechanics, Springer Verlag, Berlin 1983.

[Viv72]   Vivoli, J.: Vibrations de Plaques et Potentiels de Couches, Acustica 26 (1972) 305-314.

[Viv74]   Vivoli, J. and Filippi, P.: Eigenfrequencies of Thin Plates and Layer Potentials, J. Acoust. Soc. Am. 55 (1974) 562-567.

[Wal]   Wallin, H.: Interpolating and Orthogonal Polinomials on Fractals, Constr. Approx. 5 (1989) 137-150.

[Wal89]   Wallin, H.: The Trace to the Boundary of Sobolev Spaces on a Snowflake, Rep. Dept. Math. Univ. of Umea, Sweden 1989.

[Was]   Washizu, K.: Variational Methods in Elasticity and Plasticity, Pergamon Press, Oxford 1968.

[We82a]   Weeen, Van der F.: Application of the Boundary Integral Equation Method to Reissner's Plate Model, Int. J. Numer. Meth. Eng. 18 (1982) 1-10.

[We82b]   Weeen, Van der F.: Applications of the Direct Boundary Element to Reissner's Plate Model. In: "Proc. 4th Int. Conf. Boundary Element Methods in Engineering", (ed. by C.A. Brebbia), Springer Verlag, Southampton, N.York 1982.

[Wen65a]   Wendland, W.: Lösung der ersten und zweiten Randwertaufgaben des Innen und Aubengebietes für die Potentialgleichung im $\mathbb{R}_3$ durch Randbelegungen, Doct. Dissertation, Berlin 1965.

[Wen65b]   Wendland, W.: Die Methode der Randbelegungen bei der Lösung der ersten und zweiten Randwertaufgabe der Potentialgleichung für Ränder mit Kanten und Ecken, ZAMM 45 (1965), T84-T87.

[Wen68]   Wendland, W.: Die Behandlung von Randwertaufgaben im $\mathbb{R}_3$ mit Hilfe von Einfach - und Doppelschicht - potentialen, Num. Math. 11 (1968) 380-404.

[Wes]   Westphal, W.: Physik, 24. Anflage, Springer Verlag, Berlin 1963.

[West]   Westergaard, H.M.: Water Pressure on Dams During Earthquakes, Trans. ASCE 98 (1933) 418-433.

[Wid]   Widrow, B. and Hoff, M.: Adaptive Switching Circuits, 1960 IRE WESCON Convention Record, New York IRE, 96-104.

[Wil]   Williams, J.G. and Rhodes, M.D.: Effect of Resin on Impact Damage Tolerance of Graphite/Epoxy Laminates. In "Proceedings of the 6th International Conference on Composite Materials, Testing and Design", (ed. by I.M. Daniel). ASTM Special Technical Publication 787, ASTM Philadelphia 1982.

[Wri]   Wriggers, P., Simo, J.C. and Taylor, R.L.: Penalty and Augmented Lagrangian Formulations for Contact Problems. In "Proceedings of the International Conference on Mumerical Methods in Engineering - Theory and Applications, NUMETA '85", (ed. by J. Middleton and G.N. Pande), Vol. I, 1985 97-106.

[Wu73]   Wu, J.C. and Thompson, J.F.: Numerical Solution of Time-dependent Incompressible Navier-Stokes Equations Using an Integrodifferential Formulation, Computers and Fluids 1 (1973), 197-215.

[Wu74]   Wu, J.C., Spring, A.H. and Sankar, N.L.: A Flow Field Segmentation Method for Numerical Solution of Viscous Flows Problems, In "Lecture Notes in Physics, Vol. 35", Springer Verlag, Berlin 1974.

[Yv]   Yvon, J.P.: Etude de quelques problemes de controle pour des systèmes distribués, Thèse de Doctorat D' État. Université Paris VI, 1973.

[Zas82]   Zastrow, U.: Solution of Plane Anisotropic Elastostatical Boundary Value Problems by Singular Integral Equations, Acta Mechanica 44 (1982), 59-71.

[Za85a]   Zastrow, U.: On the Basic Geometrical Singularities in Plane Elasticity and Plate Bending Problems, Int. J. Solids and Struct. 21 (1985) 1047-1067.

[Za85b]   Zastrow, U.: Numerical Plane Stress Analysis by Integral Equations Based on the Singularity Method, Solid. Mech. Arch. 10 (1985) 187-221.

[Zer]   Zervas, P.A.: Seismic Behaviour of Frame Structures with Unilateral Contact Conditions. Doct. Dissertation, Dept. of Civil Eng., Aristotle Univ. Thessaloniki (1992).

[Zh88]   Zhong, W.X. and Sun, S.M.: A Finite Element Method for Elasto-Plastic Structures and Contact Problems by Parametric Quadratic Programming, Int. J. for Num. Meth. in Eng., 26 (1988) 2723-2738.

[Zh89]   Zhong, W.X. and Sun, S.M.: A Parametric Quadratic Programming Approach to Elastic Contact Probems with Friction, Comp. and Struct. 32 (1989) 37-43.

[Zie77a] Zienkiewicz, O.C., Kelly, D.W. and Pettess, P.: The Coupling of the Finite Element Method and Boundary Solution Procedures, Int. J. for Num. Meth. in Eng. 11 (1977), 355-375.

[Zie77b] Zienkiewicz, O.C.: The Finite Element Method, 3rd Edition, McGraw-Hill, London 1977.

[Zie80]   Zienkiewicz, O., Wood, W. L. and Taylor, R. L.: An Alternative Single-Step Algorithm for Dynamic Problems, Earthq. Eng. and Struct. Dyn., 8 (1980) 31-40.

Subject Index